AF773942

Mapping Degree Theory

Mapping Degree Theory

Enrique Outerelo
Jesús M. Ruiz

Graduate Studies in Mathematics
Volume 108

American Mathematical Society
Providence, Rhode Island

Real Sociedad Matemática Española
Madrid, Spain

Editorial Board of Graduate Studies in Mathematics

David Cox, Chair

Steven G. Krantz   Rafe Mazzeo   Martin Scharlemann

Editorial Committee of the Real Sociedad Matemática Española

Guillermo P. Curbera, Director

Luis Alías
Emilio Carrizosa
Bernardo Cascales
Javier Duoandikoetxea
Alberto Elduque
Pablo Pedregal
Rosa María Miró-Roig
Juan Soler

2000 *Mathematics Subject Classification.* Primary 01A55, 01A60, 47H11, 55M25, 57R35, 58A12, 58J20.

For additional information and updates on this book, visit
www.ams.org/bookpages/gsm-108

Library of Congress Cataloging-in-Publication Data

Outerelo, Enrique, 1939–
Mapping degree theory / Enrique Outerelo and Jesús M. Ruiz.
p. cm. — (Graduate studies in mathematics ; v. 108)
Includes bibliographical references and index.
ISBN 978-0-8218-4915-6 (alk. paper)
1. Topological degree. 2. Mappings (Mathematics). I. Ruiz, Jesús M. II. Title.

QA612.O98 2009
515′.7248—dc22

2009026383

Copying and reprinting. Individual readers of this publication, and nonprofit libraries acting for them, are permitted to make fair use of the material, such as to copy a chapter for use in teaching or research. Permission is granted to quote brief passages from this publication in reviews, provided the customary acknowledgment of the source is given.

Republication, systematic copying, or multiple reproduction of any material in this publication is permitted only under license from the American Mathematical Society. Requests for such permission should be addressed to the Acquisitions Department, American Mathematical Society, 201 Charles Street, Providence, Rhode Island 02904-2294 USA. Requests can also be made by e-mail to reprint-permission@ams.org.

© 2009 by the authors.
Printed in the United States of America.

∞ The paper used in this book is acid-free and falls within the guidelines established to ensure permanence and durability.
Visit the AMS home page at http://www.ams.org/

10 9 8 7 6 5 4 3 2 1   14 13 12 11 10 09

To Jesús María Ruiz Amestoy

Contents

Preface

This book springs from lectures on degree theory given by the authors during many years at the Departamento de Geometría y Topología at the Universidad Complutense de Madrid, and its definitive form corresponds to a three-month course given at the Dipartimento di Matematica at the Università di Pisa during the spring of 2006. Today mapping degree is a somewhat classical topic that appeals to geometers and topologists for its beauty and ample range of relevant applications. Our purpose here is to present both the history and the mathematics.

The notion of degree was discovered by the great mathematicians of the decades around 1900: Cauchy, Poincaré, Hadamard, Brouwer, Hopf, etc. It was then brought to maturity in the 1930s by Hopf and also by Leray and Schauder. The theory was fully burnished between 1950 and 1970. This process is described in Chapter I. As a complement, at the end of the book there is included an index of names of the mathematicians who played their part in the development of mapping degree theory, many of whom stand tallest in the history of mathematics. After the first historical chapter, Chapters II, III, IV, and V are devoted to a more formal proposition-proof discourse to define and study the concept of degree and its applications. Chapter II gives a quick presentation of manifolds, with special emphasis on aspects relevant to degree theory, namely regular values of differentiable mappings, tubular neighborhoods, approximation, and orientation. Although this chapter is primarily intended to provide a review for the reader, it includes some not so standard details, for instance concerning tubular neighborhoods. The main topic, degree theory, is presented in Chapters III and IV. In a simplified manner we can distinguish two approaches to the theory: the Brouwer-Kronecker degree and the Euclidean degree. The first is developed in Chapter III by differential means, with a quick diversion into the de Rham computation in cohomological terms. We cannot help this diversion: cohomology is too appealing to skip. Among other applications, we obtain in this chapter a differential version of the Jordan Separation Theorem. Then, we construct the Euclidean degree in

Chapter IV. This is mainly analytic and astonishingly simple, especially in view of its extraordinary power. We hope this partisan claim will be acknowledged readily, once we obtain quite freely two very deep theorems: the Invariance of Domain Theorem and the Jordan Separation Theorem, the latter in its utmost topological generality. Finally, Chapter V is devoted to some of those special results in mathematics that justify a theory by their depth and perfection: the Hopf and the Poincaré-Hopf Theorems, with their accompaniment of consequences and comments. We state and prove these theorems, which gives us the perfect occasion to take a glance at tangent vector fields.

We have included an assorted collection of some 180 problems and exercises distributed among the sections of Chapters II to V, none for Chapter I due to its nature. Those problems and exercises, of various difficulty, fall into three different classes: (i) suitable examples that help to seize the ideas behind the theory, (ii) complements to that theory, such as variations for different settings, additional applications, or unexpected connections with different topics, and (iii) guides for the reader to produce complete proofs of the classical results presented in Chapter I, once the proper machinery is developed.

We have tried to make internal cross-references clearer by adding the Roman chapter number to the reference, either the current chapter number or that of a different chapter. For example, III.6.4 refers to Proposition 6.4 in Chapter III; similarly, the reference IV.2 means Section 2 in Chapter IV. We have also added the page number of the reference in most cases.

One essential goal of ours must be noted here: we attempt the simplest possible presentation at the lowest technical cost. This means we restrict ourselves to elementary methods, whatever meaning is accepted for elementary. More explicitly, we only assume the reader is acquainted with basic ideas of differential topology, such as can be found in any text on the calculus on manifolds.

We only hope that this book succeeds in presenting degree theory as it deserves to be presented: we view the theory as a genuine masterpiece, joining brilliant invention with deep understanding, all in the most accomplished attire of clarity. We have tried to share that view of ours with the reader.

Los Molinos and Majadahonda *E. Outerelo and J.M. Ruiz*
June 2009

Chapter I

History

In the body of mathematics, the notion of degree stands as a beautiful achievement of topology and one of the main contributions of the twentieth century, which has been called the century of topology. In Chapter I we try to outline how the ideas that led to this fundamental notion of degree were sparked and came to light. It is only natural that such a task is biased by our personal opinions and preferences. Thus, it is likely that a specialist in, say, partial differential equations would present the tale in a somewhat different way. All in all, a choice must be made and ours is this:

§1. *Prehistory:* Gauss, Cauchy, Liouville, Sturm, Kronecker, Poincaré, Picard, Bohl (1799–1910).

§2. *Inception and formation:* Hadamard, Brouwer (1910–1912).

§3. *Accomplishment:* Hopf, Leray, Schauder (1925–1934).

§4. *Renaissance and reformation:* Nagumo, de Rham, Heinz (1950–1970).

§5. *Axiomatization:* Führer, Amann, Weiss (1970–1972).

§6. *Further developments:* Equivariant theory, infinite dimensions.

The presentation of these topics is mainly discursive and descriptive, rigorous proofs being deferred to Chapters II through V where there will be complete arguments for all the most classical results presented here.

1. Prehistory

Rougly speaking, degree theory can be defined as the study of those techniques that give information *on the existence of solutions of an equation of the form* $y = f(x)$, where x and y dwell in suitable spaces and f is a map from one to the other. The theory also gives clues for *the number of solutions and their nature.* An important particular case is that of an equation $x = f(x)$, where f is a map from a domain D of a linear space into D itself: this is the so-called *Fixed Point Problem.*

By its very nature, it is clear that the origins of degree theory should be traced back to the first attempts to solve algebraic equations such as

$$z^n + a_1 z^{n-1} + \cdots + a_n = 0,$$

where the coefficients a_i are complex numbers, $a_n \neq 0$. That such an equation always has some solution is the *Fundamental Theorem of Algebra.* This result was most beloved by KARL-FRIEDRICH GAUSS (1777–1855), who found at least four different proofs, in 1799, 1815, 1816, and 1849. It is precisely in the first and fourth proofs where we find what can be properly considered the first ideas of topological degree. By some properties of algebraic curves (which were formalized only in 1933 by ALEXANDER OSTROWSKI (1893–1986)), Gauss was able to prove that inside a circle of big enough radius, the algebraic curve corresponding to the real part of the polynomial shares some point with the algebraic curve corresponding to the imaginary part. In this way the following two lines of research were born:

Problem I. Find the common solutions of the equations

$$\begin{cases} f(x,y) = 0, \\ F(x,y) = 0 \end{cases}$$

inside a given closed planar domain, on whose border the two functions $f(x,y)$ and $F(x,y)$ do not vanish simultaneously.

Problem II. Find the number of real roots of a polynomial in one variable with real coefficients, in a given closed interval $[a,b]$ of the real line.

* * *

The first contributions to **Problem I** are due to AUGUSTIN LOUIS CAUCHY (1789–1857). In a memoir presented before the Academy of Turin, on November 17, 1831, and in the paper [Cauchy 1837a], Cauchy introduces a new calculus that, in its own words, can be used to solve equations of any degree.

Some parts of Cauchy's arguments are not completely precise, and the way these parts were made rigorous by JACQUES CHARLES FRANÇOIS STURM (1803–1855) and JOSEPH LIOUVILLE (1809–1882) is quite relevant in the history of the analytic definition of the topological degree of a continuous mapping.

Let us describe this. The definition of the *index of a function* given by Cauchy in [Cauchy 1837a] is the following:

Let x be a real variable and $f(x)$ a function that becomes infinite at $x = a$. If the variable x increases through a, the function will either change from negative to positive or change from positive to negative or not change sign at all. We will say that the index of f at a *is* -1 *in the first case,* $+1$ *in the second, and* 0 *in the third. We define the* integral index of f *between two given limits $x = x_0$ and $x = x_1$, denoted by $J_{x_0}^{x_1}(f)$, as the sum of the indices of f corresponding to the values of x in the interval $[x_0, x_1]$ at which f becomes infinite. If f is a function in two variables, we define the* integral index of f *between the limits $x_0, x_1; y_0, y_1$ to be the number*

$$ {}_{x_0}^{x_1}J_{y_0}^{y_1}(f) = \tfrac{1}{2}\left[J_{x_0}^{x_1}(f(\cdot, y_1)) - J_{x_0}^{x_1}(f(\cdot, y_0)) - J_{y_0}^{y_1}(f(x_1, \cdot)) + J_{y_0}^{y_1}(f(x_0, \cdot))\right]. $$

In his 1831 memoir, Cauchy obtained the index of a function by integral techniques and residues and proved the following result:

Theorem. *Let Γ be a closed curve that is the contour of an area S, and let $Z(z) = X(x, y) + iY(x, y)$ be an entire function. Then*

$$ \frac{1}{2\pi i}\int_\Gamma \frac{Z'(z)}{Z(z)}dz = \tfrac{1}{2}\, J_{s=s'}^{s=s''}(X/Y) $$

is the number of zeros of $Z(z)$ in S; here s stands for the arc length along Γ, and $s'' - s'$ is the length of Γ.

Cauchy generalized this result in a memoir published June 16, 1833, in Turin. The generalization follows:

Theorem. *Let $F(x, y)$ and $f(x, y)$ be two functions of the variables x, y, continuous between the limits $x = x_0, x = x_1, y = y_0, y = y_1$. We denote by $\Phi(x, y), \phi(x, y)$ the derivatives of the functions with respect to x, and by $\Psi(x, y), \psi(x, y)$ their derivatives with respect to y. Finally, let N be the number of the different systems of values x, y, between the above limits, verifying simultaneously the equations $F(x, y) = 0, f(x, y) = 0$. Then*

$$ N = {}_{x_0}^{x_1}J_{y_0}^{y_1}(\Delta), $$

where

$$ \begin{aligned} \Delta(x, y) &= \frac{f(x, y)}{F(x, y)}\left(\Phi(x, y)\psi(x, y) - \Psi(x, y)\phi(x, y)\right) \\ &= \frac{f(x, y)}{F(x, y)}\left(\frac{\partial F(x, y)}{\partial x}\frac{\partial f(x, y)}{\partial y} - \frac{\partial F(x, y)}{\partial y}\frac{\partial f(x, y)}{\partial x}\right). \end{aligned} $$

An elementary "proof" of this theorem appears in [Cauchy 1837b]. However, Liouville and Sturm in [Liouville-Sturm 1837] give three examples showing that the second theorem above above can fail. The first example is

$$\begin{cases} F(x,y) = x^2 + y^2 - 1, \\ f(x,y) = y. \end{cases}$$

In this example

$$\Delta(x,y) = \frac{2xy}{x^2 + y^2 - 1},$$

and drawing around the origin a rectangle containing the circle $x^2+y^2 = 1$, one sees that

$${}_{x_0}^{x_1}J_{y_0}^{y_1}(\Delta) = 0,$$

because Δ never becomes infinity on the sides of the rectangle. However, the system

$$\begin{cases} x^2 + y^2 - 1 = 0, \\ y = 0 \end{cases}$$

has the two solutions $(1,0), (-1,0)$ inside the rectangle. Liouville and Sturm conclude their note with the following remark:

There is a theorem that can replace Cauchy's. Let us consider a closed contour Γ on which $F(x,y)$ and $f(x,y)$ do not vanish simultaneously, and let us also assume that inside this contour the function

$$\begin{aligned} w &= \Phi(x,y)\psi(x,y) - \Psi(x,y)\phi(x,y) \\ &= \frac{\partial F(x,y)}{\partial x}\frac{\partial f(x,y)}{\partial y} - \frac{\partial F(x,y)}{\partial y}\frac{\partial f(x,y)}{\partial x} \end{aligned}$$

does not vanish at the values (x,y) at which $f(x,y)$ and $F(x,y)$ vanish.

In this situation, among the solutions (x,y) of the equations $F(x,y) = 0$, $f(x,y) = 0$, inside Γ, some correspond to positive values of w and others to negative values of w. We denote by μ_1 the number of solutions of the first kind, and by μ_2 the number of solutions of the second kind. With this notation we have

$$\frac{1}{2}\delta = \mu_1 - \mu_2,$$

where δ stands for how many more times the function $\frac{f(x,y)}{F(x,y)}$ changes from positive to negative than from negative to positive, at those points in the contour Γ at which that function becomes infinite, when the contour is traced in the positive direction.

We see that the function w is the *Jacobian of the mapping* (F, f) (Liouville and Sturm always consider entire functions). Consequently, we find displayed here for the first time the importance of the sign of the functional determinant

$$w = \begin{vmatrix} \dfrac{\partial F}{\partial x} & \dfrac{\partial F}{\partial y} \\ \dfrac{\partial f}{\partial x} & \dfrac{\partial f}{\partial y} \end{vmatrix}$$

when dealing with the computation of the number of solutions of the system

$$\begin{cases} F(x,y) = 0, \\ f(x,y) = 0 \end{cases}$$

in a planar region.

Today, in the hypotheses of the Liouville-Sturm Theorem, the number $\mu_1 - \mu_2$ is called the *topological degree of the mapping* (F, f) *at the origin,* and this is the starting point for the analytic definition of degree. But this will not take full shape until 1951.

* * *

In the later paper [Cauchy 1855], Cauchy states the *Argument Principle,* which is another way to compute the indices he has defined earlier. These results, translated into more modern terminology, read as follows.

Winding number (or index) of a planar curve around a point. *Let* $\Gamma \subset \mathbb{C}$ *be a closed oriented curve with a* $\mathcal{C}^1$ *parametrization:*

$$z(t) = x(t) + iy(t) + a, \quad 0 \le t \le 1, \quad z(0) = z(1), \quad a \in \mathbb{C} \setminus \Gamma.$$

Then,

$$w(\Gamma, a) = \frac{1}{2\pi i} \int_\Gamma \frac{dz}{z-a} = \frac{1}{2\pi} \int_0^1 \frac{x(t)y'(t) - x'(t)y(t)}{x^2(t) + y^2(t)} dt$$

is an integer.

This integer is called the *winding number (or index) of* Γ *around* a.

Geometrically, the winding number tells us *how many times the curve wraps around the point.* In case Γ is only continuous, the winding number is defined through a $\mathcal{C}^1$ approximation Γ_1 of Γ, because $w(\Gamma_1, a)$ remains constant for Γ_1 close enough to Γ.

The following example illustrates this notion:

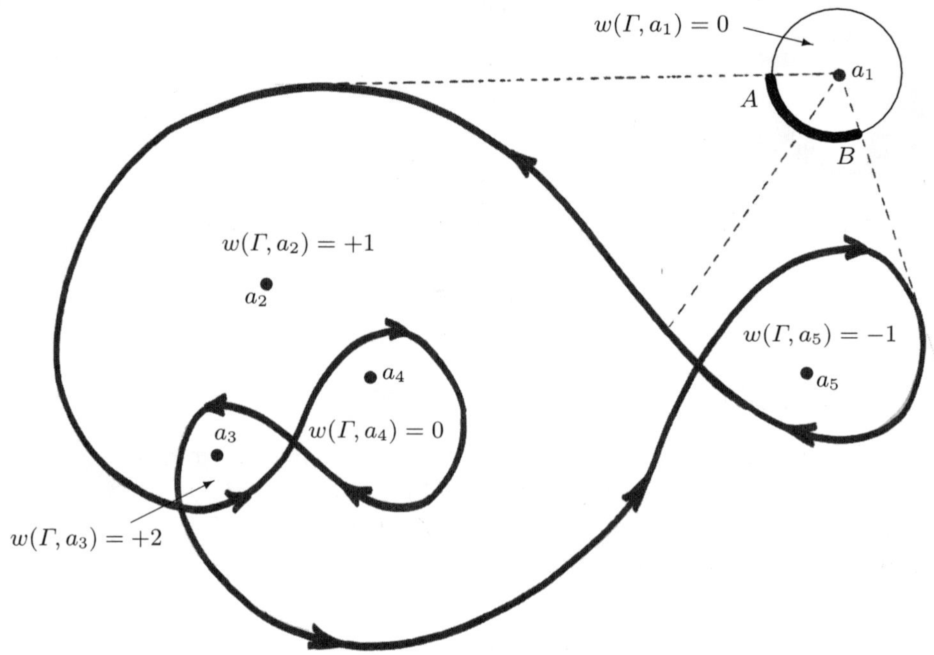

To proceed one step further, Cauchy considers a simply connected domain $G \subset \mathbb{C}$ (that is, G has no holes), a holomorphic function $f : G \to \mathbb{C}$, $\zeta = f(z)$, and a $\mathcal{C}^1$ closed curve $\Gamma \subset G$, on which f has no zeros. Then:

Argument Principle. *The following formula holds:*

$$w(f(\Gamma), 0) = \frac{1}{2\pi i} \int_{f(\Gamma)} \frac{d\zeta}{\zeta} = \frac{1}{2\pi i} \int_{\Gamma} \frac{f'(z)}{f(z)} dz = \sum_k w(\Gamma, a_k)\alpha_k,$$

where the a_k's are the zeros of f in the domain D bounded by Γ and the α_k's are their respective multiplicities.

Suppose next that Γ has no self-intersection and that it has the positive (counterclockwise) orientation. Then D is a connected domain (this is the *Jordan Separation Theorem,* which we will discuss later), and $w(\Gamma, a) = +1$ for all $a \in D$, so that the last formula becomes

$$w(f(\Gamma), 0) = \sum_k \alpha_k,$$

that is:

Theorem. *The total number of zeros (counted with multiplicities) that f has in D is the winding number of the curve $f(\Gamma)$ around the origin.*

In general, the winding number can be negative, but we can still say that f has at least $|w(f(\Gamma), 0)|$ zeros in the domain bounded by Γ.

* * *

Let us now turn to **Problem II.** The first full solution is due to Sturm. In 1829 and 1835 he gave an algorithm to find the exact number of distinct real roots of a polynomial. The theorem was later generalized by CARL GUSTAV JACOB JACOBI (1804–1851), CHARLES HERMITE (1822–1901), and JAMES JOSEPH SYLVESTER (1814–1897).

Exploring the topological content of Sylvester's article [Sylvester 1853], LEOPOLD KRONECKER (1823–1891) introduces in his papers [Kronecker 1869a] and [Kronecker 1869b] a method that extends Sturm's. Indeed, at the end of his work Kronecker writes:

In my research developed in this article, I started from a theorem by Sturm. A generalization of that result was found by Hermite some time ago, but I have been able to extend the continued fraction algorithm developed by Sylvester to further widen Sturm's theorem.

Let us describe Kronecker's contribution. He starts with the following definition:

Regular function systems. A *regular function system* consists of $n+1$ real functions $F_0, F_1, \dots, F_n$ in n real variables $x_1, \dots, x_n$, such that

(a) $F_0, F_1, \dots, F_n$ are continuous and have no common zeros. They admit partial derivatives with respect to all n variables, and those derivatives take finite values.

(b) The functions $F_0, F_1, \dots, F_n$ take positive and negative values. Moreover, each function takes positive (resp., negative) values infinitely often.

(c) The domains $\{F_i < 0\}$, $i = 0, \dots, n$, represent n-dimensional varieties that only contain finite values of the variables $x_1, \dots, x_n$.

(d) No functional determinant

$$\left| \frac{\partial F_i}{\partial x_j} \right|_{\substack{k \neq i = 0, 1, \dots, n \\ j = 1, \dots, n}}, \qquad k = 0, 1, \dots, n,$$

vanishes at any zero of the system $F_k \neq 0, F_0 = F_1 = \dots = F_n = 0$.

(e) The common zero set of any chosen $n-1$ functions among $F_0, F_1, \ldots,$ F_n is a $\mathcal{C}^1$ curve.

Then Kronecker looks at the orientations of the $\mathcal{C}^1$ curve involved in this definition (condition (e) above). He considers this part basic in his research on systems of functions in several variables:

Orientation Principle. Kronecker chooses for every pair (k, ℓ), $k < \ell$, an orientation of the $\mathcal{C}^1$ curve (recall (e) above)

$$F(k,\ell) = \{x \in \mathbb{R}^n : F_i(x) = 0 \text{ for } i \neq k, \ell\}.$$

This orientation is denoted by $|k\ell|$; he then puts $|\ell k| = -|k\ell|$.

Next, he defines:

(a) A point $e \in F(k,\ell) \cap \{F_k = 0\}$ is called an *incoming (eingang)* point of $F(k,\ell)$ (in $\{x \in \mathbb{R}^n : F_k(x) \cdot F_\ell(x) < 0\}$) if the following condition holds true: *walking the curve $F(k,\ell)$ as oriented by $|k\ell|$, we leave the set $\{x \in \mathbb{R}^n : F_k(x) \cdot F_\ell(x) > 0\}$ at the point e and enter $\{x \in \mathbb{R}^n : F_k(x) \cdot F_\ell(x) < 0\}$.*

The set of all these incoming points e is denoted by $E(k,\ell)$.

(b) A point $a \in F(k,\ell) \cap \{F_k = 0\}$ is called an *outgoing (ausgang)* point of $F(k,\ell)$ (off $\{x \in \mathbb{R}^n : F_k(x) \cdot F_\ell(x) < 0\}$) if the following condition holds true: *walking the curve $F(k,\ell)$ as oriented by $|k\ell|$, we leave the set $\{x \in \mathbb{R}^n : F_k(x) \cdot F_\ell(x) < 0\}$ at the point a and enter $\{x \in \mathbb{R}^n : F_k(x) \cdot F_\ell(x) > 0\}$.*

The set of all these outgoing points a is denoted by $A(k,\ell)$.

After the preceding preparation, Kronecker shows that the number

$$\#E(k,\ell) - \#A(k,\ell)$$

is *even* and does not depend on the indices k, ℓ, and then he defines:

Kronecker characteristic. The *characteristic of the regular function system* $F_0, F_1, \ldots, F_n$ is the integer

$$\chi(F_0, F_1, \ldots, F_n) = \tfrac{1}{2}\big(\#E(k,\ell) - \#A(k,\ell)\big).$$

It is convenient to stress that in the course of his proof of this fact

Kronecker obtains the following very modern description of his invariant:

$$\chi(F_0, F_1, \ldots, F_n) = (-1)^j \cdot \sum_{\substack{(F_0,\ldots,F_n)(x)=0 \\ F_j(x)<0}} \operatorname{sign} J_x(F_0, \ldots, \widehat{F}_j, \ldots, F_n)$$
$$= (-1)^j \cdot \sum_{\substack{(F_0,\ldots,F_n)(x)=0 \\ F_j(x)>0}} \operatorname{sign} J_x(F_0, \ldots, \widehat{F}_j, \ldots, F_n)$$

(here J stands for the Jacobian determinant) for any $j = 0, \ldots, n$.

Once this invariant is defined, Kronecker shows how it detects solutions of the given regular system:

Kronecker Existence Theorem. *Let $F_0, F_1, \ldots, F_n$ be a regular function system. If $\chi(F_0, F_1, \ldots, F_n) \neq 0$, then for every $i = 0, 1, \ldots, n$, the system*

$$F_0(x) = 0, \ldots, F_{i-1}(x) = 0,\ F_i(x) < 0,\ F_{i+1}(x) = 0, \ldots, F_n(x) = 0$$

has some solution $x \in \mathbb{R}^n$.

This result and the above remark extend to a system of n functions with n unknowns what Liouville and Sturm had done thirty years earlier, as has already been mentioned.

We illustrate all of this with a simple example:

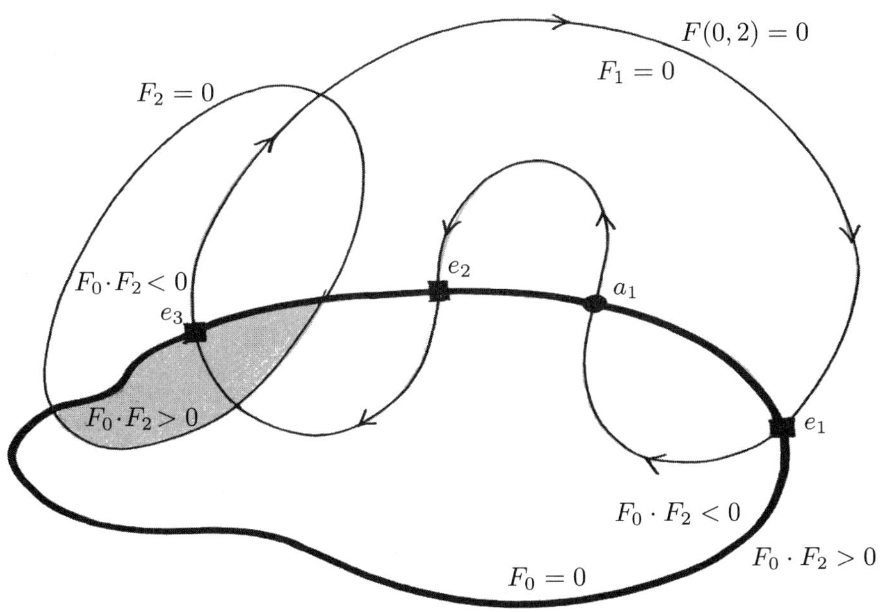

Here, we depict a regular system consisting of three functions $F_0, F_1, F_2 : \mathbb{R}^2 \to \mathbb{R}$. Each set

$$F(k,\ell) = \{(x_1,x_2) \in \mathbb{R}^2 : F_i(x_1,x_2) = 0 \text{ for } i \neq k,\ell\} = \{F_i = 0\}, i \neq k,\ell,$$

is a $\mathcal{C}^1$ curve, and we have the incoming points $E(0,2) = \{e_1,e_2,e_3\}$ and the outgoing points $A(0,2) = \{a_1\}$. We get

$$\chi(F_0,F_1,F_2) = \frac{1}{2}(\#E(0,2) - \#A(0,2)) = \frac{1}{2}(3-1) = 1 \neq 0,$$

and we see immediately in the picture that each of the three systems

$$\begin{array}{ll} (0) & F_0 < 0,\ F_1 = 0,\ F_2 = 0, \\ (1) & F_0 = 0,\ F_1 < 0,\ F_2 = 0, \\ (2) & F_0 = 0,\ F_1 = 0,\ F_2 < 0 \end{array}$$

indeed has solutions.

In 1877, influenced by some discussions with his friend KARL THEODOR WILHELM WEIERSTRASS (1815–1897) on the matter of complex analysis and potential theory, Kronecker gave a representation of his characteristic by means of an integral, today known as the *Kronecker integral.* Indeed, he shows this:

Kronecker Integral Theorem. *Let $F_0, F_1, \ldots, F_n$ be a regular system. Then for every $j = 0,1,\ldots,n$,*

$$\chi(F_0,F_1,\ldots,F_n) = -\frac{1}{\operatorname{vol}(\mathbb{S}^{n-1})}\int_{S_j} \frac{\Delta}{\Phi^n}\, du$$

where

$$\Delta = \begin{vmatrix} F_0(w) & F_1(w) & \ldots & \widehat{F_j(w)} & \ldots & F_n(w) \\ \frac{\partial F_0}{\partial x_1} & \frac{\partial F_1}{\partial x_1} & \ldots & \widehat{\frac{\partial F_j}{\partial x_1}} & \ldots & \frac{\partial F_n}{\partial x_1} \\ \ldots & \ldots & \ldots & \ldots & \ldots & \ldots \\ \frac{\partial F_0}{\partial x_{n-1}} & \frac{\partial F_1}{\partial x_{n-1}} & \ldots & \widehat{\frac{\partial F_j}{\partial x_{n-1}}} & \ldots & \frac{\partial F_n}{\partial x_{n-1}} \end{vmatrix}$$

and $S_j = \{x \in \mathbb{R}^n : F_j(x) = 0\}$, $\Phi = \sqrt{F_0^2 + F_1^2 + \cdots + \widehat{F_j^2} + \cdots + F_n^2}$.

Note that in case $n = 2$, and supposing that F_1 and F_2 are the components of a holomorphic function $f(z)$ on $F_0 \leq 0$, the above integral for $j = 0$ becomes

$$\frac{1}{2\pi i}\int_\Gamma \frac{f'(z)}{f(z)} dz.$$

Thus, the Kronecker Existence Theorem is in fact a generalization of Cauchy's results.

There are antecedents to the Kronecker integral in dimension > 2. Even Kronecker remarks in [Kronecker 1869a] that a special case of his integral can be found in a previous paper by Gauss on potential theory [Gauss 1813]. In this case $n = 3$,

$$F_1 = 4\pi(x_1 - x_1^0),\ F_2 = 4\pi(x_2 - x_2^0),\ F_3 = 4\pi(x_3 - x_3^0),$$

and the Kronecker integral gives the flow through the surface $F_0 = 0$ of the electric field

$$V(z) = \frac{x - x^0}{\|x - x^0\|^3}$$

created by the unit charge placed at $x^0 = (x_1^0, x_2^0, x_3^0)$. Gauss establishes in this paper that the flow is equal to 4π or 0, according to whether the point x^0 is interior or exterior to the surface. In modern terms, the flow through the surface $F_0 = 0$ is expressed by

$$\int_{F_0=0} \langle V, \nu\rangle dS,$$

where ν is the outward normal vector field on the surface and dS is the area element.

* * *

After Kronecker's pioneering work on the characteristic and the integral of a regular system, there were a number of papers on the subject, with important applications to geometric and topological questions. Following the chronological development, me must mention first the contributions by JULES HENRI POINCARÉ (1854–1912), who used the Kronecker integral in the qualitative theory of autonomous (independent of time) ordinary differential equations. Already in 1881, Poincaré uses the index of a planar curve to study these differential equations, but in 1883 he uses Kronecker's theorem in a note in Comptes Rendus [Poincaré 1883]. There he writes:

Mr. Kronecker has presented to the Berlin Academy, in 1869, a memoir on functions of several variables, including an important theorem from which the following result follows easily:

Let $\xi_1, \xi_2, \dots, \xi_n$ be n continuous functions in the n variables $x_1, x_2, \dots, x_n$, the variable x_i restricted to range among the limits $-a_i$ and $+a_i$. Let us suppose that for $x_i = a_i$ the function ξ_i is always positive, and for

$x_i = -a_i$ the function ξ_i is always negative. Then I say that there is a system of values for the x at which all the ξ vanish. This result can be applied to the three-body problem to prove it has infinitely many special solutions with important properties that we are to present.

This is an *Intermediate Value Theorem in arbitrary dimension.* Three years later Poincaré published the paper [Poincaré 1886] in which he studies the curves defined by differential equations of the second order. There he looks for the singular points of those equations and discusses their distribution using the Kronecker integral. This contains the argument for the invariance of the characteristic under continuous deformations, which is used in the proof of the result stated above. Kronecker had himself considered this invariance, and actually he could prove it in some particular cases.

* * *

Another important question implicit in Kronecker's work is the determination of the *exact* number of solutions of a system of equations. But it is CHARLES EMILE PICARD (1856–1941) who in 1891 published a note in Comptes Rendus [Picard 1891] with the precise formulation of the problem whose resolution was to be his main contribution to degree theory. Picard states:

Let us consider n equations $f_i(x_1, x_2, \ldots, x_n) = 0$, $i = 1, 2, \ldots, n$, where we suppose the f_i's represent continuous functions in n real variables $x_1, x_2, \ldots, x_n$ defining a point in some domain D. The question of finding the number of common roots of these equations in that domain has held the attention of geometers for a long time. A formula has been given in this sense by Mr. Kronecker, in his famous investigations of the characteristic of function systems. Unfortunately, the Kronecker integral, a multiple integral of order $n-1$ on the surface of the domain D, does not give the number of roots we are looking for. The functional determinant of the system plays a fundamental role in this theory, and one only obtains the difference of the numbers of roots at which the determinant is positive and the number of roots at which the determinant is negative.

I will show here, in a few lines, that it is possible to represent the exact number of roots by a suitable integral.

Picard's method consists of attaching new equations to get a new system that has the same roots and whose functional determinant is always positive and applying Kronecker's theory afterwards. Picard presented his results in full in classes at La Sorbonne, Paris, and finally published them in a two-volume treatise on analysis [Picard 1891/1905]. After this, Kronecker's theory became a classic.

In 1904 PIERS BOHL (1865–1921) published a paper [Bohl 1904] where he proves the following result:

Theorem. *Let* (G) *be the domain defined by* $-a_i \leq x_i \leq a_i$ $(a_i > 0, i = 1, 2, \ldots, n)$. *There do not exist continuous real-valued functions* $F_1, \ldots, F_n$, *without common zeros in* (G), *such that* $F_i = x_i$ $(i = 1, 2, \ldots, n)$ *on the boundary of* (G).

For the proof, Bohl uses the Kronecker integral and *Stokes' Theorem* (proving along the way the Kronecker Existence Theorem). From the above proposition, and by continuous deformation techniques, Bohl deduces the following two statements:

Proposition. (a) *Let* (G) *be the domain defined by* $-a_i \leq x_i \leq a_i$ $(a_i > 0,\ i = 1, 2, \ldots, n)$, *and let* $F_1, \ldots, F_n$ *be continuous real-valued functions such that* $F_i = x_i$ $(i = 1, 2, \ldots, n)$ *on the boundary of* (G). *Then, for every point* $(\gamma_1, \gamma_2, \ldots, \gamma_n)$ *in* (G) *there is another* $(x_1, x_2, \ldots, x_n)$ *such that* $F_i(x_1, x_2, \ldots, x_n) = \gamma_i$ $(i = 1, 2, \ldots, n)$.

(b) *Let* (G) *be the domain defined by* $-a_i \leq x_i \leq a_i$ $(a_i > 0,\ i = 1, 2, \ldots, n)$, *and let* $f_1, \ldots, f_n$ *be continuous real-valued functions that do not vanish simultaneously in* (G). *Then there are a constant* $N < 0$ *and a point* $(u_1, \ldots, u_n)$ *in the boundary of* (G) *such that* $f_i(u_1, u_2, \ldots, u_n) = N u_i$ $(i = 1, 2, \ldots, n)$.

It is possible that these results were already in the Ph.D. thesis defended by Bohl in 1900 at Dorpat University. Note here that (b) implies what is today the well-known fact that the boundary of (G) is not a retract of (G), which in turn implies a fixed point result. Thus we can say that Bohl was really close to discovering the *Brouwer Fixed Point Theorem.*

On the other hand, in 1895 Poincaré published his famous memoir *Analysis Situs* [Poincaré 1895b], where, starting from the attempts by GEORG FRIEDRICH BERNHARD RIEMANN (1826–1866) and ENRICO BETTI (1823–1892), he starts the development of what will later be called *combinatorial topology* and *simplicial homology.* Poincaré refined these ideas in two complements to his *Analysis Situs,* [Poincaré 1899, Poincaré 1900], where we find the first notion of *(abstract) polyhedron,* which will later be called *triangulation of a compact manifold* V *of dimension* n. This was a finite set $\mathcal{T}$ of disjoint cells of different dimensions $\leq n$ such that:

(1) Every cell of dimension d, or d-cell, is the image in V of an open ball $B \subset \mathbb{R}^d$ via a diffeomorphism from an open neighborhood of the closure $\overline{B}$ into a submanifold W of V of dimension d.

(2) The boundary in W of such a d-cell is a union of cells of dimensions $\leq d-1$.

(3) V is a union of cells of $\mathcal{T}$.

We will come back to this, when ten years later the notion of manifold comes to full development.

2. Inception and formation

In 1910 Jacques Salomon Hadamard (1865–1963) published a remarkable paper [Hadamard 1910], which is in fact an appendix to the book [Tannery 1910]. This paper marks the transition from the origins of degree theory to the construction of a more elaborate and systematic theory.

Hadamard says:

The proof given by Mr. Ames for the Jordan Theorem in the plane *is based on the index (or variation of the argument) of a planar curve with respect to a point (the Cauchy concept). The generalization of this concept to higher dimensions is provided by the Kronecker index. This concept is now classic, mainly after the* Traité d'Analyse *by Mr. Picard (vol. I, p. 123; vol. II, p. 193) of 1891 and 1893, and after several contemporary papers that give new applications of that index.*

Later Hadamard mentions Poincaré, Bohl, and Luitzen Egbertus Jan Brouwer (1881–1966).

Let us now describe things more explicitly. The result is the following well-known statement:

Jordan Separation Theorem. *Let Γ be a Jordan curve in $\mathbb{R}^2$, that is, a curve homeomorphic to a circle. Then $\mathbb{R}^2 \setminus \Gamma$ has exactly two connected components whose common boundary is Γ.*

The argument that Hadamard refers to in his comment runs as follows:

(1) First, by some geometric and topological means, one sees that $\mathbb{R}^2 \setminus \Gamma$ has at most two connected components and only one unbounded.

This reduces the result to its most essential part, that Γ *disconnects the plane.*

(2) Second, one computes the index with respect to Γ. This gives:

(a) Points at which the index vanishes. These belong to the unbounded component of $\mathbb{R}^2 \setminus \Gamma$, which is called the *exterior of* Γ.

(b) Points at which the index is ± 1. These are in a second (bounded) component, the *interior of* Γ.

The figure below illustrates the second (and most relevant) part of this proof.

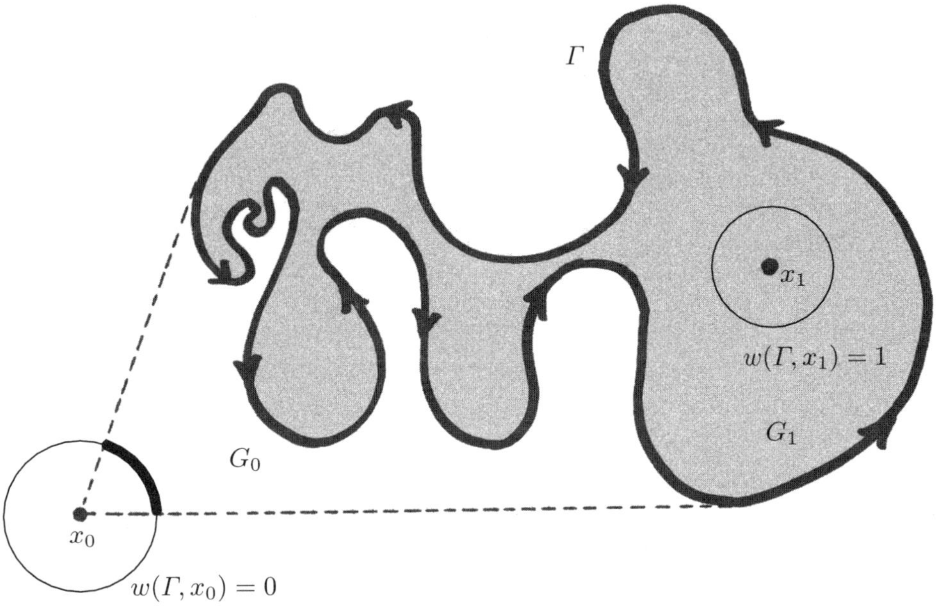

Camille Jordan (1838–1922) stated and tried to prove this theorem in [Jordan 1893]. As mentioned above, the essential content of the result is that *a Jordan curve divides the plane,* which jointly with the fact that *a simple arc does not divide the plane* is the oldest proposition of *set topology* in Euclidean spaces. The first complete proof of the Jordan Theorem was given by Oswald Veblen (1880–1960) in [Veblen 1905].

In [Schönflies 1902], Arthur Moritz Schönflies (1853–1928) gives the following additional information concerning this theorem:

Schönflies Theorem. *Let Γ be a Jordan curve in $\mathbb{R}^2$ and G_1 its interior. Then, for all $x \in G_1$ and all $a \in \Gamma$, there is a simple arc from x to a whose points other than a are all in G_1.*

We say that a is *accessible from* G_1. This finally led to the following characterization of Jordan curves:

If a compact set in $\mathbb{R}^2$ has two complementary domains, from which every point of the set is accessible, then the set is a Jordan curve.

Here we see the typical topological beast which is not a Jordan curve:

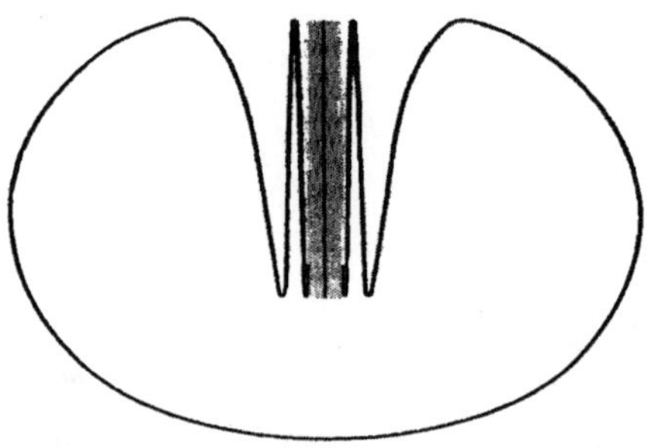

* * *

Let us now take a closer look at Hadamard's paper. In the first paragraph, Hadamard analyzes the proof above of the Jordan Separation Theorem, focusing on the part involving the order (= index) of a point with respect to the curve. He thus presents clearly what he wants to generalize. To do that, he defines, in the second paragraph, what he means by *a surface in the Euclidean space.* His notions are based on ideas introduced by Poincaré, at the beginning of the century, to treat polyhedra. We will not go into detail here, but in a sketchy way, we can think of an $(n-1)$-*surface in* $\mathbb{R}^n$ as a subspace that decomposes into pieces homeomorphic to the standard $(n-1)$-simplex, which glue in a suitable way along their faces. Hadamard concludes this paragraph by stating *Green's Theorem,* which reduces a volume (triple) integral to a surface (double) one.

In the third paragraph, Hadamard defines the order of a point with respect to a closed oriented surface in $\mathbb{R}^n$. Let us suppose we are given a hypersurface S, with coordinates $(x_1, \dots, x_n) \in \mathbb{R}^n$ with respect to parameters $(u_1, \dots, u_{n-1})$, such that the x_i's have continuous partial derivatives with respect to the u_j's (later, this condition is weakened to the mere existence of partial derivatives); suppose also that the origin is not a point of S. Then, Hadamard shows this key fact:

Hadamard Integral Theorem. *The following integral is an integer:*

$$w = -\frac{1}{\operatorname{vol}(\mathbb{S}^{n-1})} \int_S \frac{\Delta}{\gamma^n}\, du_1 \cdots du_{n-1}$$

where

$$\Delta = \begin{vmatrix} x_1 & \cdots & x_n \\ \frac{\partial x_1}{\partial u_1} & \cdots & \frac{\partial x_n}{\partial u_1} \\ \cdots & \cdots & \\ \frac{\partial x_n}{\partial u_{n-1}} & \cdots & \frac{\partial x_n}{\partial u_{n-1}} \end{vmatrix} \quad \textit{and} \quad \gamma = \sqrt{x_1^2 + \cdots + x_n^2}\,.$$

Hadamard calls this integer w, the *order of the origin with respect to* S. The order with respect to an arbitrary point $a = (a_1, \ldots, a_n) \notin S$ is obtained by translation, that is, by replacing x_i with $x_i - a_i$ everywhere in the formula above.

The proof of this theorem proceeds by induction, using Green's Theorem, the starting step being the order (= index) of a point with respect to a planar curve. It is a long proof: seven pages in the paper! Along the way, Hadamard obtains several important byproducts. For instance:

Proposition. *Let S be a hypersurface, and consider a point $a \notin S$. If there exists a half-line starting at a and not meeting S, then the order of S with respect to a is zero.*

This of course corresponds to the unbounded component of $\mathbb{R}^n \setminus S$ (if we advance the Jordan Separation Theorem in dimension n). But among other things, Hadamard explains how the order can be computed:

Proposition. *Let S be a hypersurface, which is a closed polyhedron. Let w stand for the order with respect to S of a point $a \notin S$. Then*

$$w = N_1 - N_2,$$

where N_1, N_2 are computed as follows. Pick a half-line ℓ starting at a and meeting S only at points lying in $(n-1)$-simplices (not in faces). Then $N_1 + N_2$ is the number of points in $\ell \cap S$, and such a point is counted in N_1 (resp., N_2) if the n vertices of the simplex that contains it, ordered in the corresponding orientation and preceded by the point a, determine an n-simplex oriented according to the orientation of the coordinate system.

This is much simpler than it reads. Consider the following polygonal curve in the plane:

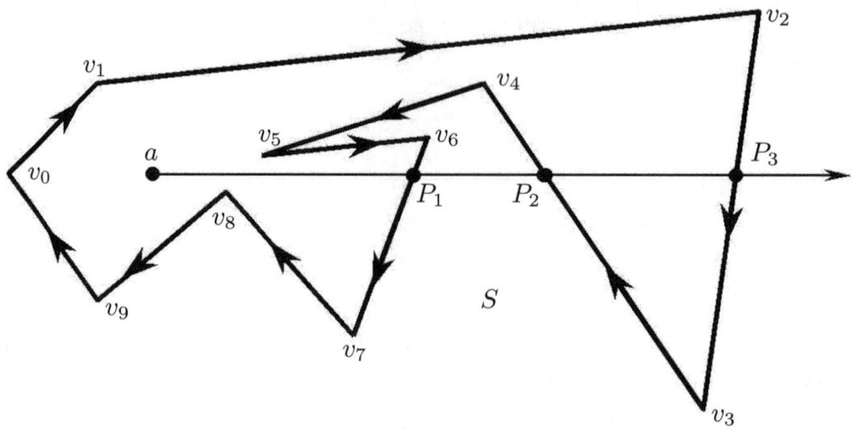

We have three points P_1, P_2, P_3 in $\ell \cap S$, and to decide which is in N_1 and which in N_2, we construct the three simplices

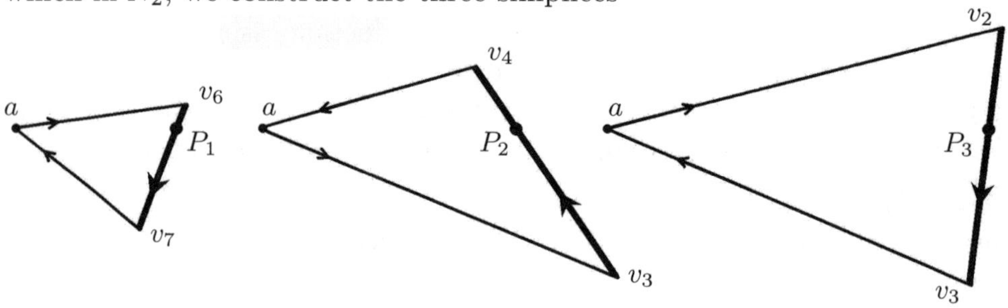

Comparing with the standard orientation of the coordinate system

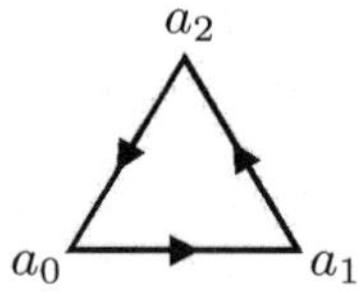

we immediately see that $P_1 \in N_2$, $P_2 \in N_1$, and $P_3 \in N_2$. Hence, $w(S, a) = 1 - 2 = -1$.

Furthermore, Hadamard analyzes the variation of w as a moves and concludes:

Proposition. (a) *The order is constant if the point varies without crossing the surface.*

(b) *The order is ± 1 if S is a convex polyhedron and the point is an interior point.*

In the fourth paragraph, Hadamard defines the Kronecker index of a function system for functions defined on a closed hypersurface:

Definition. *Let S be a hypersurface of the Euclidean space $\mathbb{R}^n$, and let $f_1, \ldots, f_n$ be a system of n continuous functions defined on S, which do not vanish simultaneously at any point of S. The* index of the system *is the order of the origin with respect to the hypersurface $(f_1, \ldots, f_n)(S)$ generated by the functions.*

Note that this definition has meaning by *the non-vanishing condition,* which guarantees that the origin does not belong to the hypersurface $(f_1, \ldots, f_n)(S)$.

The main property of this index is given in the following theorem:

Boundary Theorem. *Let S be the boundary of a domain $V \subset \mathbb{R}^n$, and let $f_1, \ldots, f_n$ be defined and continuous on the whole domain V. If $f_1, \ldots, f_n$ do not vanish simultaneously at any point of V, then the index w of the system $f_1, \ldots, f_n$ on S is zero.*

This can be trivially rewritten as a solutions existence statement:

Corollary. *Let S be the boundary of a domain $V \subset \mathbb{R}^n$, and let $f_1, \ldots, f_n$ be defined and continuous on S, such that they do not vanish simultaneously at any point of S. If the index of $f_1, \ldots, f_n$ on S is not zero, then the system $f_1 = 0, \ldots, f_n = 0$ has some solution in V.*

On the other hand, the computation of the index of a function system on a surface depends on the following important result:

Poincaré-Bohl Theorem. *Let us consider, on the same closed hypersurface $S \subset \mathbb{R}^n$, two function systems $f_1, \ldots, f_n$ and $g_1, \ldots, g_n$, both satisfying the non-vanishing condition on S. Then:*

(i) *If the systems have different indices, then there is at least one point x in S such that*

$$\frac{f_1(x)}{g_1(x)} = \cdots = \frac{f_n(x)}{g_n(x)} < 0.$$

(ii) *If the indices of the two systems are not in the ratio $(-1)^n$, then there is at least one point x in S such that*

$$\frac{f_1(x)}{g_1(x)} = \cdots = \frac{f_n(x)}{g_n(x)} > 0.$$

Using this, Hadamard obtains:

Schönflies Theorem. *Consider the closed disc*

$$\overline{D}^2 = \{(x,y) \in \mathbb{R}^2 : x^2 + y^2 \leq 1\}$$

and its boundary $\mathbb{S}^1$. *Let* $f, g : \overline{D}^2 \to \mathbb{R}$ *be two continuous functions such that the mapping* $(f,g) : \overline{D}^2 \to \mathbb{R}^2$ *is injective, and let* C *be the closed curve defined by the parametrization* $(f|_{\mathbb{S}^1}, g|_{\mathbb{S}^1})$. *Then, for every point* (X,Y) *interior to* C, *there is a point* $(x,y) \in \overline{D}^2$ *such that* $(f(x,y), g(x,y)) = (X,Y)$.

Note how the Jordan Separation Theorem for plane curves is used here. Moreover, Hadamard states the Jordan Separation Theorem *in* $\mathbb{R}^n$ without proof and then deduces the Schönflies theorem *in* $\mathbb{R}^n$:

Schönflies Theorem in $\mathbb{R}^n$. *Let* V *be a domain with boundary a surface* S *in one piece (this hypothesis is not essential), and let* $f_1, \ldots, f_n$ *be continuous functions from* V *into* $\mathbb{R}$ *such that the mapping* $f = (f_1, \ldots, f_n) : V \to \mathbb{R}^n$ *is injective. Consider the surface* S' *parametrized by* $f|_S = (f_1|_S, \ldots, f_n|_S)$. *Then for every point* $(X_1, \ldots, X_n)$ *interior to* S', *there is a point* $(x_1, \ldots, x_n)$ *in* V *such that* $f(x_1, \ldots, x_n) = (X_1, \ldots, X_n)$.

As a consequence, Hadamard gets:

Proposition. *If* $f : V \to V'$ *is a homeomorphism between two domains, the interior points of* V *map to interior points of* V', *and the boundary points of* V *map to boundary points of* V'.

This is a particular case of the *Invariance of Domain Theorem.* Concerning the Jordan Separation Theorem in $\mathbb{R}^n$, it is very likely, as we will see later, that Hadamard had known it from some private communication with Brouwer.

Another very important result that Hadamard proves follows:

Brouwer Fixed Point Theorem. *Every continuous mapping from the closed ball* $\overline{D}^n \subset \mathbb{R}^n$ *into itself has at least one fixed point.*

Proof. The argument runs as follows. Let $f : \overline{D}^n \to \overline{D}^n$ be a continuous mapping. We can suppose $f(x) \neq x$ for all $x \in \partial\overline{D}^n = \mathbb{S}^{n-1}$ (otherwise we are done). Then, by the Poincaré-Bohl Theorem $\mathrm{Id}_{\mathbb{S}^{n-1}} - f|_{\mathbb{S}^{n-1}}$ and $\mathrm{Id}_{\mathbb{S}^{n-1}}$ have the same index: were it

$$\frac{x_1 - f_1(x)}{x_1} = \cdots = \frac{x_n - f_n(x)}{x_n} = \mu < 0,$$

then $f(x) = (1-\mu)x \notin \overline{D}^n$. Consequently, the index of $\mathrm{Id}_{\mathbb{S}^{n-1}} - f|_{\mathbb{S}^{n-1}}$ is 1, and by the basic properties of indices, there is an $x_0 \in \overline{D}^n$ such that $x_0 - f(x_0) = 0$, that is, $f(x_0) = x_0$. □

Another remarkable statement (without proof) by Hadamard is the famous *Poincaré-Hopf Index Theorem,* relating the Poincaré index (characteristic) of a domain to the zeros of a vector field. In particular, Hadamard proves:

Vanishing of tangent fields. *Let V be a domain that admits a tangent space at every point, depending continuously on the point. If the index σ of V is not zero, then it is not possible to attach continuously a tangent line to V at every point (that is, the tangent line must be undefined somewhere).*

This is the case, for instance, for the sphere in $\mathbb{R}^3$. The fact that Hadamard formulates this theorem seems to result from a misunderstanding with Brouwer, who had proved it for spheres.

To conclude, Hadamard states (again without proof) the following theorem due to Brouwer:

Proposition. *Every continuous mapping from a sphere into itself that preserves orientations has at least one fixed point.*

The result is false if the mapping reverses orientation (think of the antipodal isometry)

* * *

One should mention here a footnote (p. 476 of [Hadamard 1910]) where Hadamard explains that the method for the proof of the previous result was communicated by Brouwer. Thus, it becomes more and more evident that there was a quite fluent exchange of ideas between Hadamard and Brouwer. What follows is a letter from Brouwer to Hadamard that spares any further comment on the matter. We translate from [Brouwer 1976]:

Paris, January 4, 1910
6 Rue de l'Abbé de l'Epée

Dear Sir,

I can now communicate some extensions of the fixed point theorem for bijective continuous transformations of the sphere. They are reduced to arbitrary continuous transformations of the sphere. To such a transformation one can attach a finite number n as its *degree.* From a degree n transformation one can construct by continuous variations any other degree n transformation, but no more than those. In particular, one can always construct in this way a degree n rational transformation of the complex sphere.

To determine this degree, we use homogeneous coordinates (in the double sense), write x, y, z for the initial sphere, and ξ, η, ζ for the image, split the sphere into a finite number of regions, and consider firstly the transformations defined by relations

$$\xi : \eta : \zeta = f_1(x, y, z) : f_2(x, y, z) : f_3(x, y, z),$$

where f_1, f_2, f_3 are polynomials, which on the other hand can vary for the different regions of the sphere. We call this transformation *a polynomial transformation.* We choose an orientation on the sphere: then every point P in the image, *in general position,* will occur r_P times with the positive orientation, and s_P times with the negative orientation. In this situation, one can prove that $r_P - s_P$ is a constant: it is the degree of the polynomial transformation.

Let us come back to an arbitrary continuous transformation. It can be approximated by a sequence of polynomial transformations: one proves then that the latter have all the same degree: it is again the degree of the limit transformation.

The degree is always a finite integer, either positive or negative. The degree of a bijective transformation is $+1$, if the orientation is preserved, and -1, if the orientation is reversed.

Now the generalized fixed point theorem becomes what follows: Every continuous transformation of the sphere, whose degree is not -1, has at least one fixed point.

Moreover, I have extended this theorem to m-dimensional spheres. It reads then in the following way: Every continuous transformation of the m-dimensional sphere has at least one fixed point, except a) when m is odd and the degree n is $+1$, b) when m is even and the degree n is -1.

In particular, if the transformation is *bijective* [in margin, strenger formuleeren], there is at least one fixed point a) if m is odd and the orientation is reversed, b) if m is even and the orientation is preserved.

For the volume of an m-dimensional sphere [sic] in the space of dimension $m+1$ (if we include there the sphere itself) I was able just recently to establish a still more general theorem, namely: Every (possibly not bijective) continuous transformation of the volume of the m-dimensional sphere has at least one fixed point.

Concerning general continuous tangent distributions on the sphere, two articles of mine will appear soon, where I study certain questions that refer to the Dirichlet principle and to the decomposition of a field into a "quellenfrei" part and a "wirbelfrei" part. To that end, I determine first the most general form that tangent curves (characteristic curves after Poincaré) can have. As the main result in the first article, one should take the property that a characteristic curve that does not tend to a singular point must be a spiral, whose two limit cycles are also characteristic. The property that there is at least one singular point is necessary; it is not in the end an accessory corollary, on which I have insisted because it was the first easy to formulate result and because there seemed to be some close relation between this theorem and that of the fixed point in the sphere, a relation that has been clarified only through your correspondence. In the second article I have included your beautiful, direct, and more complete proof of the existence of at least one singular point.

My address will be in Paris till January 15. Maybe there will be the occasion for us to meet?

Yours sincerely,

L.E.J. Brouwer

* * *

Now we turn to Brouwer's fundamental work. We can agree that the greatest contribution due to Brouwer is the definition of the degree of a continuous mapping of n-manifolds solely by geometric and topological means. The manifolds he considers come from the ideas concerning polyhedra that, as was already mentioned, Poincaré introduced at the end of the nineteenth century and the beginning of the twentieth century. These ideas were the germ of what we know today as *combinatorial topology.* It is remarkable that Brouwer never mentions Poincaré's writtings on the topic.

The notion of a manifold used by Brouwer coincides with what nowadays we call a *pseudomanifold.* A manifold of dimension n, or n-manifold, is built up from simplices of dimension d, or d-simplices, of smaller dimensions d. Skipping strict formalisms, we can put it as follows:

Definition. *A subspace M of a Euclidean space is an n-*manifold *when it is homeomorphic to the polyhedron $|K|$ associated to a* simplicial complex K of dimension n. *That is, $|K|$ is a subspace of a Euclidean space consisting of a union of disjoint simplices of dimension $\leq n$ (points, open segments, open triangles, open tetrahedrons, etc.) such that:*

(i) *Every simplex of $|K|$ is a face of some n-simplex of $|K|$.*

(ii) *Every $(n-1)$-simplex of $|K|$ is a face of at most two n-simplices of $|K|$.*

(iii) *For every pair of n-simplices S and S' of $|K|$, there is a finite sequence $S = S_1, S_2, \ldots, S_m = S'$ of n-simplices of $|K|$ such that S_i and S_{i+1} have a common $(n-1)$-face for $i = 1, \ldots, m-1$.*

The boundary of $|K|$, *which will be denoted by $|\dot{K}|$, is the union of all $(n-1)$-simplices that are faces of exactly one n-simplex. Hence, if $|\dot{K}| = \emptyset$, every $(n-1)$-simplex is a face of exactly two n-simplices of $|K|$.*

The following examples are very simple illustrations of these notions:

(a) *A 2-manifold with boundary.*

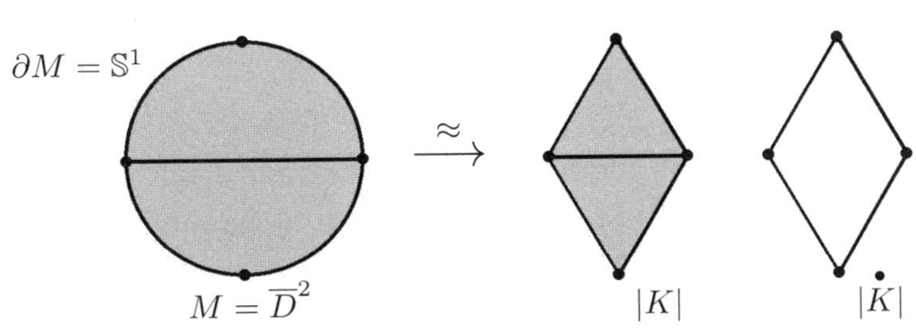

(b) *A* 1-*manifold without boundary.*

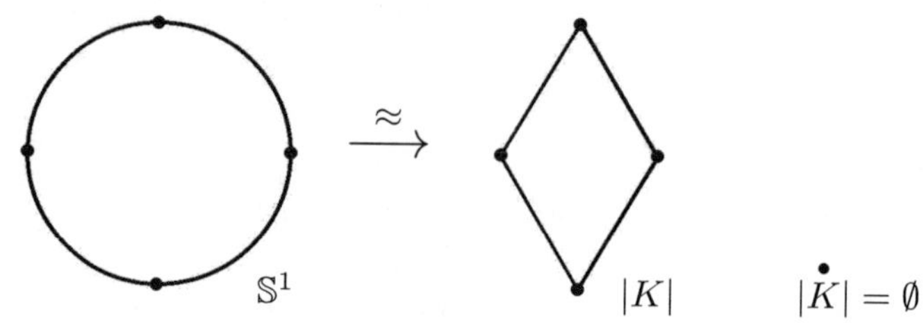

(c) *A space which is not a polyhedron.*

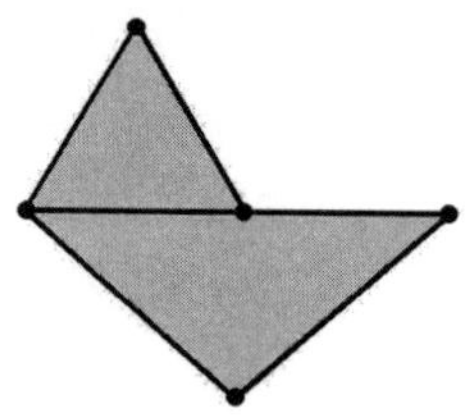

Once the notion of a manifold is fixed in this form, Brouwer constructs the topological degree of a continuous mapping of manifolds in [Brouwer 1912a], article dated in Amsterdam, July 1910. In a footnote in the first page we read:

> *While this paper was in print, the note by J. Hadamard,* Sur quelques applications de l'indice de Kronecker, *has appeared in the second volume of J. Tannery's* Introduction à la théorie des fonctions d'une variable. *In that note some aspects of the theory we present here are anticipatedly developed.*

This once again confirms the mutual influence between the two mathematicians.

Let now M and N be two n-manifolds, which we assume to be connected, compact, boundaryless, and oriented. For the definition of degree, Brouwer proceeds in two steps:

Step I: The simplicial case. Consider a *simplicial continuous mapping* $g : M \to N$. This means that g is continuous, and there are triangulations $\varphi : |K| \equiv M$ and $\psi : |L| \equiv N$ of M and N such that the *localization*

$$\bar{g} = \psi \circ g \circ \varphi^{-1} : |K| \to |L|$$

restricts to an affine map on each simplex of $|K|$. For such a g, Brouwer shows that there is a dense connected open set Ω in $|L|$ such that if $y \in \Omega$ is not in the union S of the images by $\bar{g}$ of the $(n-1)$-simplices of $|K|$, then $g^{-1}(y)$ is a finite set $\{x_1, \ldots, x_r\}$ each of whose points belongs to an (open) n-simplex of $|K|$. These simplices σ are disjoint, and the restriction of $\bar{g}$ to every σ is a homeomorphism onto $\bar{g}(\sigma)$. Let p (resp., q) be the number of simplices on which the restriction $\bar{g}|_\sigma$ preserves (resp., reverses) the orientation, and consider the difference $p - q$. Then Brouwer shows that the number $p - q$ is the same for all $y \in \Omega$ off S. To that end, he joins any two such y's by a polygonal contained in Ω and moves y along: p and q increase or decrease simultaneously by the same number when the polygonal crosses S; hence $p-q$ does not change. This happens because each $(n-1)$-simplex is a face of exactly two n-simplices of $|K|$. Consequently, the integer $d = p - q$ is well defined and is called the *degree* $d(g)$ *of* g.

Step II: The general case. To deal with arbitrary non-simplicial mappings, Brouwer introduces two constructions that have become fundamental in combinatorial topology: *barycentric subdivision* of a simplicial complex (subdivision of each simplex by taking its barycenter as a new vertex) and *simplicial approximation* of continuous mappings. By means of these refined geometric techniques, he defines the degree $d(f)$ of an arbitrary continuous mapping $f : M \to N$ through a *good* simplicial approximation of f. In fact, he sees that any two close enough simplicial approximations of f have the same degree as defined in Step I. In fact, Brouwer shows that close approximations are homotopic by a piecewise linear homotopy, which implies the equality of degrees.

Let us make some pictures for this construction. Suppose we have the following simplicial mapping:

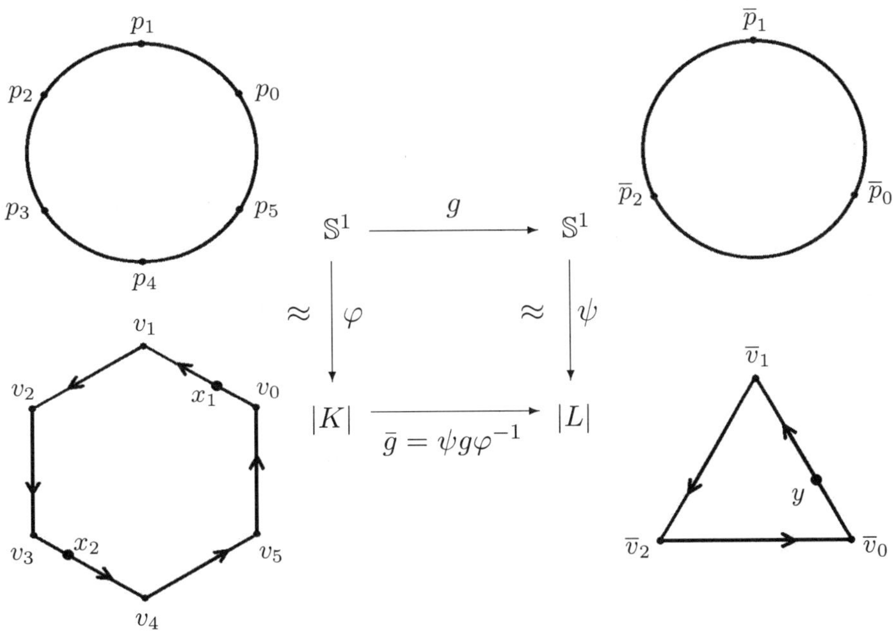

Here the localization $\bar{g}$ is affine on every 1-simplex of $|K|$; hence it is determined by the images of the vertices: in this case we suppose

$$\bar{g}(v_0) = \overline{v}_0, \quad \bar{g}(v_1) = \overline{v}_1, \quad \bar{g}(v_2) = \overline{v}_2,$$

$$\bar{g}(v_3) = \overline{v}_0, \quad \bar{g}(v_4) = \overline{v}_1, \quad \bar{g}(v_5) = \overline{v}_2.$$

Then take a point

$$y \in |L| \setminus \bar{g}(\{v_0, v_1, v_2, v_3, v_4, v_5\}),$$

and let $\bar{g}^{-1}(y) = \{x_1, x_2\}$. In this situation, the restrictions

$$\bar{g}|_{(v_0,v_1)} : (v_0, v_1) \to (\overline{v}_0, \overline{v}_1) \;\text{ and }\; \bar{g}|_{(v_3,v_4)} : (v_3, v_4) \to (\overline{v}_0, \overline{v}_1)$$

are homeomorphisms that preserve the orientation; hence the degree of f is

$$d(f) = d(\bar{g}) = p - q = 2 - 0 = 2.$$

Next, we look at the general case: a non-affine map like $f(z) = z^2$. The picture is the following:

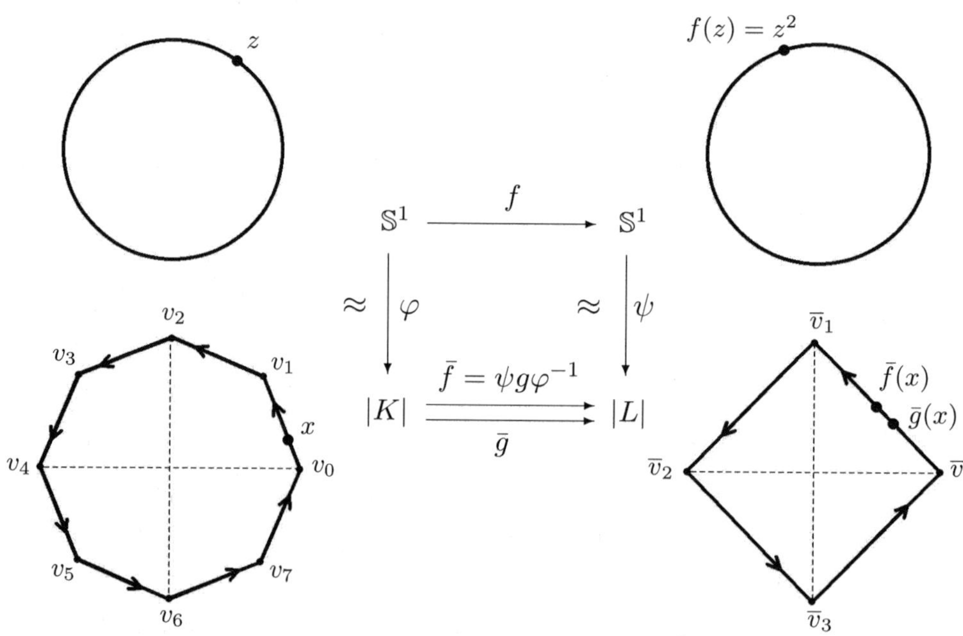

Here $\bar{g}$ is the simplicial approximation of $\bar{f}$, and we have

$$\bar{g}(v_0) = \overline{v}_0, \quad \bar{g}(v_1) = \overline{v}_1, \quad \bar{g}(v_2) = \overline{v}_2, \quad \bar{g}(v_3) = \overline{v}_3,$$

$$\bar{g}(v_4) = \overline{v}_0, \quad \bar{g}(v_5) = \overline{v}_1, \quad \bar{g}(v_6) = \overline{v}_2, \quad \bar{g}(v_7) = \overline{v}_3.$$

Computing as above, $d(\bar{g}) = 2$, and we conclude $d(f) = 2$.

For his notion of degree, Brouwer obtains the following essential properties:

Proposition. *The following statements hold:*

(a) $d(f) = 0$ *if* f *is not surjective.*

(b) $d(f) = \pm 1$ *if* f *is a homeomorphism.*

(c) $d(g \circ f) = d(g) \cdot d(f)$ *for the composition* $g \circ f$ *of two continuous mappings* $f : M \to N$, $g : N \to P$.

(d) $d(f_1) = d(f_2)$ *if* $f_1, f_2 : M \to N$ *are homotopic.*

Furthermore, Brouwer proves the following two important theorems:

Theorem. *A continuous vector field tangent to an* n*-sphere of even dimension always has singular points (= zeros).*

Theorem. *A continuous mapping without fixed points from an* n*-sphere into itself has degree* -1 *if* n *is even and degree* $+1$ *if* n *is odd.*

Proof. The proof of the second result follows. First, the homotopy

$$F(t, x) = \frac{tf(x) - (1-t)x}{\|tf(x) - (1-t)x\|}$$

is well defined because f has no fixed point. Thus, the antipodal mapping F_0 is homotopic to $f = F_1$, and thus F_0 and f have the same degree, which is ± 1 according to the dimension. □

From this, Brouwer deduces:

Corollary. (a) *Every non-surjective continuous mapping of an* n*-sphere must have some fixed point.*

(b) *Every continuous mapping of a sphere of even dimension that is homotopic to the identity must have some fixed point.*

(c) *Every continuous mapping of a sphere of odd dimension that is homotopic to the antipodal mapping must have some fixed point.*

Proof. For (a), a non-surjective mapping has degree $0 \neq \pm 1$, and the preceding theorem applies. For (b), the degree of the given mapping is that of the identity; hence it is $1 \neq -1$. As the dimension is even, again the preceding theorem gives the conclusion. Assertion (c) is proven in the same way, since the degree of the antipodal map in odd dimension is -1. □

Another very important result due to Brouwer (and as mentioned before, revisited by Hadamard) is the following:

Brouwer Fixed Point Theorem. *A continuous mapping from a closed n-ball $\overline{D}^n$ into itself must have some fixed point.*

Proof. The proof runs as follows. Given a continuous mapping $f : \overline{D}^n \to \overline{D}^n$, Brouwer identifies $\overline{D}^n$ with the upper hemisphere $\mathbb{S}^n_+$ of the n-sphere $\mathbb{S}^n$ and defines

$$g : \mathbb{S}^n \to \mathbb{S}^n : x \mapsto g(x) = \begin{cases} f(x) & \text{if } x \in \mathbb{S}^n_+, \\ f(x_1, \ldots, x_n, -x_{n+1}) & \text{if } x \notin \mathbb{S}^n_+. \end{cases}$$

This mapping is continuous, but not surjective, hence has some fixed point that must be a fixed point of f. □

With this powerful tool in hand, Brouwer proved in later articles many other important results: the Invariance of Domain Theorem, the Jordan Separation Theorem in arbitrary dimension, and various dimension properties. Moreover, in the paper [Brouwer 1912b], he defines the *link coefficient* $\ell(K_1, K_2)$ of two disjoint oriented compact manifolds K_1, K_2 in $\mathbb{R}^n$, with complementary dimensions h and $n - h - 1$. This link is the degree of the mapping

$$f : K_1 \times K_2 \to \mathbb{S}^{n-1} : (x, y) \mapsto \frac{y - x}{\|y - x\|},$$

that is,

$$\ell(K_1, K_2) = d(f).$$

We must mention here a contribution made 80 years earlier by Gauss, who in [Gauss 1833] actually computed the link coefficient of two curves.

The Italian mathematician Carlo Miranda proved in [Miranda 1940] that the Brouwer Fixed Point Theorem is equivalent to Poincaré's Intermediate Value Theorem of 1883 (I.1, p. 12). For this reason, some authors call this theorem the *Poincaré-Miranda Theorem.*

3. Accomplishment

This is the place to mention Heinz Hopf (1894–1971). Hopf's interest in degree theory came from the lectures given by Erhard Schmidt (1876–1959) at Breslau in 1917 and at Berlín in 1920 on the invariance of dimension and the proof of the Jordan Theorem in arbitrary dimension using

Brouwer's degree. Hopf himself says in [Hopf 1966] that those results appealed to him deeply:

I was fascinated; this fascination—of the power of the method of the mapping degree—has never left me since, but has influenced major parts of my work.

Hopf's essential contribution to degree theory springs from his Ph.D. thesis, which he defended at Berlin University in 1926. Among other as important geometric questions, he completed Brouwer's results concerning continuous mappings of closed oriented manifolds of the same dimension, with special attention to the case when the target manifold is a sphere. We explain it in the following paragraphs.

Let G and G' be two domains in the Euclidean n-space, and let f, g be two continuous mappings from G into G'. Let $a \in G$ be an *isolated coincidence point of* f *and* g (that is, $f(a) = g(a)$ and this does not hold true for any other point in a small enough neighborhood $U \subset G$ of a). The *coincidence index of* f *and* g *at* a is the degree of the continuous mapping

$$\varphi : S_r(a) \to \mathbb{S}^{n-1} : x \mapsto \frac{f(x) - g(x)}{\|f(x) - g(x)\|},$$

where $S_r(a)$ is the spherical surface of center a and radius $r > 0$ contained in U. By its local nature, this definition extends to mappings of arbitrary manifolds. Modifying Brouwer's proof of the Fixed Point Theorem for spherical surfaces, Hopf proved ([Hopf 1925]):

Theorem. *Let* f *and* g *be continuous mappings from a compact oriented manifold* M *of dimension* n *into the sphere* $\mathbb{S}^n$*, and denote by* d, d' *their respective degrees. Then, if* f *and* g *have finitely many concidence points, the sum of the coincidence indices of* f *and* g *at those points is* $(-1)^n d + d'$.

Brouwer had proved in [Brouwer 1912a] that continuous homotopic mappings have the same degree and in [Brouwer 1912c] that continuous mappings of $\mathbb{S}^2$ into itself with the same degree are homotopic. Hopf, using the theorem above, proved in [Hopf 1926a] the most general form of this fact:

Hopf Theorem. *Two mappings from a compact oriented manifold* M *of dimension* n *into the sphere* $\mathbb{S}^n$ *that have the same degree are homotopic.*

Later, in [Hopf 1933], Hopf extended this result to the case when the source manifold M is a polyhedron. A further generalization to Banach spaces was to be published by Erich Rothe in [Rothe 1936].

Now, the dual problem that arises in a natural way is the homotopic classification of continuous mappings from a sphere into a topological space. This led to the discovery of *homotopy groups* of higher order by EDUARD ČECH (1893–1960), who defined them in a very short paper presented at the Zurich ICM, in 1932 [Čech 1932]. However, these groups did not attract much attention, as they are always commutative, in sharp contrast with the behavior of the fundamental group. But in 1935 WITOLD HUREWICZ (1904–1956) rediscovered and studied systematically these homotopy groups ([Hurewicz 1935a], [Hurewicz 1935b], [Hurewicz 1936a], and [Hurewicz 1936b]). The two basic ingredients from which the theory develops are:

(1) the *fundamental group* (now *first homotopy group*), which Poincaré introduced in [Poincaré 1895b], and

(2) the *Hopf invariant,* the means used by Hopf to distinguish infinitely many homotopy classes of mappings $\mathbb{S}^3 \to \mathbb{S}^2$. This he achieves by combinatorial topology techniques in [Hopf 1931]. Some years later, in [Hopf 1935], Hopf extends his result to mappings $\mathbb{S}^{2n-1} \to \mathbb{S}^n$.

On the other hand, in [Poincaré 1885a], Poincaré had proved that if a continuous vector field on a compact orientable surface of genus g has finitely many critical points, the sum of their indices (Poincaré indices) is the topological invariant $2 - 2g$, which is the Euler characteristic of the surface. This was generalized by Brouwer in [Brouwer 1912a] to spherical surfaces of arbitrary dimension. As we have said before, Hadamard announced without proof the same result for compact orientable manifolds of arbitrary dimension [Hadamard 1910]. But it is Hopf who, finally, proved in [Hopf 1926b] the following result:

Poincaré-Hopf Theorem. *Let M be a compact orientable manifold of dimension n, and let X be a continuous vector field with finitely many critical points. Then the sum of the indices of those critical points is the Euler characteristic $\chi(M)$ of M.*

But there is more to tell. In his path to these important results, Hopf gained deeper insight into the topological meaning of the various invariants involved. In fact, he analyzed Kronecker's work on the characteristic of a system of several variables and defined the so-called *integral curvature* of a closed oriented hypersurface M in the Euclidean $(n+1)$-space. That invariant is the degree of the Gauss mapping ν of the hypersurface ($\nu(x)$ is the unitary vector perpendicular to the hypersurface at x). If M is described by an equation $F = 0$, the integral curvature of M is the Kronecker characteristic of the system

$$\left(F, \frac{\partial F}{\partial x_1}, \ldots, \frac{\partial F}{\partial x_{n+1}}\right).$$

In the articles mentioned, Hopf proved:

Theorem. *The integral curvature of a Jordan hypersurface in $\mathbb{R}^{n+1}$ (one that bounds a domain) coincides with the Euler characteristic of the hypersurface.*

To refine this statement, Hopf introduced the *models* of a closed oriented n-manifold. These are the hypersurfaces in $\mathbb{R}^{n+1}$ that are homeomorphic to M. Then, he proved:

Gauss-Bonnet Theorem in arbitrary dimension. *Let M be a closed oriented manifold of dimension n.*

(1) *If n is odd, the integral curvature of the models of M is not a topological invariant, not even for Jordan models.*

(2) *If n is even, the integral curvature of the models of M is a topological invariant, namely, half the Euler characteristic of M.*

If N is a model of M and $\kappa_1, \ldots, \kappa_n$ are its principal curvatures, then

$$\tfrac{1}{2}\chi(N) = \frac{1}{\mathrm{vol}(\mathbb{S}^n)} \int_N \kappa_1 \cdots \kappa_n dw.$$

This was obtained by Walter Franz Anton Dyck (1856-1934) in [Dyck 1888] and [Dyck 1890].

In particular, we see that all models of even-dimensional manifolds have even Euler characteristic, and thus we find manifolds which cannot be embedded in codimension 1:

Corollary. *There are closed oriented n-manifolds which have no model in $\mathbb{R}^{n+1}$, even allowing self-intersections.*

An example is the complex projective plane, which has Euler characteristic 3 (as it can be computed with a vector field using the Poincaré-Hopf Theorem).

* * *

Brouwer-Kronecker degree theory and its applications were systematically developed by Hopf in the topology text [Aleksandrov-Hopf 1935] written jointly with PAVEL SERGEEVICH ALEKSANDROV (1896–1982). The subject is covered in the final three chapters, XII, XIII, and XIV, of that book. The titles of the sections in those chapters give a detailed account of the contents of the theory. We translate them as follows:

CHAPTER XII. The Brouwer degree. The Kronecker characteristic

§1. The order of a point with respect to a cycle. Recall results in link theory. The Poincaré-Bohl theorem. The Rouché theorem. Homology of $\mathbb{R}^n \setminus \{p\}$. The global degree. Relation between order and degree. The winding number in the plane. The Kronecker integral.

§2. The Kronecker characteristic. Local degree of mapping in $\mathbb{R}^n$. The existence theorem for a point. Application: The fundamental theorem of algebra. The characteristic of a function system. The index of a point. The algebraic number at a point. The local degree in $\mathbb{R}^n$. Topological applications. The invariance theorem. The functional determinant.

§3. Special theorems and applications. Relations among vector fields on $\mathbb{R}^n$ and mappings. Vector fields on the sphere. The Brouwer fixed point theorem on an n-dimensional element. Vector fields on a spherical surface of even dimension: A fixed point theorem. Symmetries on the spherical surface. Another fixed point theorem. Borsuk theorem on antipodal mappings. Analytic corollary. Mappings from $\mathbb{S}^n$ to $\mathbb{R}^n$. The covering theorem for the spherical surface $\mathbb{S}^n$.

§4. The degree of mappings between polyhedra. Definition of local degree. Continuous deformations of mappings (homotopies). Cycle mappings. Cycle mappings on acyclic irreducible closed complexes. Coincidence of the local and global degrees. Determination of the global degree by homological invariants. Essential properties of mappings.

Appendix. Comment on the Brouwer link number as a characteristic. The Gauss integral. The number of cuts as a degree. The link number as an order. The Gauss integral.

CHAPTER XIII. Homotopy and mapping extension theorems

§1. More on the Kronecker existence theorem. An extension problem for mappings into $\mathbb{R}^n$. Reduction to mappings into $\mathbb{S}^n$. Elementary lemma on extension and homotopy of mappings. Solution to the extension problem for a simplex. Reduction of Theorem II to a lemma. Algebraic part of the proof of the lemma. Geometric part of the proof of the lemma. Criterion for the local essentiality of mappings on $\mathbb{R}^n$.

§2. Mappings from n-dimensional polyhedra into the spherical n-dimensional surface. Equivalence between homology and homotopy. Listing of the types of mappings in the simplest case. Essential mappings. Resolution theorem.

§3. Mappings from n-dimensional polyhedra into the circle. Mappings from the n-dimensional sphere into the circle. The extension theorem. Types of mappings. Essentiality.

§4. Characterization of the identity and polyhedra borders by deformation properties. A deformation theorem from a polyhedron into itself. The essentiality of a polyhedron into itself. The particular case of polygons. Characterization of the identity. Characterization of the border. Stability of polyhedra. Examples.

Appendix. Mappings that are homologous but not homotopic.

CHAPTER XIV. Fixed points

§1. A fixed point existence theorem. Fixed simplex. A generalization of the Poincaré-Euler formula. The Lefschetz number of a continuous mapping of a polyhedron into itself. Fixed point existence theorem. Examples. Remarks.

§2. The index of fixed points. The index of a zero of a vector field and of a singular point of an oriented field. The index of a fixed point. Index properties. Normal fixed points. Fixed points of affine mappings. Topological invariance of the index of fixed points. The invariance of the index at a singular point of an oriented field.

§3. Algebraic number of fixed points of a continuous mapping from a polyhedron into itself. The algebraic number of regular fixed points. General fixed points theorem. Regular fixed points of simplicial mappings. Reduction to the approximation theorem. Proof of the approximation theorem. Remarks on the concept of number of fixed points.

§4. Oriented fields on closed manifolds. Preliminaries on smooth manifolds. Oriented fields and their singularities in manifolds.

* * *

After this, the next step in degree theory was taken by JEAN LERAY (1906–1998) and JULIUSZ PAWEL SCHAUDER (1899–1943), who developed degree theory for *completely continuous mappings in Banach spaces.* They published their results in 1934 in the paper [Leray-Schauder 1934]. Previously, a summary of the paper had been presented at the Paris Academy of Sciences [Leray-Schauder 1933] on July 10, 1933. This summary starts as follows:

Let $\mathcal{E}$ be an abstract space, normed, linear, and complete (in the sense of Mr. Banach [Banach 1922]*); let W be an open set of $\mathcal{E}$ (whose boundary we denote by W', and its closure by $\overline{W} = W + W'$); finally, let*

$$y = x - \mathcal{F}(x) = \Phi(x)$$

be a functional transformation defined on $\overline{W}$. We do not suppose $\mathcal{F}(x)$ linear, but only completely continuous *(or* compact, vollstetig*) (that is, $\mathcal{F}(x)$ is continuous and transforms every bounded set into a relatively compact one); furthermore, $\mathcal{F}(x)$ takes values only in $\mathcal{E}$. We have been able to define, after the famous work by Mr. Brouwer* [Brouwer 1912a], *the* topological degree $d[\Phi, W, b]$ *of the transformation Φ at a point b off $\Phi(W')$, in such a way that the well-known properties of this degree are still valid. We describe briefly our definition.*

The Leray-Schauder construction is this. Let us suppose the point b is the origin $0 \in \mathcal{E}$, and set

$$h = \min\{d(0, \Phi(x)) : x \in W'\}.$$

Consider a new functional transformation $\mathcal{F}_h(x)$ at a distance $< h$ from $\mathcal{F}(x)$ and such that all its values belong to a common linear subset M of $\mathcal{E}$ of finite dimension; denote by W_M the intersection of W with that subset M. Let $\Phi_h(x) = x - \mathcal{F}_h(x)$. We set

$$d(\Phi, W, 0) = d(\Phi_h, W_M, 0),$$

where the right-hand side degree comes from Brouwer's theory.

Then, the authors prove that the value $d(\Phi, W, 0)$ does not depend on the choice of Φ_h and W_M, after which they develop the main properties of this degree. They introduce the following:

Definition. *The degree at 0 of the transformation*

$$y = x - \mathcal{F}(x) = \Phi(x) \tag{1}$$

is called the total index *of the solutions in W of the equation*

$$x - \mathcal{F}(x) = 0. \tag{2}$$

With this terminology, they prove:

Theorem. *Suppose that equation* (2) *depends continuously on a parameter k varying in a segment K of the real line. Let Ω be an open bounded set of the space $\mathcal{E} \times K$ such that* (i) *$\mathcal{F}$ is defined on $\overline{\Omega}$ and* (ii) *the boundary Ω' of Ω does not contain solutions to* (1)*. Then, the total index of the solutions in the interior of Ω is the same for all values of k.*

Thus, we see that the existence of at least a solution to (2) is guaranteed for every value k as soon as one knows a value k_0 for which the total index is not zero. The authors conclude by stressing that their theorem has a wide range of applications. In particular, it applies to limit (boundary) problems concerning *elliptic second-order partial differential equations,* and they can use it to generalize the well-known existence theorems due to Sergei Natanovich Bernstein (1880–1968).

These important applications to differential equations of the Leray-Schauder methods lead to the consideration of more general operators, in particular linear and non-linear Fredholm or Noether operators. The first investigations in this line are due to Renato Caccioppoli, who in [Caccioppoli 1936] obtained a mod 2 degree theory for continuously differentiable Fredholm mappings of index 0.

4. Renaissance and reformation

After the publication of Brouwer's fundamental papers, much effort was devoted to establishing the main properties of the degree of a continuous mapping, and in particular the Fixed Point Theorem, by analytic methods that did not involve the concepts of combinatorial topology. It is clear that the first satisfactory approach to this question was Hadamard's use of the Kronecker integral. The second attempt is based on the paper [Sard 1942] by ARTHUR SARD (1909–1980). There, we find the famous:

Sard Theorem. *Let G be an open set in $\mathbb{R}^m$, and let $f : \overline{G} \to \mathbb{R}^n$ be a $\mathcal{C}^p$ mapping ($p \geq 1$), with $p \geq m - n + 1$. Let C_f stand for the set of* critical values *of f, that is, $y = f(x) \in C_f$ when the Jacobian matrix of f at x has rank $<$ $\min(m, n)$ (which is the maximum possible rank). Then, C_f has Lebesgue measure zero.*

In particular, C_f has empty interior, hence dense complement, in $\mathbb{R}^n$.

The final conclusion comes from the fact that *a subset of $\mathbb{R}^n$ whose Lebesgue measure is zero does not contain any open set,* a result proved previously by ARTHUR BARTON BROWN in [Brown 1935]. Due to this, the theorem above is often called the *Sard-Brown Theorem.*

Using this theorem and the classical Weierstrass Theorem on polynomial uniform approximation of continuous functions on compact sets in Euclidean spaces, in 1951 MITIO NAGUMO constructed Brouwer's degree theory by elementary analytic means. He published the results in [Nagumo 1951a]; the journal had received the paper on March 6, 1950. In the introduction, the author says:

This paper establishes a theory of degree of mapping for open sets in a Euclidean space of finite dimension, based on the theory of infinitesimal analysis, which is free of the notion of simplicial mapping. Although the results are not new, I hope in this way to make it possible to incorporate the theory of degree of mapping into a course in infinitesimal analysis.

The basic ideas of Nagumo's construction are these:

Step I: Let G be an open bounded set in $\mathbb{R}^n$, $f : \overline{G} \to \mathbb{R}^n$ a $\mathcal{C}^1$ mapping, and $a \in \mathbb{R}^n \setminus \big(f(\overline{G} \setminus G) \cup C_f\big)$. Then, by the Inverse Function Theorem, the derivative $df(x)$ is a linear isomorphism for every $x \in f^{-1}(a)$, and consequently $f^{-1}(a)$ is a finite set, say $f^{-1}(a) = \{x_1, \ldots, x_p\}$. Thus, Nagumo can define

$$d(f,G,a) = \begin{cases} \sum_{i=1}^{p} \operatorname{sign} \det df(x_i) & \text{if } f^{-1}(a) = \{x_1, \ldots, x_p\}, \\ 0 & \text{if } f^{-1}(a) = \emptyset. \end{cases}$$

Step II: Let G be an open bounded set in $\mathbb{R}^n$, $f : \overline{G} \to \mathbb{R}^n$ a continuous mapping, and $a \in \mathbb{R}^n \setminus f(\overline{G} \setminus G)$. Then $d(g, G, a)$ has the same value for all $\mathcal{C}^2$ mappings $g : \overline{G} \to \mathbb{R}^n$ such that

(1) $|g(x) - f(x)| < \operatorname{dist}\big(a, f(\overline{G} \setminus G)\big)$ for all $x \in \overline{G}$, and

(2) $a \in \mathbb{R}^n \setminus C_g$.

Notice here that condition (1) implies $a \notin g(\overline{G} \setminus G)$; hence Step I applies to g, and Nagumo can also define the *degree of f at a* by setting

$$d(f, G, a) = d(g, G, a)$$

for any g as above.

We remark that Step II is where Nagumo used the Sard-Brown and Weierstrass Theorems. On the other hand, it is worth stressing that the degree $d(f, G, a)$ is the Hadamard index of $f|_{\partial G}$, in case $\partial G = \overline{G} \setminus G$ is a hypersurface.

This same method leads to the construction of degree for continuous mappings of boundaryless compact oriented manifolds, once one extends the Sard and Weierstrass Theorems to those manifolds. It is easy to see that the degree obtained this way coincides with Brouwer's.

Nagumo also wrote a second important article, [Nagumo 1951b], where we find the first axiomatic approach to degree theory. This second article was also received by the journal on March 6, 1950, and published in the same volume right after the first one. In the introduction, the author explains that the goal of the paper is to complete the details of Leray-Schauder's construction of degree and that following their ideas, he will describe a degree theory for locally convex topological spaces and prove the Invariance of Domain Theorem for locally convex complete linear spaces. From this paper, we extract paragraph 1.1, relevant for the axiomatic developement of degree theory. Nagumo starts:

First we shall explain the notion of degree of mapping in a finite dimensional Euclidean space.

And he presents the following existence statement:

Nagumo axiomatization. *Let G be an open set in $\mathbb{R}^m$ and let $f : \overline{G} \to \mathbb{R}^m$ be such that $f(x) - x$ is bounded for $x \in \overline{G}$. Let $a \notin f(\overline{G} \setminus G)$. Then there will be determined an integer $d(f, G, a)$, called the* degree of mapping *of G at a by f, with the following properties:*

(i) $d(\mathrm{Id}_{\overline{G}}, G, a) = 1$ *for* $a \in G$; $d(\mathrm{Id}_{\overline{G}}, G, a) = 0$ *for* $a \notin G$.

(ii) *If* $d(f, G, a) \neq 0$, *then there is an* $x \in G$ *such that* $f(x) = a$.

(iii) *If* $G \supset \bigcup_i G_i$, $\overline{G} \supset \bigcup_i \overline{G_i}$, *for finitely many disjoint open sets* G_i, *and* $a \notin f(\overline{G_i} \setminus G_i)$ *for any* i, *then*

$$d(f, G, a) = \sum_i d(f_i, G, a).$$

(iv) *If* $f_t(x) - x$ *is a bounded continuous function of* (t, x) *for* $0 \leq t \leq 1$, $x \in \overline{G}$, *if* $a(t) \in \mathbb{R}^n$ *is continuous, and if* $a(t) \notin f_t(\overline{G} \setminus G)$ *for* $0 \leq t \leq 1$, *then* $d(f_t, G, a(t))$ *is constant for* $0 \leq t \leq 1$.

(v) *Let* X *be the set of all roots of the equation* $f(x) = a$ *in* G, *and let* G_0 *be any open set such that* $X \subset G_0 \subset G$. *Then*

$$d(f, G, a) = d(f, G_0, a).$$

Then, Nagumo remarks that (v) follows from (ii) and (iii) and concludes:

The existence and uniqueness of $d(f, G, a)$ satysfying the above conditions can be verified, if we use simplicial mappings for approximations of f (at first for bounded G and then for general G). But I may refer to [Nagumo 1951a] *in which the existence of $d(f, G, a)$ is given, based on infinitesimal analysis but free from the notion of simplicial mapping.*

Only one objection can be made here: Nagumo does not prove uniqueness!

* * *

We close this section by recalling two additional ways to define the Brouwer-Kronecker degree.

De Rham cohomology. On one hand, GEORGES DE RHAM (1903–1990) published the book [de Rham 1955], where he developed homology and degree theory from a completely different viewpoint. He used flows in connection with the theory of distributions due to LAURENT SCHWARTZ (1915–2002). We can describe this method very roughly as follows.

Every differentiable mapping $f : M \to N$ of boundaryless compact connected oriented n-manifolds induces in (de Rham) cohomology a linear mapping $f^* : H^n(N, \mathbb{R}) \to H^n(M, \mathbb{R})$. Since $H^n(\,\cdot\,, \mathbb{R}) \equiv \mathbb{R}$, that linear mapping is just multiplication times some number, *but this number is an integer d, which we call the degree of f.* Among other things, two important results are behind this approach: Stokes' Theorem and the Change of Variables Formula for Integrals.

Heinz Integral Formula. On the other hand, and related to the above method, in 1959 ERHARD HEINZ presented a more elementary construction, which appears in [Heinz 1959] and runs as follows.

Let $y = y(x)$ be a $\mathcal{C}^1$ mapping defined on a bounded open set $\Omega \subset \mathbb{R}^m$, which is continuous on $\overline{\Omega}$. Furthermore, assume we are given a point $z \in \mathbb{R}^m$ such that $y(x) \neq z$ for all $x \in \overline{\Omega} \setminus \Omega$, and let $\Phi(r)$ be a real-valued function such that the following hold:

(1) $\Phi(r)$ is continuous on the interval $0 \leq r < +\infty$ and vanishes on a neighborhood of $r = 0$ and on $\varepsilon \leq r < +\infty$, where

$$0 < \varepsilon < \min\{y(x) - z : x \in \overline{\Omega} \setminus \Omega\}.$$

(2) $\int_{\mathbb{R}^m} \Phi(|x|)dx = 1$.

Then the Brouwer degree $d(y(x), \Omega, z)$ is uniquely defined by

$$d(y(x), \Omega, z) = \int_{\Omega} \Phi(|y(x) - z|)J(y(x))dx.$$

5. Axiomatization

The development of degree theory was fully completed in 1971, with the axiomatic characterization of the Brouwer-Kronecker topological degree. Such a characterization was obtained by LUTZ FÜHRER in his Ph.D. thesis [Führer 1971], presented at the Freie Universität Berlin. These results were published in the paper [Führer 1972], received by the journal on October 28, 1971. Let us also mention that Führer quoted Nagumo.

However, the axiomatic characterization of degree is often attributed to HERBERT AMANN and STANLEY A. WEISS. In fact, in the paper [Amann-Weiss 1973] (received by the journal on August 21, 1972) they characterize axiomatically not only the Brouwer-Kronecker degree, but also the Leray-Schauder degree. The authors conclude their introduction with the following acknowledgement:

After having finished the first draft of this paper the authors learned of the thesis of Führer [Führer 1971] *in which the uniqueness of the Brouwer degree has been proved also.*

Thus, priorities are clear: the axiomatization of the Brouwer-Kronecker degree is due to Führer, that of the Leray-Schauder degree to Amann-Weiss.

* * *

Let us start with the following:

Führer Characterization. *There exists a unique mapping*

$$d : \left\{ (f, D, y) : \begin{array}{l} D \subset \mathbb{R}^n \ \ \textit{bounded open}, \\ f : \overline{D} \to \mathbb{R}^n \ \ \textit{continuous}, \\ y \in \mathbb{R}^n \setminus f(\partial D) \end{array} \right\} \longrightarrow \mathbb{Z}$$

such that we have the following:

(1) Homotopy invariance: *For every bounded open set $D \subset \mathbb{R}^n$ and all continuous mappings $F : [0,1] \times \overline{D} \to \mathbb{R}^n$ and $\gamma : [0,1] \to \mathbb{R}^n$ such that*

$$\gamma(t) \in \mathbb{R}^n \setminus F(\{t\} \times \partial D) \ \ \textit{for} \ \ 0 \le t \le 1,$$

the following formula holds:

$$d(F(t,\cdot), D, \gamma(t)) = d(F(0,\cdot), D, \gamma(0)) \ \ \textit{for} \ \ 0 \le t \le 1.$$

(2) Normality: *For every bounded open set $D \subset \mathbb{R}^n$ and every point $p \in D$,*

$$d(\mathrm{Id}_{\overline{D}}, D, p) = 1.$$

(3) Additivity: *For every bounded open set $D \subset \mathbb{R}^n$, every disjoint union $D = D_1 \cup D_2$ of two open sets, every continuous mapping $f : \overline{D} \to \mathbb{R}^n$, and every point $p \in \mathbb{R}^n \setminus f(\partial D \cup \partial D_1 \cup \partial D_2)$,*

$$d(f, D, p) = d(f, D_1, p) + d(f, D_2, p).$$

(4) Existence of solutions: *For every bounded open set* $D \subset \mathbb{R}^n$*, every continuous mapping* $f : \overline{D} \to \mathbb{R}^n$*, and every point* $p \in \mathbb{R}^n \setminus f(\partial D)$*, such that* $d(f, D, p) \neq 0$*, the equation* $f(x) = p$ *has some solution in* D*.*

This is the full description of the Brouwer-Kronecker degree. This on one hand has the virtue that the theory becomes clearer and systematic, but on the other hand, and this is very important, this also shows that all different definitions of degree invented by the 1960s coincide, no matter which topological or analytical means were involved.

As often happens, once a collection of complete axioms is found, there is a search for simplicity that aims to distill the collection as much as possible. In this respect, it is worth quoting here the following result obtained by Klaus Deimling in [Deimling 1985]:

Deimling Characterization. *There exists a unique mapping*

$$d : \left\{ (f, D, y) : \begin{array}{l} D \subset \mathbb{R}^n \ \textit{bounded open,} \\ f : \overline{D} \to \mathbb{R}^n \ \textit{continuous,} \\ y \in \mathbb{R}^n \setminus f(\partial D) \end{array} \right\} \longrightarrow \mathbb{Z}$$

such that we have the following:

(1) Normality: *For every bounded open set* $D \subset \mathbb{R}^n$ *and every point* $p \in D$*,*

$$d(\mathrm{Id}_{\overline{D}}, D, p) = 1.$$

(2) Additivity: *For every bounded open set* $D \subset \mathbb{R}^n$*, every pair of two disjoint open sets* $D_1, D_2 \subset D$*, every continuous mapping* $f : \overline{D} \to \mathbb{R}^n$*, and every point* $p \notin f(\overline{D} \setminus D_1 \cup D_2)$*,*

$$d(f, D, p) = d(f, D_1, p) + d(f, D_2, p).$$

(3) Homotopy invariance: *For every bounded open set* $D \subset \mathbb{R}^n$ *and all continuous mappings* $H : [0,1] \times \overline{D} \to \mathbb{R}^n$ *and* $y : [0,1] \to \mathbb{R}^n$ *such that*

$$y(t) \notin H(\{t\} \times \partial D) \ \textit{for} \ 0 \leq t \leq 1,$$

the following formula holds:

$$d(H(t, \cdot), D, y(t)) = d(H(0, \cdot), D, y(0)) \ \textit{for} \ 0 \leq t \leq 1.$$

* * *

Next we describe the axiomatization of the Leray-Schauder degree according to Amann-Weiss.

Let $\mathcal{E}$ be a separated topological linear space, and consider a family $\mathcal{W}$ of open subsets including $\emptyset$ (but not only $\emptyset$). To each $\Omega \in \mathcal{W}$, attach a set $\mathcal{M}(\Omega)$ of continuous mappings $\overline{\Omega} \to \mathcal{E}$. We equip this set $\mathcal{M}(\Omega)$ with the uniform convergency topology.

Definition 1. *The family*

$$\mathbb{M}(\mathcal{W}) = \{\mathcal{M}(\Omega) : \Omega \in \mathcal{W}\}$$

is called an admisible class *of mappings on $\mathcal{E}$ when the following properties hold true:*

(1) $\mathrm{Id}_{\overline{\Omega}} \in \mathcal{M}(\Omega)$ *for every non-empty open set $\Omega \in \mathcal{W}$.*

(2) *For every pair of two open sets $\Omega_1 \subset \Omega$ in $\mathcal{W}$ and every f in $\mathcal{M}(\Omega)$, the restriction $f|_{\overline{\Omega_1}}$ belongs to $\mathcal{M}(\Omega_1)$.*

Now, fix such an admissible class $\mathbb{M}(\mathcal{W})$ for the given family $\mathcal{W}$ of open sets of $\mathcal{E}$. For every $\Omega \in \mathcal{W}$, consider the following subspace of $\mathcal{M}(\Omega)$:

$$\mathcal{M}_0(\Omega) = \{f \in \mathcal{M}(\Omega) : 0 \notin \overline{f(\partial\Omega)}\},$$

where as usual $\partial\Omega = \overline{\Omega} \setminus \Omega$ stands for the boundary of Ω. Then:

Definition 2. *A* topological degree *d for $\mathbb{M}(\mathcal{W})$ is a family of mappings*

$$d = \{d(\cdot, \Omega) : \mathcal{M}_0(\Omega) \to \mathbb{Z} : \Omega \in \mathcal{W}\}$$

for which the following conditions hold true:

(1) Normality: $d(\mathrm{Id}_{\overline{\Omega}}, \Omega) = 1$ *for every $\Omega \in \mathcal{W}$ with $0 \in \Omega$.*

(2) Additivity: *For every non-empty open set $\Omega \in \mathcal{W}$, every pair of two disjoint open subsets of Ω, $\Omega_1, \Omega_2 \in \mathcal{W}$, and every $f \in \mathcal{M}(\Omega)$ such that $0 \notin f(\overline{\Omega} \setminus \Omega_1 \cup \Omega_2)$,*

$$d(f, \Omega) = d(f|_{\overline{\Omega_1}}, \Omega_1) + d(f|_{\overline{\Omega_2}}, \Omega_2).$$

(3) Homotopy invariance: *For every non-empty open set $\Omega \in \mathcal{W}$ and every continuous mapping $h : [0,1] \to \mathcal{M}_0(\Omega)$,*

$$d(h(t), \Omega) = d(h(0), \Omega) \quad \text{for} \quad 0 \le t \le 1.$$

Once the setting is fixed in this way, the authors prove their main result:

Theorem. *Let $\mathcal{E}$ be a locally convex linear space, and let $\mathcal{W}$ be either* (i) *the family of all open sets of $\mathcal{E}$ or* (ii) *the family of all bounded open sets of $\mathcal{E}$. Then, there exists a* unique *topological degree for the admissible class*

$$\mathbb{M}(\mathcal{W}) = \{K(\Omega) : \Omega \in \mathcal{W}\},$$

where $K(\Omega)$ consists of all continuous mappings $f : \overline{\Omega} \to \mathcal{E}$ such that the image $(\mathrm{Id}_{\overline{\Omega}} - f)(\overline{\Omega})$ is relatively compact (f is a compact vector field*).*

Of course, this implies the uniqueness of both the Brouwer-Kronecker and the Leray-Schauder degrees.

6. Further developments

There are several lines along which generalizations have been developed, with various aims and scope. We discuss here: (i) the case of spaces of *infinite dimension,* (ii) the case when *source and targent have different dimensions,* and (iii) the *equivariant* case.

* * *

Degree theory in infinite dimension. This line of research comes from Leray-Schauder theory. Its main purpose is to construct degrees for vector fields other than the compact ones, i.e., Fredholm, monotonous, contractive,..., and to find the corresponding axiomatizations. The resulting theories apply to problems on partial differential equations and bifurcation in functional equations. As we mentioned before (I.3, p. 34), this was started by Caccioppoli in 1936. Years later, the idea was rediscovered and presented in a more rigorous general way by Stephen Smale in [Smale 1965], making use of the non-oriented cobordism rings invented by René Thom in his outstanding foundational paper [Thom 1954]. Smale's definition can be summarized as follows.

Let $f : M \to V$ be a proper Fredholm mapping of index $p \geq 0$ and class $q > p+1$. Then the non-oriented cobordism class of the inverse image $f^{-1}(a)$ of a regular value a is a well-defined invariant $\gamma(f)$, which vanishes if f is not surjective. In case the index p is zero, then $f^{-1}(a)$ is a finite set and $\gamma(f)$ coincides with Caccioppoli's mod 2 degree.

When orientations are taken into account for index 0 Fredholm operators, the result is a degree theory with integral values, a construction made by K. DAVID ELWORTHY and ANTHONY J. TROMBA in [Elworthy-Tromba 1970a] and [Elworthy-Tromba 1970b]. In the first of these two references, the authors present an oriented degree theory for proper Fredholm mappings of index n and class $r = n + 2$ using framed cobordism as introduced by LEV SEMENOVICH PONTRYAGIN in [Pontryagin 1955] (a gem of topology).

Then, in [Nirenberg 1971], LOUIS NIRENBERG produced a generalized topological degree theory for compact perturbations of Fredholm operators, using stable homotopy groups and their analogous version in infinite dimension. These results are extended by E. NORMAN DANCER in [Dancer 1983]. Another interesting contribution to this topic is due to JORGE IZE, who used cohomotopy groups in [Ize 1981].

* * *

Mapping degree for source and target of distinct dimensions. From the preceding discussion, we see how bifurcation theory motivates the introduction of oriented degree theories for Fredholm mappings of positive index and therefore brings in the notion of topological degree for *mappings between spaces of different dimensions.* In such theories the so-called degree is not an integer anymore, but some homotopy class in a suitable homotopy group (of a sphere, because spheres are compactifications of Euclidean spaces). An important development of this theory is presented by KAZIMIERZ GEBA, IVAR MASSABÒ, and ALFONSO VIGNOLI in [Geba *et al.* 1986].

The goal is to construct a generalized degree (the *Geba-Massabò-Vignoli degree*) for continuous mappings $f : \overline{U} \to \mathbb{R}^n$, where U is a bounded open subset of $\mathbb{R}^m$, with $m \geq n$, under the assumption that f does not vanish on $\partial U = \overline{U} \setminus U$. The authors support their theory by exploring the $m = n$ case, that is, by reformulating the Brouwer-Kronecker degree in a way suitable for generalization. Indeed, let f be given as above, with $m = n$, and denote by $f_0 : \partial U \to \mathbb{R}^n \setminus \{0\}$ the restriction of f to ∂U. We look at $\mathbb{R}^n$ and $\mathbb{R}^n \setminus \{0\}$ inside the *Alexandroff compactification* $(\mathbb{R}^n)^*$ of $\mathbb{R}^n$. Since $(\mathbb{R}^n)^* \setminus \{0\}$ is homeomorphic to $\mathbb{R}^n$, by the Tietze Extension Theorem, f_0 extends to a continuous mapping $f_\infty : (\mathbb{R}^n)^* \setminus U \to (\mathbb{R}^n)^* \setminus \{0\}$. Thus one gets a continuous mapping

$$f^* : (\mathbb{R}^n)^* \to (\mathbb{R}^n)^* : x \mapsto \begin{cases} f(x) & \text{if } x \in \overline{U}, \\ f_\infty(x) & \text{if } x \in (\mathbb{R}^n)^* \setminus U. \end{cases}$$

Such an extension is called *admissible,* but note that it is not unique. Next, we consider a homeomorphism $\sigma_n : (\mathbb{R}^n)^* \to \mathbb{S}^n$ such that

$$\begin{aligned}
&\sigma_n(\infty) = (1, 0, \ldots, 0),\\
&\sigma_n((\mathbb{R}^n)^*_+) = \{x \in \mathbb{S}^n : x_{n+1} \geq 0\},\\
&\sigma_n((\mathbb{R}^n)^*_-) = \{x \in \mathbb{S}^n : x_{n+1} \leq 0\},\\
&\sigma_{n+1}(x) = \sigma_n(x) \quad \text{for all } x \in \mathbb{R}^n.
\end{aligned}$$

It can then be shown that the homotopy class of $\sigma_n \circ f^* \circ \sigma_n^{-1}$ does not depend on the admissible extension f^*, and thus we have a well-defined element $[\sigma_n \circ f^* \circ \sigma_n^{-1}]$ in the n-th homotopy group $\pi_n(\mathbb{S}^n)$. Next, via an algebraic isomorphism $\varphi_n : \pi_n(\mathbb{S}^n) \to \mathbb{Z}$, set

$$d^*(f, U, 0) = \varphi_n([\sigma_n \circ f^* \circ \sigma_n^{-1}]).$$

Once d^* is thus constructed, it is checked that the axioms of degree theory hold true, and by the axiomatic characterization (I.5, p. 38), d^* is indeed the Brouwer-Kronecker degree.

Now, this construction can be mimicked for arbitrary $m \geq n$ to obtain a generalized degree:

$$d^*(f, U) = d^*(f, U, 0) = [\sigma_n \circ f^* \circ \sigma_m^{-1}] \in \pi_m(\mathbb{S}^n),$$

which, for $m > n$, is not an integer any more. Moreover, we see why the case $m < n$ was neglected: in that case $\pi_m(\mathbb{S}^n) = 0$.

In the article mentioned, the authors prove for this degree the basic properties, namely:

(1) Homotopy invariance. *If $h : [0, 1] \times (U, \partial U) \to (\mathbb{R}^n, \mathbb{R}^n \setminus \{0\})$ is a continuous mapping, then $d^*(h_t, U)$ is well defined for all t and does not depend on t.*

(2) Excision. *Let $f : (U, \partial U) \to (\mathbb{R}^n, \mathbb{R}^n \setminus \{0\})$ be a continuous mapping. Then, for every open set $V \subset U$ such that f has no zeros in $U \setminus V$, we have $d^*(f, V) = d^*(f, U)$.*

(3) Existence of solutions. *Let $f : (U, \partial U) \to (\mathbb{R}^n, \mathbb{R}^n \setminus \{0\})$ be a continuous mapping with $d^*(f, U) \neq 0 \in \pi_m(\mathbb{S}^n)$. Then, there is an $x \in U$ such that $f(x) = 0$.*

(4) Suspension. *Let U be an open subset of $\mathbb{R}^{m+1}$ and let*

$$f : (U, \partial U) \to (\mathbb{R}^{n+1}, \mathbb{R}^{n+1} \setminus \{0\})$$

be a continuous mapping such that

$$f(\overline{U} \cap \mathbb{R}_+^{m+1}) \subset \mathbb{R}_+^{n+1}, \quad f(\overline{U} \cap \mathbb{R}_-^{m+1}) \subset \mathbb{R}_-^{n+1}.$$

Then, setting $U_0 = U \cap \mathbb{R}^m \equiv U \cap (\mathbb{R}^m \times \{0\})$ *and* $f_0 = f|_{\overline{U}_0}$, *we have* $f_0 : (U_0, \partial U_0) \to (\mathbb{R}^n, \mathbb{R}^n \setminus \{0\})$ *and*

$$d^*(f, U) = \Sigma(d^*(f_0, U_0)),$$

where $\Sigma : \pi_m(\mathbb{S}^n) \to \pi_{m+1}(\mathbb{S}^{n+1})$ *is the suspension homomorphism (an isomorphism for* $m < 2n-1$ *and an epimorphism for* $m = 2n-1$*).*

(5) Additivity. *Let* $f : (U, \partial U) \to (\mathbb{R}^n, \mathbb{R}^n \setminus \{0\})$ *be a continuous mapping, and let* $U_1, U_2 \subset U$ *be two open disjoint sets such that* f *has no zero in* $U \setminus U_1 \cup U_2$. *Then* $d^*(f, U) = d^*(f, U_1) + d^*(f, U_2)$, *whenever* $m - n \leq n - 4$.

Later Francisco Romero Ruiz del Portal showed in his Ph.D. thesis [Ruiz del Portal 1991] that additivity also holds for $m - n \leq n - 2$, and this is definitive: there is a counterexample for $m = 2n + 1$. This appeared in [Ruiz del Portal 1992]. Another counterexample was published afterwards by Ize, Massabò, and Vignoli in [Ize *et al.* 1992], a paper that deals with equivariant degree as explained below.

To conclude, it must be noted that one major problem of this theory is the computation of the homotopy groups of the spheres, a question which is wide open today.

* * *

Equivariant degree theory. The purpose here is to define a suitable topological degree for mappings that are *invariant under the action of a Lie group* G on the given spaces. The case most studied is that of $G = \mathbb{S}^1$, that is, the so-called $\mathbb{S}^1$*-equivariant topological degree.*

Let us recall that Poincaré used what later would be called the Brouwer-Kronecker degree to study the critical points of a differential equation. But, as is well known, other very important elements for the understanding of differential equations are periodic orbits. It was to count their number that in 1965 F. Brock Fuller introduced an invariant of flows that today we call the *Fuller index.* The definition and properties of this index are given in detail in [Fuller 1967]. A careful analysis of the constructions behind the Fuller index and the generalized Geba-Massabò-Vignoli degree for $m = n+1$ led the above-mentioned Geba and GRZEGORZ DYLAWERSKI, JERZY JODEL, and WACŁAW MARZANTOWICZ, from Gdansk University, to define in a preprint in 1987 (later published as [Dylawerski *et al.* 1991]) a new homotopy invariant for $\mathbb{S}^1$-equivariant continuous mappings, which they called $\mathbb{S}^1$-*degree*. Let us describe this briefly.

Let ρ be a *finite representation of* $\mathbb{S}^1$, that is, a continuous homomorphism $\rho : \mathbb{S}^1 \to GL(V)$ into the linear group $GL(V)$ of a real linear space V of finite dimension; ρ determines an $\mathbb{S}^1$-*action on* V by $(g, v) \mapsto \rho(g)(v)$. For such a pair (V, ρ), we say the following:

(1) A set $X \subset V \times \mathbb{R}$ is *invariant* if

$$(\rho(g)(v), \lambda) \in X \quad \text{for all } g \in \mathbb{S}^1,\ (v, \lambda) \in X.$$

(2) A continuous mapping $f : X \to V$ with invariant domain X is an $\mathbb{S}^1$-*mapping* if

$$f(\rho(g)(v), \lambda) = \rho(g)(f(v, \lambda)) \quad \text{for all } g \in \mathbb{S}^1,\ (v, \lambda) \in X.$$

We will denote by $\mathfrak{A}$ the abelian group of all finite sequences $\alpha = (\alpha_r)_{r \geq 0}$, with $\alpha_0 \in \mathbb{Z}_2$ and $\alpha_r \in \mathbb{Z}$ for $r \geq 1$ (sum defined componentwise).

With this terminology fixed, the authors prove:

Theorem. *Let* (V, ρ) *run through finite representations of* $\mathbb{S}^1$, Ω *through the family of all bounded, invariant open subsets of* $V \times \mathbb{R}$, *and* $f : X \to V$ *through* $\mathbb{S}^1$*-mappings such that* X *is invariant,* $\overline{\Omega} \subset X$, *and* $f(\partial \Omega) \subset V \setminus \{0\}$. *Then there exists an* $\mathfrak{A}$*-valued function* $\mathrm{Deg}(f, \Omega)$, *called the* $\mathbb{S}^1$-degree, *satisfying the following conditions:*

(a) *If* $\mathrm{Deg}(f, \Omega) \neq 0$, *then* $f^{-1}(0) \cap \Omega \neq \emptyset$.

(b) *If* $\Omega_0 \subset \Omega$ *is open and invariant and* $f^{-1}(0) \cap \Omega \subset \Omega_0$, *then* $\mathrm{Deg}(f, \Omega) = \mathrm{Deg}(f, \Omega_0)$.

(c) *If* Ω_1, Ω_2 *are two open invariant subsets of* Ω *such that* $\Omega_1 \cap \Omega_2 = \emptyset$ *and* $f^{-1}(0) \cap \Omega \subset \Omega_1 \cup \Omega_2$, *then* $\mathrm{Deg}(f, \Omega) = \mathrm{Deg}(f, \Omega_1) + \mathrm{Deg}(f, \Omega_2)$.

(d) *If* $h : (\overline{\Omega} \times [0, 1], \partial \Omega \times [0, 1]) \to (V, V \setminus \{0\})$ *is an* $\mathbb{S}^1$*-homotopy, then* $\mathrm{Deg}(h_0, \Omega) = \mathrm{Deg}(h_1, \Omega)$.

(e) *Suppose* (W, η) *is another representation of* $\mathbb{S}^1$ *and let* U *be an open bounded, invariant subset of* W *such that* $0 \in U$. *Define* $F : \overline{U} \times \overline{\Omega} \to W \times V$ *by* $F(x, y) = (x, f(y))$. *Then* $\mathrm{Deg}(F, U \times \Omega) = \mathrm{Deg}(f, \Omega)$.

The construction of $\mathrm{Deg}(f, \Omega)$ begins with two particular cases. The first case is when we have the trivial representation $V = \mathbb{R}^n$ of $\mathbb{S}^1$. Then the $\mathbb{S}^1$-mappings in the theorem are the continuous mappings $f : (\Omega, \partial\Omega) \to (\mathbb{R}^n, \mathbb{R}^n \setminus \{0\})$, where Ω is a bounded open subset of $\mathbb{R}^{n+1}$. In this case we have a homomorphism $\Sigma : \pi_{n+1}(\mathbb{S}^n) \to \mathbb{Z}_2$ (*suspension*) that is an isomorphism for $n \geq 3$, and the authors define

$$\mathrm{Deg}(f, \Omega) = \Sigma(d^*(f, \Omega)) \in \mathbb{Z}_2,$$

where d^* is the Geba-Massabò-Vignoli generalized degree. In this case, the proof of the theorem above follows from the results in [Geba *et al.* 1986].

Secondly, suppose there exists a point $a \in \Omega$ such that

$$f^{-1}(0) = \{\rho(g)(a) : g \in \mathbb{S}^1\};$$

that is, $f^{-1}(0)$ is an *orbit* of the action of $\mathbb{S}^1$ over V. Then the set

$$\mathbb{S}^1 * a = \{g \in \mathbb{S}^1 : \rho(g)(a) = a\}$$

is finite, hence a subgroup of $\mathbb{S}^1$ consisting of k-th roots of unit. We can furthermore assume without loss of generality that (V, ρ) is orthogonal (with respect to some inner product on V) and decompose

$$V = W \oplus W^{\perp}, \quad W = \{x \in V : \rho(g)(x) = x\}.$$

Then if in that decomposition f takes the form $f(x, y, \lambda) = (f_1(x, y, \lambda), y)$, the authors define

$$\mathrm{Deg}(f, \Omega) = (\alpha_r), \quad \alpha_r = \begin{cases} \deg(f, D) & \text{for } r = k, \\ 0 & \text{otherwise,} \end{cases}$$

where $\deg(f, D)$ is the Brouwer-Kronecker degree and D is a closed disc in the linear space W, contained in $\Omega \cap (W \times \mathbb{R})$, transversal to $\mathbb{S}^1 * a = f^{-1}(0)$, and oriented so that it can be identified with the unit disc in W.

In the general case, the construction of Deg requires (i) the classical theorem that gives the full classification of all finite-dimensional representations of $\mathbb{S}^1$ ([Adams 1969]) and (ii) a quite non-trivial homotopy argument.

Another version of this degree for mappings defined on spheres has been studied using cohomological obstruction theory by Ize, Massabò, and Vignoli, in [Ize *et al.* 1986]. Afterwards, the same authors presented in the two papers [Ize *et al.* 1989] and [Ize *et al.* 1992] an equivariant degree theory for mappings defined on the closure of an arbitrary open set of the ambient space and replaced the circle group $\mathbb{S}^1$ by an arbitrary compact Lie group. In addition, they avoided the use of obstruction theory. This is different, in the non-equivariant setting, from the Geba-Massabò-Vignoli degree described earlier.

Namely, let U be a bounded open subset of $\mathbb{R}^m$ and let $f : \overline{U} \to \mathbb{R}^n$ $(m \geq n)$ be a continuous mapping such that $f(x) \neq 0$ for $x \in \partial U$. Let $\widehat{f} : B \to \mathbb{R}^n$ be a continuous extension of f to a closed ball $\overline{D}$ containing U. Let V be a bounded open neighborhood of ∂U with $f(x) \neq 0$ for $x \in \overline{V}$. Consider a Uryshon function $\varphi : \overline{D} \to [0,1]$ which is $\equiv 1$ off $U \cup V$ and $\equiv 0$ on $\overline{U}$. Define a mapping

$$F : [0,1] \times \overline{D} \to \mathbb{R}^{n+1} : (t,x) \mapsto F(t,x) = (2t + 2\varphi(x) - 1, \widehat{f}(x)).$$

It is easy to see that $F(t,x) = 0$ only if $x \in \overline{U}, f(x) = 0$, and $t = \frac{1}{2}$. Thus F maps $\partial([0,1] \times \overline{D})$ into $\mathbb{R}^{n+1} \setminus \{0\}$, which defines an element of the homotopy group $\pi_m(\mathbb{S}^n)$: this is the generalized degree of f with respect to U. When the action of a compact Lie group G is present, this construction extends without difficulty, and the equivariant degree is an element of the equivariant homotopy group $\pi_m^G(\mathbb{S}^n)$. Then, in [Ize *et al.* 1992] the authors proved that this formulation for $G = \mathbb{S}^1$ is the $\mathbb{S}^1$-degree of [Dylawerski *et al.* 1991].

All of this is revised in systematic form in the book [Ize-Vignoli 2003].

Also, we mention that Dancer defined an $\mathbb{S}^1$-degree for gradient mappings in [Dancer 1985]. In this respect, the paper [Geba *et al.* 1990], by Geba, Massabò, and Vignoli, is remarkable. Later, Sławomir Rybicki discussed an $\mathbb{S}^1$-degree for orthogonal mappings (which include gradient mappings, [Rybicki 1994]), drawing upon the earlier work in [Dylawerski *et al.* 1991]. To end these comments, let us add that Dancer, Geba, and Rybicki obtained in [Dancer *et al.* 2005] a complete classification of equivariant gradient mappings up to homotopy, and the corresponding equivariant homotopy classes can be seen as equivariant degrees.

Chapter II

Manifolds

Here we describe the objects of our theory: manifolds and mappings. We will be dealing with differentiable manifolds: our methods are those of differential topology. The general definitions and some basic facts are gathered in §§1–2. In particular, we stress the distinction between differentiable and smooth structures. In §3 we state the essential *Sard-Brown Theorem* for differentiable mappings but prove it only for smooth mappings. This brings in the subtleties concerning finite or infinite differentiability, but we have chosen a most dramatic approach: to use the fact that every differentiable structure is in fact *smooth*. In §4 we discuss in depth the existence and two different constructions of *tubular neighborhoods and differentiable retractions,* both of which will be required later. Then, in §5 we prove a key result for degree theory: differentiable mappings that are homotopic are *differentiably* homotopic. This is essential, because even for smooth manifolds, we use our methods to study continuous mappings, neither smooth nor even differentiable. In §6 we consider a special type of homotopies: the so-called *diffeotopies,* often needed to move points in manifolds at will. Finally, in §7 we recall the basics of *orientation,* which play a crucial role in degree theory.

1. Differentiable mappings

In this section we recall the basic notions concerning differentiability. In particular we discuss partitions of unity and bump functions.

(1.1) Differentiable mappings. Let $U \subset \mathbb{R}^p$ be an open set and let $r = 1, \dots, \infty$. A differentiable $\mathcal{C}^r$ function $f : U \to \mathbb{R}$ is a function whose partial derivatives $\frac{\partial^k f}{\partial x_{i_1} \cdots \partial x_{i_k}}$ exist and are continuous for all $k \leq r$. We use the terminology *smooth* when $r = \infty$. A mapping $f : X \to Y$ of arbitrary sets $X \subset \mathbb{R}^p$ and $Y \subset \mathbb{R}^q$ is $\mathcal{C}^r$ if for every point $x \in X$ there are $\mathcal{C}^r$ functions $\bar{f}_j : U \to \mathbb{R}$, $1 \leq j \leq q$, defined on an open neighborhood U of x in $\mathbb{R}^p$ such that $\bar{f} = (\bar{f}_1, \dots, \bar{f}_q)$ coincides with f on $U \cap X$; we say that $\bar{f}$ is a local $\mathcal{C}^r$ extension of f.

Notice how the extension $\bar{f}$ depends on the point x and that this is a local notion. This calls in the main globalization tool in our context: *partitions of unity.*

Proposition 1.2. *Let* $\mathcal{U} = \{U_i : i \in I\}$ *be an open covering of an open set* $U \subset \mathbb{R}^p$. *Then, there is a* smooth partition of unity $\{\theta_i : i \in I\}$ for $\mathcal{U}$, *that is, a family of smooth functions* $\theta_i : U \to [0,1]$ *such that:*

(i) *Each* $x \in U$ *has a neighborhood* W *on which all but finitely many* θ_i*'s vanish* (local finiteness), *and* $\sum_i \theta_i \equiv 1$.

(ii) *Each function* θ_i *vanishes off* U_i, *namely* $\overline{\{x \in U : \theta_i(x) \neq 0\}} \subset U_i$.

This result is quite elementary in nature, and various proofs can be found in many textbooks. We include a condensed version here for the sake of the reader.

Proof. Consider a cover of U by compact sets L_k, $k \geq 0$ such that $L_k \subset \operatorname{Int}(L_{k+1})$ and L_0 is a singleton, so that the compact sets

$$K = L_{k+1} \setminus \operatorname{Int}(L_k) \subset U$$

also cover U.

Fix k. For every $a \in K$ there are (1) an open neighborhood $D \subset \operatorname{Int}(L_{k+2}) \setminus L_{k-1}$ with $D \subset U_{i(k)}$ for some $i(k)$ and (2) a diffeomorphism $\mathrm{x} : D \to \mathbb{R}^p$ with $\mathrm{x}(a) = 0$. Now define on D the smooth function

$$\eta = \frac{f(2 - \|\mathrm{x}\|^2)}{f(2 - \|\mathrm{x}\|^2) + f(\|\mathrm{x}\|^2 - 1)}$$

where $f(t) = \exp(-1/t)$ for $t > 0$ and $f(t) = 0$ for $t \leq 0$. This function is always ≥ 0 and it is $\equiv 1$ on an open neighborhood $B \subset D$ of a and extends to U by 0 off D. In particular, η vanishes on L_{k-1}.

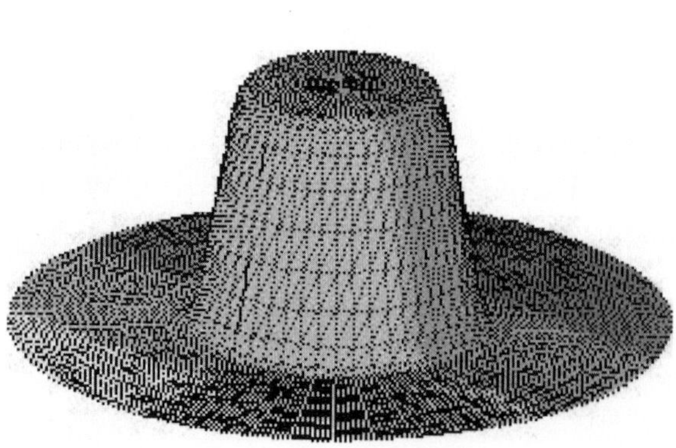

Graph of η (in $\mathbb{R}^{p+1}$)

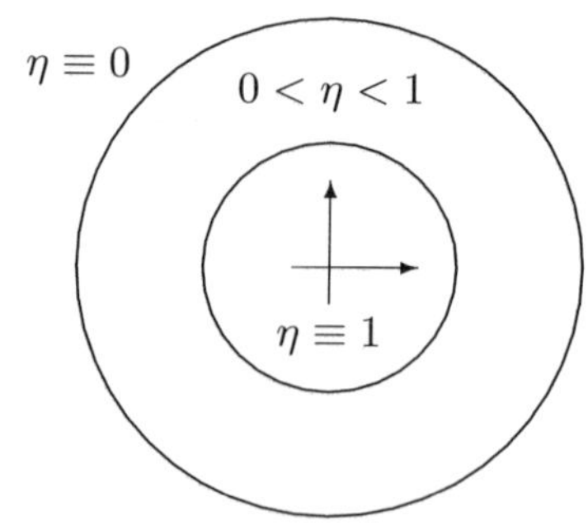

Domain of η (in $\mathbb{R}^p$)

Next, since the open sets B cover the compact K, finitely many, call them B_ℓ, already cover K, and we have the associated functions $\eta_{k\ell}$. We have thus constructed for the fixed k the finite familly of non-negative smooth functions $\{\eta_{k\ell}\}_\ell$, each vanishing off an open set $U_{i(k\ell)}$; in fact $\overline{\{\eta_{k\ell} \neq 0\}} \subset U_{i(k\ell)}$.

Now let $x \in U$, and pick the first k_0 with $x \in L_{k_0}$. By construction, each $\eta_{k\ell}$ vanishes on L_{k-1}, which for $k \geq k_0 + 2$ contains the open set $W = \mathrm{Int}(L_{k_0+1})$, which contains x. Thus, the sum $\sum_{k\ell} \eta_{k\ell}$ is actually finite on W. Moreover, again by construction, $\eta_{k\ell}(x) \neq 0$ for some k, ℓ; hence that sum is > 0. This shows that any sum of the functions $\eta_{k\ell}$ is a well-defined smooth function, and the sum h of all of them is always > 0. From these remarks, we define for each i

$$\theta_i = \frac{1}{h} \sum_{k,\ell \,|\, i(k\ell)=i} \eta_{k\ell},$$

and these are the functions we were looking for. □

As mentioned above, partitions of unity help to obtain global objects by glueing local data. Here are some useful examples of this method.

Proposition 1.3. (1) (Bump functions) *Let $A \subset \mathbb{R}^p$ be a closed set and let $U \subset \mathbb{R}^p$ be an open neighborhood of A. Then there is a smooth function $\theta : \mathbb{R}^p \to [0,1]$ that is identically 1 on A and identically 0 off U.*

(2) (Uryshon separating functions) *Let $A, B \subset \mathbb{R}^p$ be two disjoint closed sets. Then there is a smooth function $\theta : \mathbb{R}^p \to [0,1]$ that is $\equiv 1$ on A and $\equiv 0$ on B.*

Proof. Assertions (1) and (2) are equivalent, taking $U = \mathbb{R}^{n+1} \setminus B$. To prove them, take $V = \mathbb{R}^{n+1} \setminus A$ and a smooth partition of unity $\{\theta, \eta\}$ for the covering $\{U, V\}$. Since $\theta + \eta \equiv 1$ and $\eta \equiv 0$ off V, it follows that $\theta \equiv 1$ on A. Similarly, $\theta \equiv 0$ off U, that is, on B. □

Examples 1.4. We suggest that the reader represent in the one variable case the function η in the proof of II.1.2, p. 50, and find the typical bump or separating functions, whose graphs we depict below:

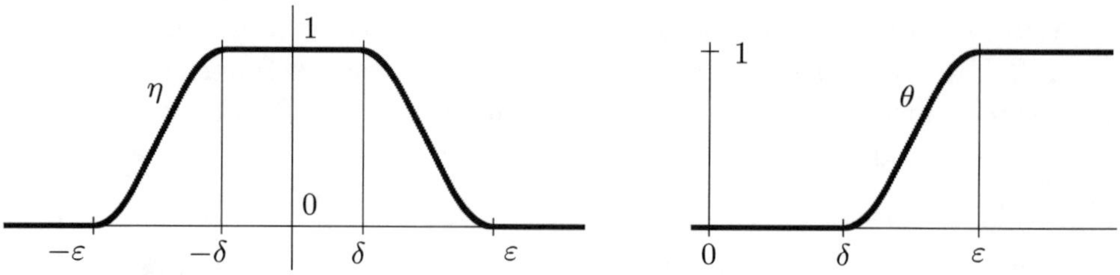

In fact, these simple examples give a precise description of the set where the value is $\equiv 1$ or $\equiv 0$ and can be used to modify other functions by multiplication. For instance, if we multiply the function $t \mapsto t$ by the separating function θ in the right-hand figure above, or by $1 - \theta$, we get

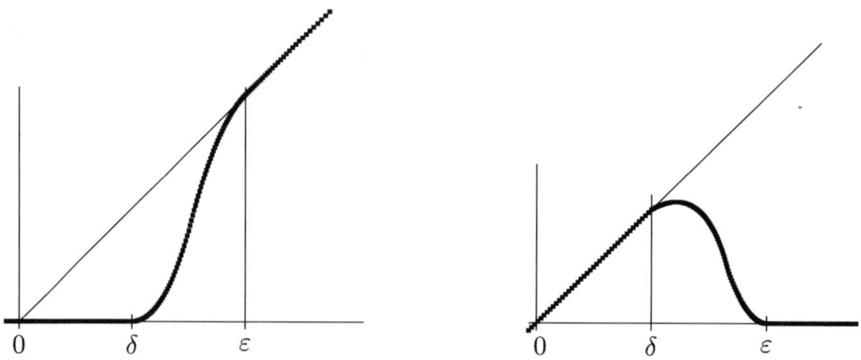

Such modifications are used in many constructions. □

Another easy construction using partitions of unity is

Proposition 1.5 (C^r Tietze Extension Theorem). *Let A be a closed subset of an open set $U \subset \mathbb{R}^p$, and let $f : A \to \mathbb{R}$ be a C^r function. Then f has a C^r extension $\bar{f} : U \to \mathbb{R}$.*

Proof. Choose local C^r extensions $\bar{f}_i : U_i \to \mathbb{R}$ of $f|A \cap U_i$ so that the U_i form an open covering of A in U. Add the open set $V = U \setminus A$ and consider a smooth partition of unity $\{\eta, \theta_i\}$ for the cover $\{V, U_i\}$. Then $\bar{f} = \sum_i \theta_i \bar{f}_i$ is well defined on the whole of U, because

(i) each term $\theta_i \bar{f}_i$ vanishes on a neighborhood of $U \setminus U_i$, hence extends by 0 off U_i, and

(ii) the sum is locally finite, as the family $\{\theta_i\}$ is.

Furthermore, $\bar{f}$ is $\mathcal{C}^r$, because the local extensions $\bar{f}_i$ are, and the θ_i's are smooth. Finally, for $x \in A \cap U_i$, $\theta_i(x)\bar{f}_i(x) = \theta_i(x)f(x)$, and the same is true for $x \in A \setminus U_i$ also, since $\theta_i(x) = 0$. Thus, for $x \in A$, $\bar{f}(x) = \big(\sum_i \theta_i(x)\big)f(x) = f(x)$. (Note that $\eta(x) = 0$.) □

Remark 1.6. Let $X \subset \mathbb{R}^p$ be an arbitrary set. The proof above can be adapted to show that every C^r mapping $f : X \to \mathbb{R}^q$ has a C^r extension to some open neighborhood U of X in $\mathbb{R}^p$. In other words, the notion of a C^r mapping, whose definition was of a local nature, is in fact global. □

Exercises and problems

Number 1. Construct a differentiable function $f : \mathbb{R} \to \mathbb{R}$ such that $f(t) = t$ for $t \leq 1$, $f(t) \geq t$ for $1 \leq t \leq 2$, and $f(t) = 2$ for $t \geq 2$.

Number 2. Exhibit two smooth functions $f, g : \mathbb{R} \to \mathbb{R}$ that do not vanish on any neighborhood of the origin but whose product does: $fg \equiv 0$.

Number 3. Exhibit two disjoint closed sets in the plane that cannot be separated by a *polynomial* function.

Number 4. Let $\{\theta_i\}$ be a smooth partition of unity for some covering $\mathcal{U}$ of an open set $U \subset \mathbb{R}^p$. Show that if all open sets of $\mathcal{U}$ have compact closure in U, then the smooth function $f = \sum_i i\theta_i : U \to \mathbb{R}$ is proper.

Number 5. (Extension by zero) Let $f : U \to \mathbb{R}$ be a smooth function defined on a neighborhood U of the origin in $\mathbb{R}^p$. Show that there are smooth functions $g : \mathbb{R}^p \to \mathbb{R}$ that (i) coincide with f in some smaller neighborhood $U' \subset U$ of the origin and (ii) vanish off U.

Number 6. Let (a_k) be a discrete sequence of pairwise distinct points in $\mathbb{R}^p$ and let (r_k) be a sequence of integers ≥ 0. For every k and $\nu = (\nu_1, \ldots, \upsilon_p)$ with $|\nu| = \nu_1 + \cdots + \upsilon_p \leq r_k$, choose $f_{k\nu} \in \mathbb{R}$. Construct a smooth function $f : \mathbb{R}^p \to \mathbb{R}$ with partial derivatives $\frac{\partial^{|\nu|} f}{\partial x_1^{\nu_1} \cdots \partial x_p^{\upsilon_p}}(a_k) = f_{k\nu}$.

2. Differentiable manifolds

Now that we have discussed differentiable mappings, we turn to the notion of manifolds. We will only consider manifolds of finite dimension.

(2.1) Local diffeomorphisms. It is clear from the definition that a composition of $\mathcal{C}^r$ mappings is again a $\mathcal{C}^r$ mapping. But in general, the inverse of a bijective $\mathcal{C}^r$ mapping need not be $\mathcal{C}^r$ again (think of $t \mapsto t^3$). Thus we say that a bijection $f : X \to Y$ is a $\mathcal{C}^r$ *diffeomorphism* when both f and f^{-1} are $\mathcal{C}^r$ mappings. The local version of this notion is essential: f is a *local $\mathcal{C}^r$ diffeomorphism at $a \in X$* when f is a $\mathcal{C}^r$ diffeomorphism from a neighborhood of a onto another neighborhood of $f(a)$.

The notion of local diffeomorphisms leads immediately to that of manifolds.

(2.2) Manifolds. A subset $M \subset \mathbb{R}^p$ is a $\mathcal{C}^r$ *manifold* when every point $x \in M$ has an open neighborhood U in M which is $\mathcal{C}^r$ diffeomorphic to an open set W in some Euclidean space $\mathbb{R}^m$.

Here we fix some notation and terminology. Of course the terms *differentiable* and *smooth* specify the class as they do for mappings. Now, a given $\mathcal{C}^r$ diffeomorphism

$$\varphi = (\varphi_1, \dots, \varphi_p) : W \to U \subset \mathbb{R}^p$$

is called a *parametrization of M at x*, or simply *of U;* the inverse diffeomorphism

$$\mathrm{x} = (\mathrm{x}_1, \dots, \mathrm{x}_m) : U \to W \subset \mathbb{R}^m$$

is called a *local coordinate system.* Given a second parametrization ψ, the composite mapping $\psi^{-1} \circ \varphi$ is the *change of coordinates,* which is a $\mathcal{C}^r$ diffeomorphism of open sets in Euclidean spaces. Hence its derivative is a linear isomorphism, and the dimension m of the domain of local coordinates is the same for the two parametrizations: this m is the *dimension of M at x: $m = \dim_x(M)$*. If the dimension is the same at all points, we just write $m = \dim(M)$; *unless otherwise stated, we always assume this.* Also, as customary, we will say that $M \subset \mathbb{R}^p$ is a *curve* when $m = 1$, a *surface* when $m = 2$, and a *hypersurface* when $p = m + 1$.

Thus, manifolds are sets locally diffeomorphic to Euclidean spaces and, in particular, locally homeomorphic to them. Consequently, manifolds share with Euclidean spaces all topological properties of a local nature: local connectedness, openness of connected components, arc-connectedness if connected, local compactness,.... Of course, since our manifolds are by definition embedded in Euclidean spaces, they inherit many other properties: metrizability, countable bases of open sets. In particular, manifolds are paracompact spaces. Moreover, combining everything, manifolds have other properties, such as exhaustion by compact sets.

(2.3) Equations. A basic property of a differentiable $\mathcal{C}^r$ manifold $M \subset \mathbb{R}^p$

is that it is *locally closed* (open in its closure) in $\mathbb{R}^p$. In fact, there are *local $\mathcal{C}^r$ equations at every point $x \in M$:* there are $\mathcal{C}^r$ functions $f_1, \dots, f_q$ defined on an open neighborhood U of x in $\mathbb{R}^p$, such that:

(i) $U \cap M = f^{-1}(0) = \{x \in U : f_1(x) = \cdots = f_q(x) = 0\}$.

(ii) The rank of the Jacobian matrix $\left(\dfrac{\partial f_i}{\partial x_j}\right)$ is q at every point $x \in U \cap M$.

(iii) $\dim_x(M) = p - q$.

In fact, starting from a parametrization at x, one constructs, using the Inverse Mapping Theorem, a $\mathcal{C}^r$ diffeomorphism $h = (h_1, \dots, h_p)$ from U onto an open set $V \subset \mathbb{R}^p$ such that $h(U \cap M) = V \cap (\mathbb{R}^m \times \{0\})$. Then the equations above are $f_i = h_{m+i}$. The existence of h simply says that the pair $M \subset \mathbb{R}^p$ is locally $\mathcal{C}^r$ diffeomorphic to the pair $\mathbb{R}^m \subset \mathbb{R}^p$.

Conversely, if a subset $M \subset \mathbb{R}^p$ can be locally described by functions verifying (i) and (ii) above, then it is a manifold and its dimension is given by (iii).

It is important to stress that a manifold M need not have *global equations,* that is, functions as above defined on $U = \mathbb{R}^p$; indeed, condition (ii) is highly restrictive. We will discuss this matter for hypersurfaces, as it is narrowly related to the Jordan Separation Theorem (II.7.5, p. 88, and III.6.4, p. 129).

Example 2.4. The first examples of manifolds are spheres:

$$\mathbb{S}^m = \{x \in \mathbb{R}^{m+1} : \sum_i x_i^2 = 1\}.$$

They can be parametrized by the so-called *stereographic projections;* in fact, two such projections suffice. Let $a_N = (0, \dots, 0, 1) \in \mathbb{R}^{m+1}$ be the north pole of $\mathbb{S}^m$, and $a_S = (0, \dots, 0, -1)$ the south pole. Then the projections from the north and south poles,

$$\pi_N : \mathbb{S}^m \setminus \{a_N\} \to \mathbb{R}^m : x \mapsto \Big(\frac{x_1}{1 - x_{m+1}}, \dots, \frac{x_m}{1 - x_{m+1}}\Big)$$

and

$$\pi_S : \mathbb{S}^m \setminus \{a_S\} \to \mathbb{R}^m : x \mapsto \Big(\frac{x_1}{1 + x_{m+1}}, \dots, \frac{x_m}{1 + x_{m+1}}\Big),$$

are local coordinate systems, corresponding to the parametrizations

$$y = (y_1, \dots, y_m) \mapsto \Big(\frac{2y_1}{1 + \|y\|^2}, \dots, \frac{2y_m}{1 + \|y\|^2}, \frac{-1 + \|y\|^2}{1 + \|y\|^2}\Big)$$

and

$$y = (y_1, \dots, y_m) \mapsto \Big(\frac{2y_1}{1+\|y\|^2}, \dots, \frac{2y_m}{1+\|y\|^2}, \frac{1-\|y\|^2}{1+\|y\|^2}\Big).$$

On the other hand, spheres do have global equations: the expression $\sum_i x_i^2 = 1$ is a global equation in the sense of II.2.3, p. 55. □

Let us point out here that the abstract approach where M is not a subset of any $\mathbb{R}^p$ is not so abstract: just assuming its topology has a countable basis, M embeds in some $\mathbb{R}^p$. This can be done by elementary means, if one disregards how big p is; we do not appeal here to the *Whitney Embedding Theorems* that control p to $2\dim(M) + 1$.

(2.5) Manifolds with boundary. The definition of manifold stems from the choice of a local model: Euclidean spaces. The choice of this model is convenient because Euclidean spaces are the natural setting for differential analysis. However one can go a little further and extend without surprises differential analysis to Euclidean half-spaces, namely to convex sets $\mathbb{H}^m \subset \mathbb{R}^m$ defined by one linear inequality $\lambda(x) \geq 0$. After linear changes of coordinates, we get the canonical $\mathbb{H}^m$ described by $x_1 \geq 0$ in coordinates $x = (x_1, \dots, x_m) \in \mathbb{R}^m$. Then, a C^r *manifold with boundary* is a set $M \subset \mathbb{R}^p$ locally C^r diffeomorphic to such half-spaces. The notions of parametrization, local coordinate system, and dimension are the same. Furthermore, since the half-spaces $\lambda \geq 0$ have quite obviously the boundary $\lambda = 0$, the more general new manifold M has its own *boundary* ∂M, which consists of the points x at which $\lambda(x) = 0$ in some coordinate system. The points in $M \setminus \partial M$ are usually called interior points of M, but this has no relation with the interior of M in $\mathbb{R}^p$.

In the end, all of this is consistent because a diffeomorphism of half-spaces preserves their boundaries. We remark here that this is true by the differentiable version of the Invariance of Domain Theorem. This differentiable version admits a thoroughly elementary proof, in dramatic contrast with the general theorem, which we will prove later twice (IV.5.6, p. 167, and IV.7.5, p. 181), by means of Euclidean degree theory.

Thus we have boundaryless manifolds and manifolds with boundary. In particular, if M is a manifold with boundary, of dimension m, then ∂M is a boundaryless manifold of dimension $m-1$, and $M \setminus \partial M$ is also a boundaryless manifold of dimension m.

Finally, we remark that the arguments of II.2.3, p. 55, applied to one single differentiable function $f : U \to \mathbb{R}$, show that if the partial derivatives

$\frac{\partial f}{\partial x_i}$ do not vanish simultaneously at any $x \in U$ with $f(x) = 0$, then the inequality $f \geq 0$ defines a manifold with boundary defined by $f = 0$.

(2.6) Product of manifolds. It is clear that the product of two manifolds is also a manifold, at least if boundaries do not interfere. In fact, we will only consider (i) the product of boundaryless manifolds, again boundaryless, and (ii) the product of a manifold N with boundary and a boundaryless manifold M, which has the boundary $\partial(N \times M) = \partial N \times M$. For instance, if $N = [0,1]$ is an interval, we get the disjoint union of two copies of M, namely

$$\partial([0,1] \times M) = M_0 \cup M_1, \quad M_0 = \{0\} \times M,\, M_1 = \{1\} \times M.$$

It is also clear that the dimension of a product is the sum of the dimensions of the factors, and the class of the product is the smallest class of a factor.

(2.7) Smoothness. The study of mappings of $\mathcal{C}^r$ manifolds is not restricted to those of class $\mathcal{C}^r$. In fact, the main interest concerns continuous mappings, not even differentiable. However, dealing with the class $\mathcal{C}^r$ involves quite a bit of difficulty, mainly due to the loss of differentiability that occurs in many constructions. This problem disappears completely for class ∞, and we will mainly take the approach of restricting our presentation to smooth manifolds.

This restriction is supported by the fact that actually *all manifolds are smooth.* This means that by looking only at smooth manifolds, we just deprive the reader of some technical diversion, but all results we prove apply to arbitrary manifolds. In any case, the consideration of finite class mappings is of relevance even on smooth manifolds and is actually a crucial matter for extending the theory to arbitrary $\mathcal{C}^r$ manifolds. We will confine this discussion to a special section in the proper place (III.4), always in a concentrated form to remain faithful to our general continence policy concerning technicalities.

The essential concept that makes manifolds the suitable object to develop calculus is that of tangency.

(2.8) Tangent space. Let M be a differentiable manifold, and let $x \in M$. Pick any parametrization $\varphi : W \to M$ at x (with $W \subset \mathbb{R}^m$ or $\mathbb{H}^m$), and let $a \in W$ be such that $\varphi(a) = x$. This φ is in fact a differentiable mapping $W \to \mathbb{R}^p$, and we have its derivative $d_a\varphi : \mathbb{R}^m \to \mathbb{R}^p$. One easily checks that the linear space $\operatorname{im}(d_a\varphi) \subset \mathbb{R}^m$ does not depend on the choice of φ:

it is the *tangent space to M at x* and is denoted by T_xM. The vectors in T_xM are called *tangent to M at x*.

It is easy to produce a basis for T_xM. Just consider the canonical basis $e_i = (0, \dots, 1, \dots, 0)$ of $\mathbb{R}^m$ and set

$$\frac{\partial}{\partial \mathbf{x}_i}\bigg|_x = d_a\varphi(e_i) = \frac{\partial \varphi}{\partial x_i}(a) = \Big(\frac{\partial \varphi_1}{\partial x_i}(a), \dots, \frac{\partial \varphi_p}{\partial x_i}(a)\Big).$$

Notice that this basis of T_xM consists of the columns of the Jacobian matrix of $d_a\varphi : \mathbb{R}^m \to \mathbb{R}^p$. The partial derivative notion comes from the fact that one can look at any vector $u \in \mathbb{R}^n$ as a *directional derivative*. We will not pursue this fruitful viewpoint here, but we keep the notation of partial derivatives anyway.

In a different way, one can use local equations to describe the tangent space. Namely, if $f_1, \dots, f_q$ are local equations of M in a neighborhood U of x (II.2.3, p. 55), then $u \in T_xM$ if and only if

$$d_xf_1(u) = 0, \dots, d_xf_q(u) = 0$$

(this is generalized later in II.3.2, p. 62). For one single equation $f = f_1$, that is, when M is a hypersurface in $\mathbb{R}^{m+1}$, we have

$$0 = d_xf(u) = \langle \operatorname{grad}_x(f), u\rangle, \quad \operatorname{grad}_x(f) = \Big(\frac{\partial f}{\partial x_1}(x), \dots, \frac{\partial f}{\partial x_p}(x)\Big);$$

that is, T_xM is the linear hyperplane perpendicular to the *gradient* $\operatorname{grad}_x(f)$.

Finally, note that for a product $M \times N$ of two manifolds we have

$$T_{(x,y)}(M \times N) = T_xM \times T_yN.$$

Example 2.9. Let us look again at spheres $\mathbb{S}^m$. One can use the stereographic projection π_N to obtain a basis $T_x\mathbb{S}^m$ at the south pole $x = a_S$:

$$\frac{\partial}{\partial \mathbf{x}_i}\bigg|_x = 2e_i \quad (1 \le i \le m).$$

Similarly, π_S gives the same basis of $T_x\mathbb{S}^m$ at the north pole $x = a_N$. We see that $T_x\mathbb{S}^m \perp x$ in both cases. This is in fact true for all $x \in \mathbb{S}^m$, and we invite the reader to pursue the computations at an arbitrary x. However, it is much easier to use the equation $f(x) = \sum_i x_i^2 - 1 = 0$: T_xM is the hyperplane perpendicular to the gradient $\operatorname{grad}_x(f) = 2x$. □

Now we can define the following:

(2.10) Derivative of a differentiable mapping. Let $f : M \to N$ be a differentiable mapping of differentiable manifolds, and consider a point $x \in M$. We can extend f to a differentiable mapping $\bar{f}$ from some open neighborhood of x in $\mathbb{R}^p$ into some open neighborhood of $y = f(x)$ in $\mathbb{R}^q$, and this extension has a derivative $d_x\bar{f} : \mathbb{R}^p \to \mathbb{R}^q$. Then, it is easy to check that by restriction we get a well-defined linear mapping

$$d_x f = d_x \bar{f}|_{T_x M} : T_x M \to T_y N$$

which does not depend on the choice of $\bar{f}$ and is called the *derivative of f at x*.

Alternatively, we can *localize,* that is, pick coordinate domains U at x and V at y such that $f(U) \subset V$, and build the following commutative diagrams:

$$\begin{array}{ccc} U & \xrightarrow{f|_U} & V \\ \varphi \uparrow \approx & & \approx \downarrow \psi^{-1} \\ A & \xrightarrow{g\,=\,\psi^{-1}\circ f\circ\varphi} & B \end{array} \qquad\qquad \begin{array}{ccc} T_x M & \xrightarrow{d_x f} & T_y N \\ d_a\varphi \uparrow \approx & & \approx \downarrow (d_b\psi)^{-1} \\ \mathbb{R}^m & \xrightarrow{d_a g} & \mathbb{R}^n \end{array}$$

(where φ is the parametrization of M at x, with $\varphi(a) = x$, and ψ is that of N at y, with $\psi(b) = y$). One sees that the matrix of the linear mapping $d_a f$ with respect to the bases of the partial derivatives in both spaces $T_x M$ and $T_y N$ is the Jacobian matrix of

$$d_a g = (d_b\psi)^{-1} \circ d_x f \circ d_a\varphi.$$

We insist that the consideration of manifolds is necessary to have derivatives in any reasonable sense: the notion of differentiable mappings was given from the very beginning with no restriction on the sets involved, but for derivatives we need tangent spaces, available only for manifolds. Then, once derivatives are defined, calculus follows: chain rule, etc. The only caution concerns the boundary sometimes. For instance, it is easy to see that *if a differentiable function $f : M \to \mathbb{R}$ has a local extreme at a point $x \in M \setminus \partial M$, then the derivative $d_x f : T_x M \to \mathbb{R}$ is identically zero.*

In the same vein, the Inverse Mapping Theorem must be modified to read: a *differentiable mapping $f : M \to N$ is a local diffeomorphism at x if and only if $d_x f$ is a linear isomorphism and f preserves boundaries locally at x.* Of course, if M is boundaryless, the latter condition is meaningless.

(2.11) Tangent vectors and curve germs. We can describe tangencies in a very geometrical way: *a tangent vector to M at x is always tangent to some curve germ in M.* This means that for every $u \in T_xM$ there are differentiable mappings $\gamma : [0, \varepsilon) \to M \subset \mathbb{R}^p$ such that $\gamma(0) = x$ and

$$u = \gamma'(0) = (\gamma_1'(0), \ldots, \gamma_p'(0)) = d_0\gamma(1) \in T_xM \subset \mathbb{R}^p.$$

Furthermore, if $f : M \to N$ is differentiable, $d_xf(u) = (f \circ \gamma)'(0)$.

Example 2.12. Here there is a simple application of tangent vectors: *if a curve and a manifold meet* transversally *at a point, then the point is isolated in their intersection.*

Let Γ be a curve and M a manifold, both in $\mathbb{R}^p$, and let $a \in \Gamma \cap M$. The hypothesis says that no tangent vector to Γ at a is tangent to M at a. For the proof we can suppose that a is the origin in $M = \mathbb{R}^m \times \{0\}$. Then choose a parametrization $\gamma : t \mapsto (\gamma_i(t))$ of Γ with $\gamma(0) = a$, and we know that $\gamma'(0) \notin T_aM = \mathbb{R}^m \times \{0\}$. Hence $\gamma_i'(0) \neq 0$ for some $i > m$, so that $\gamma_i(t)$ is strictly monotonous near 0 and 0 is an isolated zero. This means that $\gamma(t) \notin M = \mathbb{R}^m \times \{0\}$ for small enough $t \neq 0$, as claimed. $\square$

Exercises and problems

Number 1. Prove that a subset $M \subset \mathbb{R}^p$ is a differentiable manifold if and only if for every point $x \in M$ there are an open set A in some half-space $\mathbb{H}^m$ and a differentiable mapping $\psi = (\psi_1, \ldots, \psi_p) : A \to \mathbb{R}^p$ such that (i) ψ is a homeomorphism onto some open neighborhood of x in M and (ii) the Jacobian matrix $\left(\frac{\partial \psi_i}{\partial x_j}\right)$ has rank m at $\psi^{-1}(x)$.

Number 2. Show the following:

(1) The *cusp* in $\mathbb{R}^2$ given by the equation $x^2 = y^3$ is not a smooth curve, but it is homeomorphic to $\mathbb{R}$.

(2) The *semi-cone* in $\mathbb{R}^3$ defined by $x^2 + y^2 = z^2$, $z \geq 0$, is not a smooth surface, but it is homeomorphic to $\mathbb{R}^2$.

Number 3. (1) Show that the *solid* torus $M \subset \mathbb{R}^3$ generated by the disk $y = 0$, $(x-2)^2 + z^2 \leq 1$ around the z-axis is a manifold with boundary the torus generated by the circle $y = 0, (x-2)^2 + z^2 = 1$.

(2) Show that the subsets $M_1 : x_1^2 + x_2^2 \leq \frac{1}{2}$ and $M_2 : x_3^2 + x_4^2 \leq \frac{1}{2}$ of the unit sphere $\mathbb{S}^3 \subset \mathbb{R}^4 : x_1^2 + x_2^2 + x_3^2 + x_4^2 = 1$ are diffeomorphic to M.

(3) Deduce that $\mathbb{S}^3$ is the union of two solid tori along their boundaries, so that the meridians of each one are the parallels of the other.

Number 4. Let M be a Hausdorff topological space M with a countable basis of open sets. A *chart* on M is a homeomorphism $\mathtt{x} : U \mapsto \mathbb{R}^m$ from an open set U of M onto an open set $\mathtt{x}(U) \subset \mathbb{R}^m$. Two charts $\mathtt{x}$ and $\mathtt{y}$ with domains U and V are called C^r

compatible (we fix the differentiability class r henceforth) when the mapping (of open sets in Euclidean spaces)

$$\mathtt{y} \circ \mathtt{x}^{-1} : \mathtt{x}(U \cap V) \to \mathtt{y}(U \cap V)$$

is differentiable. A *differentiable atlas* in M is a collection of $\mathcal{C}^r$ compatible charts whose domains cover M; when such an atlas exists, we say that M is an *abstract* (boundaryless) manifold.

(1) Check that the *concrete* manifolds $M \subset \mathbb{R}^p$ defined in the text are also *abstract* manifolds. Check that dimension is defined well and is the same in the abstract as in the concrete setting.

(2) Show that real and complex projective spaces $\mathbb{RP}^m$ and $\mathbb{CP}^m$ are abstract smooth manifolds, using homogeneous coordinates $(x_0 : \ldots : x_m)$ to define charts with domains $x_i \neq 0$. Compute dimensions.

Number 5. A continuous mapping $f : M \to N$ of abstract manifolds is $\mathcal{C}^r$ if for every two charts $\mathtt{x}, \mathtt{y}$ on M, N, with domains U, V such that $f(U) \subset V$, the localization $\mathtt{y} \circ f \circ \mathtt{x}^{-1} : \mathtt{x}(U) \to \mathtt{y}(V)$ is $\mathcal{C}^r$. Then we have the notion of diffeomorphism, and the loop closes, because every abstract manifold M is diffeomorphic to some concrete manifold.

(1) Check that differentiable mappings between *concrete* manifolds as defined in this section are exactly differentiable mappings in the abstract sense.

(2) Prove that $\mathbb{RP}^m$ is diffeomorphic to a compact manifold in some $\mathbb{R}^p$, using the mapping

$$(x_0 : \ldots : x_m) \mapsto \left(\frac{x_i x_j}{\sum_k x_k^2} \right)_{ij}.$$

How can this be adapted to $\mathbb{CP}^m$?

(3) Show that the real projective line is diffeomorphic to $\mathbb{S}^1$ and the complex projective line is diffeomorphic to $\mathbb{S}^2$.

Number 6. Let $\rho : \mathbb{R}^{n+1} \setminus \{0\} \to \mathbb{S}^n$ be the *radial retraction* given by $x \mapsto x/\|x\|$. For every point $a \in \mathbb{R}^{n+1}, a \neq 0$, let S_a be the sphere centered at the origin with radius $\|a\|$, and denote $H_a = T_a S_a$. Show, without explicit computations, that $d_a\rho(a) = 0$ and $d_a\rho|H_a$ is the homothecy of ratio $1/\|a\|$. Deduce from this an expression for the derivative $d_a\rho(u)$.

Number 7. Let $k \geq 1$ be an integer. Show that the set $M_k \subset \mathbb{R}^3$ given by $x^{2k} + y^{2k} + z^{2k} = 1$ is a smooth surface and that the radial retraction $x \mapsto x/\|x\|$ induces a diffeomorphism from M_k onto the unit sphere $\mathbb{S}^2$.

Number 8. Let $M \subset \mathbb{R}^p$ be a differentiable $\mathcal{C}^r$ $(r \geq 2)$ manifold of dimension m. Prove that the *tangent bundle*

$$TM = \{(x, u) \in M \times \mathbb{R}^p : u \in T_x M\} \subset \mathbb{R}^p \times \mathbb{R}^p$$

is a differentiable $\mathcal{C}^{r-1}$ manifold of dimension $2m$. Show also that the *Stiefel bundle*

$$VM = \{(x, u_1, \ldots, u_m) \in M \times \mathbb{R}^p \times \cdots \times \mathbb{R}^p : u_1, \ldots, u_m \text{ is a basis of } T_x M\}$$

is a non-compact differentiable $\mathcal{C}^{r-1}$ manifold of dimension $m(m+1)$.

3. Regular values

In this section we come to the key notion behind most constructions in differential topology. We refer to *transversality*. However, we do not need it in full generality, and for this reason we restrict our discussion to *regular values,* which is the first instance of transversality.

(3.1) Regular values. Let $f : M \to N$ be a differentiable mapping. A *critical point of* f is a point $x \in M$ whose derivative $d_x f : T_x M \to T_{f(x)} N$ is not surjective. Critical points form a closed set, because the condition on their derivatives can be expressed by some equations involving the Jacobian determinant of a localization of f. The non-critical points are called *regular points,* and they form an open set.

Now we look at $a \in N$: we say that a is a *critical value of* f if there is some critical point $x \in M$ with $f(x) = a$. Hence, the set of critical values of f is the image of that of its critical points but need not be closed (unless f itself is closed, e.g., when M is compact). If $a \in N$ is not a critical value, then we call it a *regular value of* f.

We denote by $C_f \subset M$ the set of critical points of f and by $R_f \subset N$ the set of regular values of f. Hence $R_f = N \setminus f(C_f)$.

Notice that if $\dim(M) < \dim(N)$, then $C_f = M$ and $R_f = N \setminus f(M)$.

Now we recognize that when we defined equations $f_1 = \cdots = f_q = 0$ of a manifold $M \subset \mathbb{R}^p$ on an open set $U \subset \mathbb{R}^p$ (II.2.3, p. 55), we just said that 0 was a regular value of the mapping $f = (f_1, \ldots, f_q) : U \to \mathbb{R}^q$. In fact, the main feature of regular values is that their inverse images are manifolds. This is an easy consequence of the Inverse Mapping Theorem but must be stated with some care to keep track of boundaries:

Theorem 3.2. *Let* $f : M \to N$ *be a differentiable* C^r *mapping, and let* $a \in N \setminus \partial N$ *be a regular value of both* f *and* $f|_{\partial M}$. *Then*

(1) $f^{-1}(a)$ *is a manifold with boundary* $f^{-1}(a) \cap \partial M$, *whose dimension is* $\dim(M) - \dim(N)$.

(2) $T_x f^{-1}(a) = \ker(d_x f)$ *for every* $x \in f^{-1}(a)$.

Concerning the proof of this result we will at least mention the following. For every $x \in f^{-1}(a)$ there are coordinate systems $(\mathrm{x}_1, \ldots, \mathrm{x}_m)$ of M at x and $(\mathrm{y}_1, \ldots, \mathrm{y}_n)$ of N at a such that $\mathrm{y}_i \circ f = \mathrm{x}_{m-n+i}$ for $i = 1, \ldots, n$ (and

$\mathrm{x}_1 \geq 0$ where the boundary is involved). In other words, the localization of f near x is a linear projection $(x_1, \ldots, x_m) \mapsto (x_{m-n+1}, \ldots, x_m)$. Then

$$\begin{cases} d_x f\big(\frac{\partial}{\partial \mathrm{x}_{m-n+i}}\big|_x\big) = \frac{\partial}{\partial \mathrm{y}_i}\big|_a & \text{for } i = 1, \ldots, n, \text{ and} \\ \frac{\partial}{\partial \mathrm{x}_j}\big|_x,\ j = 1, \ldots, m-n, & \text{generate } T_x f^{-1}(a). \end{cases}$$

In fact, $(\mathrm{x}_1, \ldots, \mathrm{x}_{m-n})$ is a coordinate system of our inverse image at x.

We will use this result mainly when $\dim(M) - \dim(N) = 1$, so that $f^{-1}(a)$ is a curve. Then it is important to know that curves always have global parametrizations:

Theorem 3.3 (Classification of compact differentiable curves). *Let C be a compact $\mathcal{C}^1$ curve. Then C is $\mathcal{C}^1$ diffeomorphic to either the unit circle $\mathbb{S}^1 \subset \mathbb{R}^2$ or the closed interval $[0,1] \subset \mathbb{R}$.*

Of course, to use these theorems we need to find regular values, but this is the famous Sard-Brown Theorem:

Theorem 3.4 (Sard-Brown Theorem). *Let $f : M \to N$ be a differentiable $\mathcal{C}^r$ mapping, with $r > \dim(M) - \dim(N)$. Then the set of regular values of f is residual (a countable intersection of dense open sets), hence dense, in N.*

In particular, if $\dim(M) < \dim(N)$, then $N \setminus f(M)$ is residual, hence dense, in N.

The last assertion of the theorem is sometimes called the *Easy Sard Theorem*, because it can be proved directly by an easier argument.

We remark that the proof of this theorem for finite class $r < +\infty$ is technically very demanding, as it requires the *Kneser-Glaeser Theorem* to restore the differentiability class lost by partial derivation. Of course, this is no problem in the smooth case $r = +\infty$, and the proof is much easier in this case. Furthermore, this is enough for us, as we will reduce the differentiable case to the smooth one. In fact, as mentioned before (II.2.7, p. 57), *every differentiable manifold has a smooth model*, and this deep result roughly says that our limitation to smooth manifolds is only technical.

After all these considerations, we include here the following proof:

Proof of the Sard-Brown Theorem for smooth mappings. First of all notice that since $M \subset \mathbb{R}^p$ is locally compact, we can apply the Baire Theorem, so that *any countable intersection of dense open sets of M is again dense,* which is the last remark in the statement.

Since manifolds have countable bases of open sets, one easily sees that the statement is of a local nature; hence we can assume that M, N are open sets of Euclidean spaces or half-spaces and that f is the restriction of a smooth mapping $g : U \to \mathbb{R}^q$ defined on an open subset U of $\mathbb{R}^p$. Clearly, the result for f follows from the result for g; hence we can simply suppose $f = g$ (this means we can disregard boundaries). We will prove the following statement, which in fact belongs to measure theory:

($*$) *The image of the set C_f of critical points of f has measure zero.*

Recall that a set has *measure zero* when it is contained in a countable union of cubes whose total volume can be taken arbitrarily small. It is easy to see that *a countable union of measure zero sets also has measure zero,* so that being measure zero is a local question. It is also immediate that a measure zero set has empty interior. In our case, this implies that R_f is residual. Indeed, C_f is closed, hence a countable union of compact sets, so that $f(C_f) = \mathbb{R}^q \setminus R_f$ is also a countable union of compact (hence closed) sets, each one contained in $f(C_f)$, hence each one with empty interior.

We now prove the above assertion ($*$). The proof is Milnor's by induction on p. Assume the result for smooth mappings with domain in Euclidean spaces of dimension $< p$. Denote

$$C_i = \{x \in U : \text{ all partial derivatives of order } \leq i \text{ vanish at } x\}$$

and consider the sequence of closed sets

$$C_f \supset C_1 \supset C_2 \supset C_3 \supset \cdots .$$

We split the argument into several steps.

(a) *The set $f(C_f \setminus C_1)$ has measure zero.*

We will see that every point $a \in C_f \setminus C_1$ has a neighborhood V such that $f(C_f \cap V)$ has measure zero. As we can cover $C_f \setminus C_1$ by countably many such neighborhoods, this implies (a).

Fix a and choose some partial derivative of f that does not vanish at a, say $\frac{\partial f_1}{\partial x_1}(a) \neq 0$. Then $h(x) = (f_1(x), x_2, \ldots, x_p)$ is a local diffeomorphism at a, and we can replace f by $f \circ h^{-1}$ on a neighborhood V of a. In other words, we can suppose $f_1(x) = x_1$ on V, so that f preserves every hyperplane $x_1 = t$; we denote by g the restriction of f to that hyperplane. Notice that for $x = (t, x') \in V$ it is

$$Df(x) = \left(\begin{array}{c|c} 1 & 0 \\ \hline * & Dg(x') \end{array}\right)$$

so that if $Df(x)$ is not surjective, $Dg(x')$ is not either. This means that if $x \in C_f \cap V$, then $x' \in C_g$. Thus we have

$$f(C_f \cap V) \cap \{x_1 = t\} \subset f(C_f \cap \{x_1 = t\}) \subset g(C_g),$$

and the latter set has measure zero by induction. Summing, $f(C_f \cap V)$ cuts each hyperplane $x_1 = t$ on a measure zero set, and *the Fubini Theorem* (which has an elementary proof for measure zero sets) guarantees that $f(C_f \cap V)$ has measure zero, as desired.

(b) *Each set $f(C_i \setminus C_{i+1})$ has measure zero.*

As before, we see that every point $a \in C_i \setminus C_{i+1}$ has a neighborhood V such that $f(C_i \cap V)$ has measure zero. By definition, there is a partial derivative w of f of order i that vanishes at a and such that $\partial w(a)/\partial x_j \neq 0$; for ease of notation, we suppose $j = 1$. Then $h(x) = (w(x), x_2, \dots, x_p)$ is a local diffeomorphism at a, and on some neighborhood V of a the inverse mapping $h^{-1} : W \to V$ is well defined and smooth. We have $h(C_i \cap V) \subset \{x_1 = 0\}$, and every point in $h(C_i \cap V)$ is a critical point of the restriction $g = f \circ h^{-1}|W \cap \{x_1 = 0\}$. Hence, $f(C_i \cap V) = g(h(C_i \cap V)) \subset g(C_g)$, and the latter set has measure zero by induction. We have proved (b).

(c) *If $i > p/q - 1$, the set $f(C_i)$ has measure zero.*

Consider a compact cube $K \subset U$ of side $\delta > 0$, and let us see that $f(C_i \cap K)$ has measure zero. Since C_i can be covered by countably many such K's, the conclusion follows. Now, by Taylor expansion and the definition of C_i, we have

$$f(y) = f(x) + R(x, y), \quad \|R(x, y)\| \leq c\|y - x\|^{i+1}$$

for any $x, y \in C_i \cap K$. The constant c depends only on f and K. Now, subdivide K into n^p cubes L with side δ/n and diagonal $\sqrt{p}\delta/n$. Choose $x \in C_i \cap L$, so that for every $y \in L$,

$$\|f(y) - f(x)\| \leq \|R(x, y)\| \leq c\|y - x\|^{i+1} \leq \rho = c\Big(\frac{\sqrt{p}\delta}{n}\Big)^{i+1}.$$

Thus, $f(L)$ is contained in the cube L^* with center $f(x)$ and side 2ρ. Consequently, $f(C_i \cap K)$ is contained in the union of n^p cubes with side 2ρ, whose total volume is

$$c' n^{p-q(i+1)}.$$

Since $i > p/q - 1$, this volume goes to zero when $n \to \infty$. This proves (c).

It is clear that the three facts (a), (b), and (c) imply (*), and this completes the proof of the Sard-Brown Theorem for smooth mappings. □

Exercises and problems

Number 1. Let P be a homogeneous polynomial of degree d in $n+1$ indeterminates.

(1) Prove that

$$\sum_{i=1}^{n+1} x_i \frac{\partial P}{\partial x_i} = d \cdot P.$$

(2) Show that every equation $P = \varepsilon$, $\varepsilon \neq 0$, defines a smooth hypersurface M_ε in $\mathbb{R}^{n+1}$ (except if there are no zeros).

(3) Prove that M_ε is diffeomorphic to M_1 for $\varepsilon > 0$ and to M_{-1} for $\varepsilon < 0$.

(4) Also, must M_1 and M_{-1} be diffeomorphic? How does d affect this?

(5) What about the equation $P = 0$?

Number 2. Let $M \subset \mathbb{R}^n$ be a non-empty real quadric given by

$$\sum_{1 \leq i,j \leq n} a_{ij} x_i x_j = 1,$$

with $\det(a_{ij}) \neq 0$. Show that M is a smooth hypersurface diffeomorphic to a product $\mathbb{S}^k \times \mathbb{R}^\ell$, for suitable k and ℓ.

Number 3. Prove that the set $M \subset \mathbb{R}^{m\times n}$ of all matrices of rank k is a smooth manifold of codimension $(m-k)(n-k)$.

Number 4. Let $\Sigma \subset \mathbb{R}^{n\times n}$ be the linear space of all symmetric matrices of order n, and consider the smooth mapping $f : \mathbb{R}^{n\times n} \to \Sigma : A \mapsto A^t A$.

(1) Show that $d_A f(B) = A^t B + B^t A$, $A, B \in \mathbb{R}^{n\times n}$, and deduce that the *orthogonal group* $O(n) = f^{-1}(I) \subset \mathbb{R}^{n\times n}$ is a compact manifold of dimension $n(n-1)/2$.

(2) Prove that $O(n)$ has two diffeomorphic connected components, one, $SO(n)$, defined by the condition $\det = +1$.

(3) Compute the tangent space to $O(n)$ at $A = I$. At which other $A \in O(n)$ is the tangent space the same?

Number 5. Let $M \subset \mathbb{R}^p$ be a differentiable $\mathcal{C}^r$ $(r \geq 2)$ manifold of dimension m. Prove that the *(orthonormal) Stiefel bundle*

$$OM = \{(x, u_1, \ldots, u_m) \in M \times \mathbb{R}^p \times \cdots \times \mathbb{R}^p : u_1, \ldots, u_m \text{ is an orthonormal basis of } T_x M\}$$

is a differentiable $\mathcal{C}^{r-1}$ manifold of dimension $\frac{1}{2}m(m+1)$, compact if and only if M is. (Compare with Problem 8 in II.2.)

Number 6. Let $U \subset \mathbb{R}^m$ be an open set, and let $f : U \to \mathbb{R}^n$ be a smooth mapping, $m \geq n$. Suppose that 0 is a critical point and $f(0) = 0$. Consider the matrices $m \times n$ as elements of $\mathbb{R}^{m\times n}$ and consider

$$F : (U \setminus \{0\}) \times \mathbb{R}^{m\times n} \to \mathbb{R}^n : (x, A) \mapsto f(x) + Ax.$$

(1) Check that the origin is a regular value of F, hence that $M = F^{-1}(0)$ is a smooth manifold.

(2) Show that if $A \in \mathbb{R}^{m\times n}$ is a regular value of the projection $\pi : M \to \mathbb{R}^{m\times n}$, then the mapping

$$F_A : U \to \mathbb{R}^n : x \mapsto f(x) + Ax$$

has no critical point $x \neq 0$ with value $F_A(x) = 0$.

(3) Conclude that there are A's with entries arbitrarily small such that the origin is the only critical point with value 0.

Number 7. A *real projective hypersurface of degree* k is a subset $M \subset \mathbb{RP}^m$ defined by a homogeneous equation $P(x_0, \ldots, x_m) = 0$ of degree k.

(1) Under what conditions on P is M a smooth manifold?

(2) Show that if $k = 2\ell$ is even,

$$f : \mathbb{RP}^m \to \mathbb{R} : (x_0 : \ldots : x_m) \mapsto \frac{P(x_0, \ldots, x_m)}{(x_0^2 + \cdots + x_m^2)^\ell}$$

is a well-defined smooth mapping. When is 0 a regular value of f?

(3) What about *real projective hyperplanes* $H \subset \mathbb{RP}^m$, which are defined by *linear* homogeneous equations?

4. Tubular neighborhoods

Here we recall the constructions of tubular neighborhoods. These are a key ingredient in the next section for the approximation results that reduce homotopy to smooth homotopy, but, also, we will need them in a crucial way to prove the Poincaré-Hopf Index Theorem at the very end this book (V.7.2, p. 219).

The essential fact is that *every differentiable boundaryless manifold in a Euclidean space* $\mathbb{R}^p$ *is the differentiable retraction of some neighborhood of it in* $\mathbb{R}^p$. We first take a purely topological approach that preserves differentiability class well:

Proposition 4.1. *Let* $M \subset \mathbb{R}^p$ *be a boundaryless differentiable* $\mathcal{C}^r$ *manifold. Then there is an open neighborhood* U *of* M *in* $\mathbb{R}^p$ *and a* $\mathcal{C}^r$ retraction $\rho : U \to M$ *(that is, a* $\mathcal{C}^r$ *mapping with* $\rho(x) = x$ *for* $x \in M$*).*

Proof. Since the pair $M \subset \mathbb{R}^p$ is locally $\mathcal{C}^r$ diffeomorphic to the pair $\mathbb{R}^m \subset \mathbb{R}^p$ (II.2.3, p. 55) and linear projections $\mathbb{R}^p \to \mathbb{R}^m$ are smooth retractions, each point $x \in M$ has an open neighborhood U in $\mathbb{R}^p$ equipped with a

$\mathcal{C}^r$ retraction $\rho : U \to U \cap M$. Henceforth the proof consists of a careful glueing of such local retractions. We split the argument into several steps. First we note that since M is locally closed in $\mathbb{R}^m$, it is closed in some open neighborhood Ω in $\mathbb{R}^p$; *for the rest of the proof, all closures are considered inside this open set* Ω. Also, all mappings are $\mathcal{C}^r$.

Step I: Glueing two local retractions. Let $\sigma : V \to V \cap M$ and $\tau : W \to W \cap M$ be two retractions defined on open sets V, W of Ω. Then there is a retraction

$$\rho : U \to U \cap M = (V \cup W) \cap M.$$

If $T = V \cap W \cap M = \emptyset$, there are disjoint open sets $V_1, W_1 \subset \Omega$ such that $V \cap M \subset V_1 \subset V$ and $W \cap M \subset W_1 \subset W$. Indeed, in the metric space $V \cup W$, the two disjoint closed sets $V \setminus W$, $W \setminus V$ can be separated by disjoint open sets $V_1 \subset V$, $W_1 \subset W$. Now, we take $U = V_1 \cup W_1$, and ρ is the trivial glueing of two mappings defined on disjoint open sets. After this remark we assume $T \neq \emptyset$.

The idea is to glue σ and τ through a homotopy that deforms one retraction into the other over T. To that end, consider $D = \sigma^{-1}(T) \cap \tau^{-1}(T)$. This set is open in Ω, hence in $\mathbb{R}^p$, and the set

$$E = \{x \in D : D \supset [\sigma(x), \tau(x)]\} \supset T$$

is open too. Indeed, suppose D contains a segment $[\sigma(x), \tau(x)]$. Since D is open in $\mathbb{R}^p$, it contains a *convex* open neighborhood Q of that segment, and then $\sigma^{-1}(Q) \cap \tau^{-1}(Q) \subset E$ is an open neighborhood of x. Then we can define mappings H_t, $0 \leq t \leq 1$, on E by

$$H_t(x) = \sigma\big((1-t)\sigma(x) + t\tau(x)\big).$$

Next, notice that the disjoint sets $V \cap M \setminus W$ and $W \cap M \setminus V$ are closed in $V \cup W$. For instance $V \cap M$ is closed in V, and hence $V \cap M \setminus W$ is closed in $V \setminus W$, which is closed in $V \cup W$. Consequently, we can find a smooth bump function θ, which is $\equiv 0$ on a neighborhood $U_\sigma \subset V$ of $V \cap M \setminus W$ and $\equiv 1$ on a neighborhood $U_\tau \subset W$ of $W \cap M \setminus V$. Finally, we define the retraction ρ on the open set $U = U_\sigma \cup E \cup U_\tau$ by

$$\rho(x) = \begin{cases} \sigma(x) & \text{on } U_\sigma, \\ H_{\theta(x)}(x) & \text{on } E, \\ \tau(x) & \text{on } U_\tau. \end{cases}$$

Some straightforward computations show this construction is consistent.

Step II: Exhaustion of M by local retractions. There is a sequence of retractions $\rho_k : U_k \to M_k = U_k \cap M$ such that

$$M = \bigcup_k M_k \quad \text{and} \quad \overline{M}_k \subset M_{k+1}.$$

As mentioned before, we can cover M by local retractions $\tau : V \to V \cap M$, and then, since M has a countable basis of open sets, we get a sequence $\tau_k : W_k \to W_k \cap M$, with $M \subset \bigcup_k W_k$. Now, by Step I, for each k we can glue τ_k and τ_{k-1}, and τ_{k-2}, and so on, to obtain a retraction $\sigma_k : V_k \to V_k \cap M$, with

$$V_k \cap M = (W_1 \cup \cdots \cup W_k) \cap M.$$

Next, set

$$M_k = \{x \in M : \operatorname{dist}(x, M \setminus V_k) > \tfrac{1}{k}\}.$$

This is an open subset of $V_k \cap M$, and

$$\overline{M}_k \subset \{x \in M : \operatorname{dist}(x, M \setminus V_k) \geq \tfrac{1}{k} > \tfrac{1}{k+1}\} \subset M_{k+1},$$

because $V_k \subset V_{k+1}$. On the other hand, for every $x \in M$, say $x \in W_k \subset V_k$, we have

$$\begin{cases} \operatorname{dist}(x, M \setminus V_\ell) \geq \operatorname{dist}(x, M \setminus V_k) & \text{for } \ell \geq k \text{ (as } V_\ell \cap M \supset V_k \cap M), \\ \operatorname{dist}(x, M \setminus V_k) > \frac{1}{\ell} & \text{for } \ell \text{ large (as } \operatorname{dist}(x, M \setminus V_k) > 0), \end{cases}$$

and thus $x \in M_\ell$ for large ℓ. Consequently, $M = \bigcup_k M_k$. Finally, take $U_k = \sigma_k^{-1}(M_k)$ and $\rho_k = \sigma_k|_{U_k}$.

Step III: Splitting into two sequences of disjoint retractions and conclusion.

First, we shrink the domains of the retractions ρ_k of Step II as follows. Take targets

$$M'_1 = M_1, \quad M'_2 = M_2, \quad M'_k = M_k \setminus \overline{M}_{k-2} \text{ for } k \geq 3$$

and domains $U'_k = \rho_k^{-1}(M'_k)$. The virtue of this is that the M'_k's still cover M, since $\overline{M}_k \subset M_{k+1}$, but besides, we have

$$M'_k \cap M'_\ell = \emptyset \quad \text{if } k \equiv \ell \mod 2$$

(and $k \neq \ell$, of course). Indeed, suppose $\ell = k + 2n$. Then $k \leq \ell - 2$, and

$$\begin{cases} M'_k \subset M_k \subset M_{\ell-2}, \\ M'_\ell = M_\ell \setminus \overline{M}_{\ell-2}, \end{cases}$$

hence our assertion. Now put

$$V' = \bigcup_e U'_{2e-1}, \quad W' = \bigcup_e U'_{2e}.$$

These two sets are open, and $M \subset V' \cup W'$. Let us look at V' first. The sets $M'_{2e-1} = U'_{2e-1} \cap M \subset V' \cap M$ are closed in V' and form a discrete family.

Indeed, suppose that some $x \in M'_{2e_0-1}$ is the limit of a sequence of points $x_e \in M'_{2e-1} \subset M$. Then $x_e \in U'_{2e_0-1}$ for e large enough; hence $x_e \in U'_{2e_0-1} \cap M = M'_{2e_0-1}$, a contradiction.

Consequently, the sets M'_{2e-1} can be separated by disjoint open sets $A_{2e-1} \subset V'$, and ρ_{2e-1} restricts to a retraction $U'_{2e-1} \cap A_{2e-1} \to M'_{2e-1}$. All these restrictions glue trivially, because their domains are disjoint, to give another retraction

$$\rho_{\text{odd}} : \bigcup_e (U'_{2e-1} \cap A_{2e-1}) \to \bigcup_e M'_{2e-1}.$$

Similarly, starting with W', we get a second retraction

$$\rho_{\text{even}} : \bigcup_e (U'_{2e} \cap B_{2e}) \to \bigcup_e M'_{2e},$$

and we are done, because by Step I we can glue these two retractions to obtain $\rho : U \to M$, the one we sought. □

The preceding proof shows that being a retract of an open set in $\mathbb{R}^p$ is a local property. Hence, the differentiable content of the last theorem is the basic fact that manifolds are *locally* retracts. Actually, this local property characterizes manifolds:

Proposition 4.2. *Let $M \subset \mathbb{R}^p$ be a subset such that every point $x \in M$ has an open neighborhood U in $\mathbb{R}^p$ on which there is a differentiable $\mathcal{C}^r$ retraction $\rho : U \to U \cap M$. Then M is a boundaryless $\mathcal{C}^r$ manifold.*

Proof. Fix $x \in M$ and a $\mathcal{C}^r$ retraction $\rho : U \to U \cap M$, with $x = \rho(x) \in U \cap M$. After a translation, we can assume x is the origin. Denote $\pi = d_0\rho$, and define

$$h = (\mathrm{Id}_{\mathbb{R}^p} - \pi) \circ (\mathrm{Id}_{\mathbb{R}^p} - \rho) + \pi \circ \rho.$$

Since $\rho = \rho \circ \rho$, we have

$$\pi = d_0\rho = d_0(\rho \circ \rho) = \pi \circ \pi,$$

which readily gives $d_0 h = \mathrm{Id}_{\mathbb{R}^p}$. Hence h is a $\mathcal{C}^r$ diffeomorphism on an open neighborhood $V \subset U$ of the origin. Further computation shows that $h \circ \rho = \pi \circ h$; hence

$$h(W \cap M) = h(\rho(W)) = \pi(h(W))$$

for $W = V \cap \rho^{-1}(V)$. But $\pi(h(W))$ is open in $L = \pi(\mathbb{R}^p)$, because linear maps are open. Thus $W \cap M$ is $\mathcal{C}^r$ diffeomorphic to an open set of a linear space; that is, $W \cap M$ is a $\mathcal{C}^r$ manifold. □

There is a second and more popular method to construct retractions. We present this method now for *smooth manifolds*. As usual, we denote by $\perp$ perpendicularity in Euclidean spaces.

Proposition and Definition 4.3. *Let $M \subset \mathbb{R}^p$ be a boundaryless, smooth manifold of dimension m. Then the set*

$$\nu M = \{(x, u) \in \mathbb{R}^p \times \mathbb{R}^p : \ u \perp T_x M,\ x \in M\}$$

is a boundaryless, smooth manifold of dimension p, and the smooth mapping

$$\vartheta : \nu M \to \mathbb{R}^p : (x, u) \mapsto x + u.$$

induces a diffeomorphism from an open set $\Omega \supset M \times \{0\}$ onto another $U \supset M$. Consequently, we have the smooth retraction

$$\rho : U \to M : y = \vartheta(x, u) \mapsto x$$

and it follows that (maybe for a smaller U):

(1) *M is closed in U,*

(2) *$y - \rho(y) \perp T_{\rho(y)} M$ for all $y \in U$, and*

(3) $\mathrm{dist}(y, M) = \|y - \rho(y)\|$ *for all $y \in U$.*

In fact, (3) *characterizes ρ.*

We summarize the situation by saying that $\rho : U \to M$ is a tubular retraction of M in $\mathbb{R}^p$.

Proof. To start with, we prove that

(a) *νM is a boundaryless, smooth manifold of dimension p.*

Consider any local equations for M, that is, differentiable functions $f_1, \ldots, f_q$, $q = p - m$, defined on an open set $V \subset \mathbb{R}^p$, such that:

(i) $V \cap M = \{x \in V : f_1(x) = \cdots = f_q(x) = 0\}$ and

(ii) the rank of the Jacobian matrix $\left(\dfrac{\partial f_i}{\partial x_j}\right)$ is q at every point $x \in V \cap M$.

Condition (i) means that the gradients

$$\operatorname{grad}_x(f_i) = \left(\frac{\partial f_i}{\partial x_1}(x), \dots, \frac{\partial f_i}{\partial x_p}(x)\right), \quad i = 1, \dots, q,$$

are perpendicular to T_xM, and condition (ii) means that they are independent. Hence, they generate $T_xM^\perp$. Consequently we can define a bijection

$$(V \cap M) \times \mathbb{R}^q \to (V \times \mathbb{R}^p) \cap \nu M : (x, y) \mapsto (x, u)$$

by

$$(*) \qquad u = \sum_{i=1}^{q} y_i \operatorname{grad}_x(f_i).$$

This is in fact a diffeomorphism: its inverse is obtained by solving the linear system $(*)$ in the unknowns y_i. Since we can cover M by open sets like V, we see that νM is locally diffeomorphic to $M \times \mathbb{R}^q$, which is a manifold of dimension $m + q = p$.

Next we see that

(b) *ϑ is a local diffeomorphism at every point $(x, 0)$.*

Indeed, it is enough to check that $d_{(x,0)}\vartheta$ is a linear isomorphism. But ϑ induces by restriction the identifications

$$\vartheta|_{M \times \{0\}} : M \times \{0\} \equiv M \quad \text{and} \quad \vartheta|_{\{x\} \times T_xM^\perp} : \{x\} \times T_xM^\perp \equiv T_xM^\perp,$$

so that the image of $d_{(x,0)}\vartheta$ contains $T_xM + T_xM^\perp = \mathbb{R}^p$. Hence, that derivative is surjective, and since $\dim(\nu M) = \dim(\mathbb{R}^p)$, we are done.

Once we know that ϑ is a local diffeomorphism around $M \times \{0\}$, what remains to show is that

(c) *ϑ is injective on some neighborhood of $M \times \{0\}$.*

Clearly, ϑ is injective on the set

$$E = \{(x, u) \in \nu M : \|u\| < \tau(x)\},$$

where τ is defined as follows:

$\tau(x)$ is the infimum of all $s \in \mathbb{R}$ such that there are $(x,u),(y,v) \in \nu M$ with $\|v\| \leq \|u\| = s$, $x \neq y$, $\vartheta(x,u) = \vartheta(y,v)$.

Thus, we must check that E is a neighborhood of $M \times \{0\}$.

Fix $z \in M$. We already know that ϑ is injective in a neighborhood V of $(z,0)$, say $V : \|x-z\| + \|u\| < \varepsilon$ for a suitable $\varepsilon > 0$. We claim that

$$\tau(x) \geq \tfrac{1}{6}\varepsilon \quad \text{if} \quad \|x-z\| < \tfrac{1}{2}\varepsilon.$$

For, suppose $\|x-z\| < \frac{1}{2}\varepsilon$ and let (x,u), (y,v) in the definition of $\tau(x)$ verify $\|v\| \leq \|u\| < \frac{1}{6}\varepsilon$. Then $(x,u) \in V$ and $(y,v) \notin V$; moreover, $\|x-y\| = \|u-v\| \leq 2\|u\|$. Consequently:

$$\begin{aligned}\tfrac{1}{3}\varepsilon &> 2\|u\| \geq \|x-y\| \geq \|y-z\| - \|x-z\| \\ &\geq (\varepsilon - \|v\|) - \tfrac{1}{2}\varepsilon > (\varepsilon - \tfrac{1}{6}\varepsilon) - \tfrac{1}{2}\varepsilon = \tfrac{1}{3}\varepsilon,\end{aligned}$$

a contradiction.

Now, the claim implies

$$E \supset \{(z,u) \in \nu M : \ \|z-x\| < \tfrac{1}{2}\varepsilon,\ \|u\| < \tfrac{1}{6}\varepsilon\},$$

and the latter set is an open neighborhood of $(x,0)$.

Consequently, there is some open neighborhood Ω of $M \times \{0\}$ on which ϑ is an injective local diffeomorphism; hence

(d) *ϑ is a diffeomorphism from Ω onto an open neighborhood U of M.*

Now, of the three conditions on p. 71, (1) and (2) are immediate by construction; hence it only remains to show that, perhaps after shrinking U, condition (3) also holds true. But this is a local question: it suffices to prove that

(e) *every point $x \in M$ has a neighborhood $U^x \subset U$ on which* (3) *holds.*

Indeed, then one replaces U by $\bigcup_x U^x$. Thus, fix $x \in M$ to prove (e).

Choose a compact neighborhood L of x in M. Let W be the interior of L in M, and set $\varepsilon = \frac{1}{2}\operatorname{dist}(x, M \setminus W)$. Now, let $V = W \cap \{\|z-x\| < \varepsilon\}$, and choose $\delta < \frac{1}{2}\varepsilon$ such that $L \times \{\|u\| \leq \delta\} \subset \Omega$. We claim that $U^x = \vartheta(V \times \{\|u\| < \delta\})$ does the job.

Indeed, pick $y \in U^x$, that is, $\rho(y) \in V$ and $\|y - \rho(y)\| < \delta$. Let us estimate the distance $\operatorname{dist}(y, M)$. For $z \in M \setminus L$, the various choices above give the following sequence of bounds:

$$\begin{aligned}\|z-y\| &\geq \|z-x\| - \|x-y\| \geq 2\varepsilon - (\|x-\rho(y)\| + \|\rho(y)-y\|) \\ &> 2\varepsilon - (\varepsilon + \delta) = \varepsilon - \delta > \delta > \|y - \rho(y)\|,\end{aligned}$$

which show that

$$\begin{cases} \|y - z\| > \|y - \rho(y)\| & \text{for } z \in M \setminus L \text{ and} \\ \delta > \|y - \rho(y)\| \geq \operatorname{dist}(y, M) = \operatorname{dist}(y, L). \end{cases}$$

But L is compact; hence $\operatorname{dist}(y, L) = \|y - z\|$ for some $z \in L$. This means that the function $h : t \mapsto \|y - t\|^2$ has a global minimum on M at z, and consequently, its derivative vanishes on $T_z M$; equivalently, the gradient

$$\operatorname{grad}_z(h) = 2(y - z)$$

is perpendicular to $T_z M$, and $v = y - z \in T_z M^\perp$. We have $\|v\| = \|y - z\| = \operatorname{dist}(y, L) < \delta$, so that $(z, v) \in L \times \{\|u\| \leq \delta\} \subset \Omega$. Since ϑ is injective on Ω and $z + v = \rho(y) + (y - \rho(y))$, we conclude that $z = \rho(y)$ and

$$\operatorname{dist}(y, M) = \operatorname{dist}(y, L) = \|y - z\| = \|y - \rho(y)\|.$$

All of this also shows that $\rho(y)$ is the unique point $z \in M$ such that $\|y - z\| = \|y - \rho(y)\|$; hence (3) characterizes ρ. □

If one copies the above proof for differentiable manifolds of finite class $\mathcal{C}^r$, the set νM is a $\mathcal{C}^{r-1}$ manifold (the gradients of the local $\mathcal{C}^r$ equations are of class $\mathcal{C}^{r-1}$), and we end up with a $\mathcal{C}^{r-1}$ retraction. But notice the way in which we apply the Inverse Function Theorem on νM, which requires $r - 1 \geq 1$, that is, $r \geq 2$. These restrictions were avoided in II.4.1, p. 67, by purely topological means, but there no information concerning perpendicularity and distances was obtained. That information will be essential for the Poincaré-Hopf Index Theorem.

Exercises and problems

Number 1. Let $M \subset \mathbb{R}^{m+1}$ be a differentiable hypersurface and consider a point c off M. Prove that $a \in M$ is a critical point of the Euclidean distance $M \to \mathbb{R} : x \mapsto \operatorname{dist}(x, c)$ if and only if $a - c \perp T_x M$.

Number 2. Let $f : \mathbb{R}^p \to \mathbb{R}^n$ be a smooth map such that the origin is a regular value of f; hence $M = f^{-1}(0)$ is a boundaryless smooth manifold of dimension $m = p - n$. Prove that there is a diffeomorphism $\varphi : M \times \mathbb{R}^n \to \nu M$ such that $\vartheta \circ \varphi(x, u) = x$.

Number 3. Let $M \subset \mathbb{R}^p$ be a boundaryless smooth manifold of dimension m. Show that

$$N = \{(x, u, v) \in \mathbb{R}^p \times \mathbb{R}^p \times \mathbb{R}^p : x \in M, u \perp T_x M, v \in T_x M\}$$

is a boundaryless smooth manifold diffeomorphic to $M \times \mathbb{R}^p$.

Number 4. Let $M \subset \mathbb{R}^p$ be a compact boundaryless smooth manifold of dimension m. For $\varepsilon > 0$ we denote

$$\nu_\varepsilon M = \{(x, u) \in \nu M : \|u\| < \varepsilon\}.$$

Show that:

(1) For ε small enough, ϑ restricts to a diffeomorphism from $\nu_\varepsilon M$ onto an open neighborhood of M in $\mathbb{R}^p$.

(2) There is a diffeomorphism h from νM onto an open neighborhood of M in $\mathbb{R}^p$ such that $h(x,0) = x$ for every $x \in M$.

Number 5. Let $M \subset \mathbb{R}^p$ be a boundaryless smooth manifold of dimension m. For every continuous mapping $\varepsilon : M \to \mathbb{R}$ everywhere > 0 we denote

$$\nu_\varepsilon M = \{(x,u) \in \nu M : \|u\| < \varepsilon(x)\}.$$

Show that:

(1) There is an ε such that ϑ restricts to a diffeomorphism from $\nu_\varepsilon M$ onto an open neighborhood of M in $\mathbb{R}^p$.

(2) There is a diffeomorphism h from νM onto an open neighborhood of M in $\mathbb{R}^p$ such that $h(x,0) = x$ for every $x \in M$.

5. Approximation and homotopy

In this section we show that homotopies can always be made differentiable, which is essential for studying homotopies of continuous mappings using differentiable methods.

As is well known, a *continuous homotopy* is a continuous mapping $H : [0,1] \times M \to N$ (of course, one can use any closed interval), and as customary, we usually denote $H_t(x)$ instead of $H(t,x)$. Thus a homotopy is a continuous uniparametric family of mappings $H_t : M \to N$. Two mappings $f, g : M \to N$ are *homotopic* if there is a continuous homotopy H_t with $H_0 = f$ and $H_1 = g$. As is well known, this is an equivalence relation, whose set of classes is denoted $[M,N]$. Here there is other standard notation: the *k-th homotopy group* is $\pi_k(N) = [\mathbb{S}^k, N]$ for N connected, and the *k-th cohomotopy group* is $\pi^k(M) = [M, \mathbb{S}^k]$ (this is indeed a group, because M is a manifold). This notion is enough for studying continuous mappings on compact manifolds, but here we will transfer the compactness assumption from manifolds to maps, which requires some additional care.

(5.1) Proper mappings. A continuous mapping $f : M \to N$ is called *proper* when it is closed and the inverse image of every point of Y is compact. In the context of manifolds, these two conditions are equivalent to the single condition that the inverse image of every compact set is compact. In our setting, we can also formulate properness in terms of convergency: *f is proper if and only if every sequence x_k whose image $f(x_k)$ converges has some convergent subsequence.*

Note also that if M is compact, then all continuous mappings are proper.

As mentioned before, for proper mappings plain homotopy is not convenient. Instead, we consider the notion of a *proper homotopy,* which simply is a homotopy $H : [0,1] \times M \to N$ that is a proper mapping. Then, two proper mappings $f, g : M \to N$ are *properly homotopic* if there is a proper homotopy H_t with $H_0 = f$ and $H_1 = g$.

The preceding definitions can be used in the differentiable setting, just by requiring all mappings involved to be differentiable. But then, we need two facts: (i) that every proper mapping is homotopic to a proper differentiable one and (ii) that two differentiable mappings that are homotopic are homotopic by a differentiable homotopy. To prove this, we must use approximation techniques and these techniques require the use of retractions (described in the preceding section).

Theorem 5.2. *Let $X \subset \mathbb{R}^p$ be a locally closed set, $N \subset \mathbb{R}^q$ a boundaryless differentiable $\mathcal{C}^r$ manifold, and $f : X \to N$ a continuous (proper) mapping. For every positive continuous function $\varepsilon : X \to \mathbb{R}$ there is a (proper) $\mathcal{C}^r$ mapping $g : X \to N$ such that*

$$\|f(x) - g(x)\| < \varepsilon(x) \quad \textit{for every } x \in X.$$

Proof. Since X is closed in some open neighborhood $W \subset \mathbb{R}^p$, f extends to a continuous mapping, which we still denote $f : W \to \mathbb{R}^q$, but the target need not be N (Tietze Extension Theorem). In the same way, ε extends to a continuous function $\varepsilon : W \to \mathbb{R}$, and substituting $\{\varepsilon > 0\}$ for W, we can suppose ε is positive on W. We proceed now in several steps.

Step I: Approximation of $f : W \to \mathbb{R}^q$. Since f is continuous, every point $x \in W$ has a neighborhood $W^x \subset W$ such that $\|f(z) - f(x)\| < \varepsilon(z)$ for all $z \in W^x$. Let $\{\theta_x\}$ be a smooth partition of unity for the covering $\{W^x\}$ of W, and define

$$h = \sum_x f(x)\theta_x : W \to \mathbb{R}^q.$$

Clearly, h is smooth, and for every $z \in W$ we have

$$\begin{aligned}\|h(z) - f(z)\| &= \Big\|\sum_x f(x)\theta_x(z) - \Big(\sum_x \theta_x(z)\Big) f(z)\Big\| \\ &\le \sum_x \theta_x(z)\|f(x) - f(z)\| < \varepsilon(z)\end{aligned}$$

(notice that if $\|f(x) - f(z)\| \ge \varepsilon(z)$, then $z \notin W^x$, and so $\theta_x(z) = 0$).

Step II: Retraction of the approximation to N. Let $\rho : U \to N$ be a $\mathcal{C}^r$ retraction onto N from an open set $U \subset \mathbb{R}^q$ (II.4.1, p. 67). The function

$\varepsilon'(x) = \operatorname{dist}(f(x), \mathbb{R}^q \setminus U)$ is continuous and positive on X. Thus, we can replace W by $\{\varepsilon' > 0\}$, or merely assume $f(W) \subset U$. Using $\min\{\varepsilon, \varepsilon'\}$ instead of ε in Step I, we get $h(W) \subset U$.

For each $x \in X$, let K_x be a compact neighborhood of x in W and let $\varepsilon_x > 0$ be the minimum of ε on K_x. The set

$$V_x = \{y \in U : \ \|\rho(y) - f(x)\| < \tfrac{1}{2}\varepsilon_x\}$$

is an open neighborhood of $f(x) = \rho(f(x))$ in $\mathbb{R}^q$ and contains a ball centered at $f(x)$ with radius $\delta_x < \varepsilon_x$. Let B_x be the ball centered at $f(x)$ with radius $\frac{1}{2}\delta_x$. Now, by continuity, in an open neighborhood $A^x \subset f^{-1}(B_x) \cap K_x$ of x in W it follows that $\|f - f(x)\| < \frac{1}{2}\delta_x$. By replacing W with the union of the sets A^x, we can assume they cover W, and using a smooth partition of unity $\{\zeta_x\}_x$ for the cover $\{A^x\}_x$, we define

$$\varepsilon'' = \sum_x \tfrac{1}{2}\delta_x\zeta_x : W \to \mathbb{R},$$

which is positive and continuous. Next, we substitute $\min\{\varepsilon, \varepsilon', \varepsilon''\}$ for ε in the preceding step and claim that the composite mapping $g = \rho \circ h$, which is well defined and $\mathcal{C}^r$ because $h(W) \subset U$, is the approximation we sought.

Indeed, let us estimate $\|g - f\|$ on X. Pick $z \in X$, and we will have $\zeta_x(z) \neq 0$ exactly for finitely many points $x = x_1, \dots, x_s \in X$. Then

$$\varepsilon''(z) = \sum_{i=1}^{s} \tfrac{1}{2}\delta_{x_i}\zeta_{x_i}(z) \leq \tfrac{1}{2}\delta_{x_1},$$

δ_{x_1} being, to simplify notation, the biggest δ_{x_i}. As $z \in A^{x_1}$, $f(z) \in B_{x_1}$; hence

$$\|h(z) - f(x_1)\| \leq \|h(z) - f(z)\| + \|f(z) - f(x_1)\| < \varepsilon''(z) + \tfrac{1}{2}\delta_{x_1} \leq \delta_{x_1},$$

so that $h(z) \in V_{x_1}$ and $\|\rho(h(z)) - f(x_1)\| < \frac{1}{2}\varepsilon_{x_1}$. On the other hand, $z \in A^{x_1}$ also implies $\|f(z) - f(x_1)\| < \frac{1}{2}\varepsilon_{x_1}$, so that

$$\|g(z) - f(z)\| \leq \|\rho(h(z)) - f(x_1)\| + \|f(z) - f(x_1)\| < \varepsilon_{x_1}.$$

To conclude, notice that $\varepsilon_{x_1} \leq \varepsilon(z)$, because $z \in A^{x_1} \subset K_{x_1}$.

Step III: Properness. Suppose f proper. The concern then is to choose ε small enough. Pick two locally finite open coverings $\{V_i\}$ and $\{U_i\}$ of N such that $\overline{V}_i \subset U_i$ and $\overline{V}_i$ is compact. Then the compact sets $K_i = f^{-1}(\overline{V}_i)$ form a locally finite covering of X. Hence, each point $x \in X$ has a neighborhood W_x that meets finitely many K_i's, and we denote

$$\varepsilon_x = \min_i \operatorname{dist}(\overline{V}_i\,, N \setminus U_i)$$

for the indices i with $W_x \cap K_i \neq \emptyset$. There is a partition of unity $\{\eta_x\}$ for the covering $\{W_x\}$, and define the positive continuous function

$$\varepsilon^*(z) = \tfrac{1}{2} \sum_x \eta_x(z)\varepsilon_x\,.$$

We claim that if the approximation g is constructed using $\varepsilon \leq \varepsilon^*$, then g is proper.

The key fact is that for such a g we have $g(K_i) \subset U_i$ for all i's. Indeed, suppose $z \in K_i = f^{-1}(\overline{V}_i)$. Since $\eta_x(z) \neq 0$ implies $z \in W_x$, we see that i is one of the indices in the definition of ε_x, so that $\varepsilon_x \leq \operatorname{dist}(\overline{V}_i\,, N \setminus U_i)$. It follows that

$$\|f(z) - g(z)\| < \varepsilon^*(z) \leq \tfrac{1}{2}\Big(\sum_x \eta_x(z)\Big) \operatorname{dist}(\overline{V}_i\,, N \setminus U_i) = \tfrac{1}{2} \operatorname{dist}(\overline{V}_i\,, N \setminus U_i).$$

As $f(z) \in \overline{V}_i$, necessarily $g(z) \in U_i$, as desired.

Finally we deduce from this that g is proper. Let $L \subset N$ be compact. As the family $\{U_i\}$ is locally finite, L meets finitely many U_i's. But then, by the property just proved, $C = g^{-1}(L)$ meets only finitely many K_i's. As the K_i's cover X, the closed set C is contained in the union of finitely many of them. Such a union is compact; hence C is compact too. □

Approximation settles homotopy matters, because:

Proposition 5.3. *Let $X \subset \mathbb{R}^p$ be a locally closed set, $N \subset \mathbb{R}^q$ a boundaryless differentiable $\mathcal{C}^r$ manifold, and $f : X \to N$ a continuous (proper) mapping. There is a positive continuous function $\varepsilon : X \to \mathbb{R}$ such that every continuous mapping $g : X \to N$ with $\|f(x) - g(x)\| < \varepsilon(x)$ for all $x \in X$ is (properly) homotopic to f.*

Proof. To start with, note that any continuous mapping $g : X \to N \subset \mathbb{R}^q$ is homotopic to f in $\mathbb{R}^q$, by $H_t = (1-t)f + tg$. But this homotopy can map points off N. Then consider a tubular $\mathcal{C}^r$ retraction $\rho : U \to N$, and set $\varepsilon(x) = \operatorname{dist}(f(x), \mathbb{R}^q \setminus U) > 0$. If $\|f(x) - g(x)\| < \varepsilon(x)$ for all $x \in X$, we deduce

$$\begin{aligned}\|f(x) - H_t(x)\| &= \big\|f(x) - \big(tf(x) + (1-t)g(x)\big)\big\| \\ &= |1-t|\|f(x) - g(x)\| < \varepsilon(x) = \operatorname{dist}(f(x), \mathbb{R}^n \setminus U),\end{aligned}$$

so that $H_t(x) \in U$. Consequently, we can define a homotopy in N by $\rho \circ H_t$. That $\rho \circ H_t$ is proper when f is follows by applying the method of Step III in the proof of the preceding result II.5.2 to the proper mapping $[0,1] \times X \to N : (t,x) \mapsto f(x)$. □

In the last proof, if f and g are $\mathcal{C}^r$, so is the homotopy obtained. But in fact, we have the following more general result, essential in the differentiable setting:

Proposition 5.4. *Let $X \subset \mathbb{R}^p$ be a locally closed set and let $N \subset \mathbb{R}^q$ be a boundaryless differentiable $\mathcal{C}^r$ manifold. If two $\mathcal{C}^r$ mappings $f, g : X \to N$ are (properly) homotopic, then they are homotopic by a (proper) $\mathcal{C}^r$ homotopy.*

Proof. Let $H : [0,1] \times X \to N$ be a (proper) homotopy with $H_0 = f$, $H_1 = g$. By approximation (II.5.2, p. 76), there is a (proper) $\mathcal{C}^r$ mapping $H' : [0,1] \times X \to N$ with $\|H - H'\|$ arbitrarily small. We thus have a (proper) $\mathcal{C}^r$ homotopy H'_t, but we only know that $\|f - H'_0\|$ and $\|g - H'_1\|$ are small. Anyway, II.5.3, p. 78, provides (proper) $\mathcal{C}^r$ homotopies F_t from f to H'_0 and G_t from H'_1 to g. Clearly, we can paste F_t to H'_t and then with G_t to get another homotopy from f to g. The only difficulty is for the pasted homotopy to be $\mathcal{C}^r$ at the junctions. But this can be arranged easily: replace F_t, H'_t, and G_t by $F_{\theta(t)}, H'_{\theta(t)}, G_{\theta(t)}$, for any bump function $\theta : [0,1] \to [0,1]$ with

$$\begin{cases} \theta(t) = 0 & \text{for } t \leq \frac{1}{3} \text{ and} \\ \theta(t) = 1 & \text{for } t \geq \frac{2}{3}\,. \end{cases}$$

These modifications flatten the homotopies near $t = 0$ and $t = 1$ and make the pasting $\mathcal{C}^r$. □

Exercises and problems

Number 1. Let M and N be two differentiable $\mathcal{C}^r$ manifolds. Two $\mathcal{C}^r$ mappings $f, g : M \to N$ are $\mathcal{C}^r$ homotopic if there is a $\mathcal{C}^r$ homotopy $H : [0,1] \times M \to N$ such that $H_0 = f$ and $H_1 = g$. Show that this is an equivalence relation for $\mathcal{C}^r$ mappings. State and prove the analogous result for proper $\mathcal{C}^r$ mappings.

Number 2. Let $f, g : M \to \mathbb{R}$ be two continuous functions such that $f(x) < g(x)$ for every $x \in M$. Prove that there is a $\mathcal{C}^r$ function $h : M \to \mathbb{R}$ such that $f(x) < h(x) < g(x)$ for every $x \in M$.

Number 3. Let $M \subset \mathbb{R}^p$ be a differentiable $\mathcal{C}^r$ manifold and $C \subset M$ a closed subset. Let $f : M \to \mathbb{R}$ be a continuous function whose restriction $f|_C : C \to \mathbb{R}$ is $\mathcal{C}^r$ differentiable. Show that for every real number $\varepsilon > 0$, there is a $\mathcal{C}^r$ mapping $g : M \to \mathbb{R}$ such that $g|_C = f|_C$ and $\|g(x) - f(x)\| < \varepsilon$.

Use this result to prove that if M is connected, each two points in M can be connected by a differentiable $\mathcal{C}^r$ curve.

Number 4. Let $H : [0,1] \times M \to \mathbb{R}$ be a $\mathcal{C}^r$ homotopy, where M is a compact differentiable $\mathcal{C}^r$ manifold. Show that if $0 \in \mathbb{R}$ is a regular value of H_0, then it is a regular value of H_t for every $t > 0$ small enough.

Number 5. Let M be a differentiable $\mathcal{C}^r$ manifold and let $h : M \to \mathbb{R}$ be a $\mathcal{C}^r$ function, whose zero set we denote by X. Let $f : M \to \mathbb{R}$ be a continuous function that vanishes on X, and let $\varepsilon : M \to \mathbb{R}$ be a continuous strictly positive function. Show that there is a $\mathcal{C}^r$ function $g : M \to \mathbb{R}$ such that $\|f(x) - g(x)h(x)\| < \varepsilon(x)$ for every $x \in M$.

6. Diffeotopies

Here we discuss *homogeneity*, which is the property that points in manifolds can be moved around on demand. To start with, we introduce the key definition:

Definition 6.1. *Let M be a differentiable $\mathcal{C}^r$ manifold. A $\mathcal{C}^r$* diffeotopy *of M is a differentiable $\mathcal{C}^r$ homotopy $F : [0,1] \times M \to M$ such that $F_0 = \mathrm{Id}_M$ and all mappings F_t, $0 \le t \le 1$, are $\mathcal{C}^r$ diffeomorphisms of M. We say that F joins each two points $x \in M$ and $F_1(x)$.*

We say that F is the identity off *a set $A \subset M$ if $F_t(x) = x$ for $x \notin A$ and $0 \le t \le 1$.*

Remarks 6.2. (1) Every diffeotopy is a *proper* homotopy.

Indeed, the mapping $\widetilde{F} : (t,x) \mapsto (t, F_t(x))$ is a diffeomorphism of $[0,1] \times M$ (use the Inverse Mapping Theorem to check that $\widetilde{F}^{-1}$ is differentiable), hence a proper mapping. Now the projection $\pi : [0,1] \times M \to M$ is proper, because $[0,1]$ is compact, and we conclude that $F = \pi \circ \overline{F}$ is proper too.

(2) Since every point $x \in M$ is connected to $F_t(x)$ by the path $t \mapsto F_t(x)$, they lie in the same connected component. Hence, the F_t's preserve connected components. This allows many reductions to the case where M is connected.

(3) As for homotopies, the unit interval can be replaced by any other closed interval. □

Example 6.3. The mapping $(t, x) \mapsto tx$ gives a diffeotopy of $\mathbb{R}^m$ on any closed interval $[1, \lambda]$, and it can be transported to the sphere $\mathbb{S}^m \subset \mathbb{R}^{m+1}$ via the stereographic projection from the north pole to obtain a diffeotopy with equations:

$$F_t(x) = \frac{\big(2tx_1, \dots, 2tx_m, t^2(1 + x_{m+1}) - (1 - x_{m+1})\big)}{t^2(1 + x_{m+1}) + (1 - x_{m+1})}.$$

It is clear that F_t moves every parallel $x_{m+1} = h > -1$ to another $x_{m+1} = h_t$, with $\lim_{t \to \infty} h_t \to 1$. Thus we can move the set $x_{m+1} \geq h$ into $x_{m+1} \geq h'$ for any given heights h, h'. We say that F_t *collapses the sphere towards the north pole.* □

The main goal of this section is to move points around using diffeotopies. We start in Euclidean spaces.

Proposition 6.4. *There is a positive radius $\varepsilon > 0$ such that in $\mathbb{R}^m$ the origin can be joined to every point in the open ball $\|x\| < \varepsilon$ by a smooth diffeotopy that is the identity off $\|x\| \geq 2$.*

Proof. Let $\theta : \mathbb{R} \to [0, 1]$ be a bump function such that

$$\begin{cases} \theta(0) = 1 & \text{and} \\ \theta(t) \equiv 0 & \text{on } |t| \geq 1. \end{cases}$$

Let $\alpha \geq 1$ be such that $|\theta'(t)| \leq \alpha$ for all $t \in \mathbb{R}$, and take $\varepsilon = 1/\alpha \leq 1$. Let c be a point with $a = \|c\| < \varepsilon$. After a rotation we can assume $c = (a, 0, \dots, 0) \in \mathbb{R} \times \mathbb{R}^{m-1}$, and we consider coordinates $x = (y, z)$ in $\mathbb{R} \times \mathbb{R}^{m-1}$. We will produce a diffeotopy F joining the origin to c.

Let $\tau : \mathbb{R}^{m-1} \to [0, 1]$ be another bump function which is $\equiv 1$ on $\|z\| \leq 1$ and $\equiv 0$ on $\|z\| \geq \sqrt{3}$. Then define

$$F_t(x) = F_t(y, z) = (y + t\theta(y)\tau(z)a, z).$$

It is clear that F_0 is the identity and $F_1(0) = c$. Furthermore, if $\|x\| = \sqrt{y^2 + \|z\|^2} \geq 2$, it must be either $y^2 \geq 1$ or $\|z\|^2 \geq 3$; hence either $\theta(y) = 0$ or $\tau(z) = 0$, and $F_t(x) = x$. Now, we are to prove that each F_t is a diffeomorphism. To that end, we claim:

For every fixed t, z, the function $F_{t,z} : \mathbb{R} \to \mathbb{R} : y \mapsto y + t\theta(y)\tau(z)a$ is bijective.

Indeed, the function $F_{t,z}$ is strictly increasing, because its derivative is

$$1 + t\theta'(y)\tau(z)a \geq 1 - \alpha a > 0,$$

and $F_{t,z}$ is not bounded, because

$$|F_{t,z}(y)| \geq |y| - |t\theta(y)\tau(z)a| \geq |y| - \frac{1}{\alpha} \to \infty$$

when $|y| \to \infty$. This proves the claim.

It immediately follows from the claim that F_t is bijective; hence to prove it is a diffeomorphism, we can apply the Inverse Mapping Theorem and check that the derivative $d_x(F_t) : \mathbb{R}^m \to \mathbb{R}^m$ is a linear isomorphism, which can be readily seen by computing the Jacobian determinant:

$$\det\big(d_x(F_t)\big) = \left|\begin{array}{c|ccc} 1 + t\theta'(y)\tau(z)a & & * & \\ \hline & 1 & & 0 \\ 0 & & \ddots & \\ & 0 & & 1 \end{array}\right| = 1 + t\theta'(y)\tau(z)a > 0.$$

The proof is finished. □

Using this Euclidean construction, the result follows easily for manifolds:

Theorem 6.5. *Let M be a boundaryless differentiable $\mathcal{C}^r$ manifold. Let a, b be two points of M and let A be a connected open set containing both. Then there is a $\mathcal{C}^r$ diffeotopy joining a and b, which is the identity off a compact set $K \subset A$, which is a neighborhood of a and b.*

Proof. Choose for each point $x \in A$ a coordinate domain $U_x \subset A$, diffeomorphic to $\mathbb{R}^m$. By II.6.4, p. 81, we find two open relatively compact neighborhoods $W_x \subset V_x$ of x such that $K_x = \overline{V}_x \subset U_x$ and each point of W_x can be joined to x by a diffeotopy of U_x which is the identity off K_x; notice that this latter condition means that the diffeotopy can be extended to the whole manifold M by the identity. Now, since A is connected and covered by the W_x's, we find a finite chain $W_{x_1}, \dots, W_{x_p}$ with $a \in W_{x_1}$ and $b \in W_{x_r}$. Then pick points $y_k \in W_{x_k} \cap W_{x_{k+1}}$, $k = 1, \dots, p-1$, and set $y_0 = a$, $y_p = b$. By construction, for each $k = 0, \dots, p-1$, we can join y_k and y_{k+1} by a diffeotopy $F^{(k)}$ of M which is the identity off the compact set $K_{x_{k+1}} \subset A$. Clearly, if we paste those diffeotopies, we get one which joins $a = y_0$ and $b = y_p$ and which is the identity off the compact set

$K = \bigcup K_{x_{k+1}} \subset A$. The only problem, that the pasting be $\mathcal{C}^r$ at the junctions, is arranged by a bump function θ as in II.5.4, p. 79. The composite $G * F$ of two diffeotopies F and G is

$$(G * F)(t, x) = \begin{cases} F(\theta(2t), x) & \text{for } t \leq \frac{1}{2}, \\ F_1(G(\theta(2t-1), x)) & \text{for } t \geq \frac{1}{2}. \end{cases}$$

□

Corollary 6.6. *Let M be a boundaryless differentiable $\mathcal{C}^r$ manifold of dimension $m \geq 2$. Let $a_1, \dots, a_p$ and $b_1, \dots, b_p$ be two collections of points of M, and let A be a connected open set containing both. Then there is a $\mathcal{C}^r$ diffeotopy joining a_k and b_k for all $k = 1, \dots, p$, which is the identity off a compact set $K \subset A$.*

Proof. By induction on p. Assume the result for fewer that p points, and let $N = M \setminus \{a_p, b_p\}$. By assumption, there is a diffeotopy of N joining a_k and b_k for $k = 1, \dots, p-1$, which is the identity off a compact set $L \subset A \setminus \{a_p, b_p\}$ (which is connected because $\dim(A) \geq 2$); such a diffeotopy extends to M. Then, by II.6.5 above, there is a diffeotopy of M joining a_p and b_p that is the identity off a compact set K contained in the connected set $A \setminus \{a_k, b_k : 1 \leq k < p\}$. One concludes by pasting these two diffeotopies. □

Remarks 6.7. (1) On $M = \mathbb{R}$ every diffeomorphism is either decreasing or increasing, which immediately gives a restriction to joining series of points: they must be numbered in exactly the same (or the reverse) order. Then it is easy to produce by hand the diffeotopy that connects the two series (or to do the pasting as in the proof above). This is the essential restriction for manifolds of dimension 1.

(2) An easy consequence of the preceding result is that *connected manifolds are n-homogeneous (in particular homogeneous):* any two collections of n points can be transformed to each other by a diffeomorphism (in dimension 1 with the restriction explained in the above remark).

(3) Another interesting consequence that often makes life easier is that *any finite collection of points of a connected manifold has an open neighborhood diffeomorphic to $\mathbb{R}^m$.*

Indeed, given $x_1, \dots, x_p$, pick an open set U diffeomorphic to $\mathbb{R}^m$, and choose any $y_1, \dots, y_p \in U$. By remark (2), there is a diffeomorphism f, mapping $x_k \mapsto y_k$; hence $f^{-1}(U)$ is the open neighborhood we sought. □

All these results obviously work for bordered manifolds as long as the points involved are not in the boundary. We will not need an elaborate discussion.

Exercises and problems

Number 1. Let $f : \mathbb{R}^m \to \mathbb{R}^m$ be a linear isomorphism with positive determinant. Show that there is a smooth diffeotopy $F : [0,1] \times \mathbb{R}^m \to \mathbb{R}^m$ such that $F_1 = f$.

Number 2. Let $h : \mathbb{R} \to [0,1]$ be a smooth function with $h(t) = 1$ for $|t| \leq 1$ and $h(t) = 0$ for $|t| \geq 4$. Prove that there exists a positive real number $\varepsilon > 0$ such that for every $a \in \mathbb{R}^m$ with $\|a\| < \varepsilon$ and every linear mapping $u : \mathbb{R}^m \to \mathbb{R}^m$ with $\| \mathrm{Id}_{\mathbb{R}^m} - u\| < \varepsilon$, the map

$$F_t(x) = x + th(\|x\|^2)(u(x) - x + a)$$

is a smooth diffeotopy of $\mathbb{R}^m$ verifying (i) $F_t(x) = x$ for $0 \leq t \leq 1$ and $\|x\| \geq 2$ and (ii) $F_1(x) = u(x) + a$ for $\|x\| \leq 1$. (Check that each F_t is a proper local diffeomorphism whose fibers all have the same number of points.)

Number 3. Let $f : \mathbb{R}^m \to \mathbb{R}^m$ be a smooth diffeomorphism with $f(0) = 0$ and $d_0 f = \mathrm{Id}_{\mathbb{R}^m}$. Show that there is a smooth diffeotopy $F : [0,1] \times \mathbb{R}^m \to \mathbb{R}^m$ such that $F_1 = f$.

Number 4. Let U be an open neighborhood of the origin in $\mathbb{R}^m$, and let $f : U \to \mathbb{R}^m$ be a differentiable $\mathcal{C}^r$ mapping with $f(0) = 0$, $d_0 f = \mathrm{Id}_{\mathbb{R}^m}$. Consider the function h in Problem Number 2, and prove that there is a positive real number $\varepsilon > 0$ such that

$$F_t(x) = x + th\big(\tfrac{\|x\|^2}{\varepsilon^2}\big)(f(x) - x)$$

is a well-defined $\mathcal{C}^r$ diffeotopy of $\mathbb{R}^m$ verifying (i) $F_t(x) = x$ for $0 \leq t \leq 1$ and $\|x\| \geq 2\varepsilon$ and (ii) $F_1(x) = f(x)$ for $\|x\| \leq \varepsilon$.

Number 5. Let $\mathbb{S}^m \subset \mathbb{R}^{m+1}$ be a sphere of odd dimension m. Construct a diffeotopy F of $\mathbb{S}^m$ such that $F_1(x) = -x$ for all $x \in \mathbb{S}^m$.

Number 6. Let $F : [0,1] \times \mathbb{S}^m \to \mathbb{S}^m$ be a diffeotopy.

(1) Show that

$$\Phi : [0,1] \times U \to U = \mathbb{R}^{m+1} \setminus \{0\} : (t,x) \mapsto \|x\| F_t\big(\tfrac{x}{\|x\|}\big)$$

is a diffeotopy.

(2) Pick any point $a \in \mathbb{S}^m$. Show that the Jacobian determinant at a of Φ_t is positive.

Use this to prove that if m is even, there is no isotopy of $\mathbb{S}^m$ such that $F_1(x) = -x$ for all $x \in \mathbb{S}^m$.

7. Orientation

In this final section of the chapter we quickly survey the notion of orientation and some related constructions.

(7.1) Orientation of differentiable manifolds. An *orientation* ζ *on a linear space* E of finite dimension $m \geq 1$ is a choice of a basis $B = \{u_1, \dots, u_m\}$, two choices being equivalent if the determinant of the base change matrix is positive; we denote $\zeta = [u_1, \dots, u_m]$ and say that B is a *positive basis* of ζ. Clearly, there are only two orientations, which are denoted by ζ and $-\zeta$ and are called *opposite.* In $E = \mathbb{R}^m$ there is a *canonical orientation* ζ^m, corresponding to the canonical basis $\{e_1, \dots, e_m\}$. We know the two first examples of this very well: in $\mathbb{R}^2$, the *counterclockwise orientation;* in $\mathbb{R}^3$, the *screwdriver orientation.*

Now, let M be a differentiable manifold of dimension m. An *orientation on* M is a family $\zeta_M = \{\zeta_x : x \in M\}$ of orientations ζ_x in each tangent space T_xM, such that each point $a \in M$ has a coordinate system $\mathtt{x}$ on whose domain U

$$(*) \qquad \left[\frac{\partial}{\partial \mathtt{x}_1}\Big|_x, \dots, \frac{\partial}{\partial \mathtt{x}_m}\Big|_x\right] = \zeta_x \quad \text{for every } x \in U.$$

When this condition holds, the coordinate system (and the corresponding parametrization) is said to be *compatible with the orientation.*

Notice here that the base change matrix for two bases of partial derivatives is the Jacobian matrix of the corresponding change of coordinates; hence the Jacobian determinant of the change of coordinates of a pair of compatible parametrizations is positive (we just say that *the change is positive*). By definition, if M has an orientation, it has a covering by coordinate domains with positive changes, which is called a *positive atlas.* Conversely, if there is a positive atlas, we can use $(*)$ above to define an orientation ζ on M.

Suppose M is connected and oriented by ζ. Pick a coordinate domain U, with coordinate system $\mathtt{x}$. To check condition $(*)$ for $\mathtt{x}$, we must use a coordinate system compatible with ζ and check whether the change is positive. By continuity, this holds (or not) locally; hence, $(*)$ holds (or not) locally. If the coordinate domain is connected, $(*)$ holds (or not) on the whole domain. Combining this with the fact that any two points of M can be joined by a chain of connected coordinate systems, we conclude that once we know ζ at a point, then we know it everywhere. In other words, M has exactly two different orientations, ζ and its opposite: $-\zeta = \{-\zeta_x : x \in M\}$.

Finally, let us mention here that the product $M \times N$ of two oriented manifolds is immediately oriented at $(x, y) \in M \times N$ by

$$\zeta_{(x,y)} = [(u_1, 0), \dots, (u_m, 0), (0, v_1), \dots, (0, v_n)],$$

given that

$$\zeta_x = [u_1, \dots, u_m] \quad \text{and} \quad \zeta_y = [v_1, \dots, v_n].$$

In short, we write $\zeta_{M\times N} = (\zeta_M, \zeta_N)$.

(7.2) Orientation and differentiable mappings. Consider a differentiable mapping of oriented manifolds of the same dimension, $f : M \to N$. Let $x \in M$ be a regular point of f, so that $d_x f : T_x M \to T_{f(x)}N$ is a linear isomorphism. Then we say that f *preserves* (resp., *reverses*) orientation at $x \in M$ if $d_x f$ maps a positive basis of $\zeta_{M,x}$ onto a positive basis of $\zeta_{N,f(x)}$ (resp., $-\zeta_{N,f(x)}$). We denote

$$\operatorname{sign}_x(f) = +1 \quad (\text{resp., } -1)$$

and call this the *sign of f at x*.

The sign can be easily computed by localization through coordinates compatible with both orientations ζ_M and ζ_N: as the Jacobian matrix of such a localization is the matrix of $d_x f$ with respect to the bases of partial derivatives, one only has to check whether that Jacobian has positive determinant. In particular, we see that if $f : M \to N$ preserves (resp., reverses) orientation at x, it does the same on a whole neighborhood of x. In particular, if a diffeomorphism of connected manifolds preserves (resp., reverses) orientation at some point, it does so at all of them. The following is an application of this:

Proposition 7.3. *Let F_t be a diffeotopy of an oriented manifold M. Then all F_t's preserve orientation.*

Proof. Since diffeotopies preserve connected components, we can suppose M connected, and then it is enough to check that the F_t's preserve the orientation at a given point $a \in M$. To see that, let $t \in [0,1]$ and choose a local parametrization $\psi : V \to M$ compatible with the orientation of M, such that $\psi(0) = F_t(a)$. Then $F^{-1}(\psi(V))$ is a neighborhood of (t,a), and there are (i) a local parametrization $\varphi : U \to M$ compatible with the orientation of M, with $a = \varphi(0)$, and (ii) a neighborhood $J \subset I$ of t, such that $F(J \times \varphi(U)) \subset \psi(V)$. In this situation, the mapping

$$\sigma : J \to \mathbb{R} : s \mapsto \det d_0(\psi^{-1} \circ F_s \circ \varphi)$$

is well defined and continuous. Since ψ, φ, and the F_s's are diffeomorphisms, σ never vanishes and has constant sign on a neighborhood of t. In conclusion, for s close enough to t, F_s preserves or reverses orientation as F_t

does. It follows that all F_t preserve or reverse orientation simultaneously. But $F_0 = \mathrm{Id}_M$ preserves; hence all F_t preserve also. $\square$

(7.4) Orientation of hypersurfaces. Let $M \subset \mathbb{R}^{m+1}$ be a differentiable hypersurface, that is, $\dim(M) = m$. We look for a *normal vector field* $\nu : M \to \mathbb{R}^{m+1}$, which is a mapping such that $0 \neq \nu(x) \perp T_xM$ for every $x \in M$.

To start with, fix a coordinate domain U corresponding to a parametrization φ. Then for every $x \in U$, $\nu(x)$ must be perpendicular to the partial derivatives

$$\frac{\partial}{\partial \mathrm{x}_i}\Big|_x = \left(\frac{\partial \varphi_1}{\partial x_i}(\varphi^{-1}(x)), \dots, \frac{\partial \varphi_{m+1}}{\partial x_i}(\varphi^{-1}(x))\right), \quad 1 \leq i \leq m.$$

This perpendicular vector is easily obtained as a *vector product.* Take new variables $u = (u_1, \dots, u_{m+1})$ and let

$$A(u) = \begin{pmatrix} u_1 & \frac{\partial \varphi_1}{\partial x_1}(\varphi^{-1}(x)) & \cdots & \frac{\partial \varphi_1}{\partial x_m}(\varphi^{-1}(x)) \\ \vdots & \vdots & & \vdots \\ u_{m+1} & \frac{\partial \varphi_{m+1}}{\partial x_1}(\varphi^{-1}(x)) & \cdots & \frac{\partial \varphi_{m+1}}{\partial x_m}(\varphi^{-1}(x)) \end{pmatrix}.$$

Then, let the i-th component of $\nu(x)$ be the signed adjoint to u_i in $A(u)$. Computing $\det A(u)$ for $u = \frac{\partial}{\partial \mathrm{x}_i}\big|_x$ $(1 \leq i \leq m)$ through its first column, it follows immediately that this $\nu(x)$ is a normal vector field on U, which can be assumed unitary after dividing by its norm.

Now, notice that at each point $x \in M$ there are only two unitary vectors perpendicular to T_xM. This makes it reasonable to paste together the ν's obtained from different coordinate systems. For this, we must understand when $\nu(x) = \nu'(x)$ for two different parametrizations. But by construction

$$(*) \qquad \zeta^{m+1} = \left[\nu, \frac{\partial}{\partial \mathrm{x}_1}, \dots, \frac{\partial}{\partial \mathrm{x}_m}\right] = \left[\nu', \frac{\partial}{\partial \mathrm{x}'_1}, \dots, \frac{\partial}{\partial \mathrm{x}'_m}\right]$$

is the canonical orientation in $\mathbb{R}^{m+1}$, and it follows readily that $\nu(x) = \nu'(x)$ if and only if the change of coordinates is positive. From this we see that M *has a global normal vector field if and only if M is orientable.* Furthermore, in that case, there are exactly two different unitary normal vector fields, opposite each other. We can distinguish them by chosing an orientation ζ: *ν is compatible with ζ* if the coordinate systems in $(*)$ above are compatible with ζ.

Examples 7.5. (1) Let $M \subset \mathbb{R}^{m+1}$ be a *closed* differentiable hypersurface, which has a *global equation* $M : f = 0$, that is, a differentiable function $f : \mathbb{R}^{m+1} \to \mathbb{R}$ such that 0 is a regular value of f and $M = f^{-1}(0)$. Then, M disconnects $\mathbb{R}^{m+1}$: otherwise, f would have constant sign; hence every point $x \in M$ would be extremal, hence a critical point of f!

Once f is given, we know (II.2.8, p. 57) that each tangent space T_xM is perpendicular to the gradient $\text{grad}_x(f) = \big(\frac{\partial f}{\partial x_1}(x), \dots, \frac{\partial f}{\partial x_{m+1}}(x)\big)$, which is always $\neq 0$, and consequently is a global normal vector field (and $\nu = \text{grad}(f)/\|\text{grad}(f)\|$ is unitary). Thus, M is orientable.

In fact, this is another formulation of the Jordan Separation Theorem for differentiable hypersurfaces: *every closed differentiable hypersurface of a Euclidean space disconnects it, has a global equation, and is orientable.* We will prove this using degree theory (III.6.2, p. 124, III.6.4, p. 129)

(2) The first case to which we apply the above construction is the unit sphere $\mathbb{S}^m \subset \mathbb{R}^{m+1}$, whose equation is $\sum_i x_i^2 = 1$. Then the gradient $\nu(x) = x$ is already unitary and gives the canonical orientation ζ of $\mathbb{S}^m$. *We always equip* $\mathbb{S}^m$ *with this orientation;* ν is called the *outward* normal vector field.

Let us now look at two antipodal points a and $-a$ in $\mathbb{S}^m$. The two tangent spaces $T_a\mathbb{S}^m$ and $T_{-a}\mathbb{S}^m$ coincide: both are perpendicular to the line through a and $-a$. Thus we can compare the orientations ζ_a and ζ_{-a}:

$$\zeta_a = [u_1, \dots, u_m] \quad \text{if} \quad \zeta^{m+1} = [a, u_1, \dots, u_m]$$

and

$$\zeta_{-a} = [v_1, \dots, v_m] \quad \text{if} \quad \zeta^{m+1} = [-a, v_1, \dots, v_m].$$

Thus we can take $v_1 = -u_1$, $v_2 = u_2, \dots, v_m = u_m$ and conclude that $\zeta_{-a} = -\zeta_a$.

We have just seen that *the sphere has opposite orientations at antipodal points.*

(3) Now suppose $a = (0, \dots, 0, 1)$ and $-a = (0, \dots, 0, -1)$. Let π_N be the stereographic projection from a and let π_S be that from $-a$ (II.2.4, p. 55). We have

$$\zeta_a = [e_1, \dots, e_m] \quad \text{if and only if } m \text{ is even.}$$

After this remark, from II.2.9, p. 58, we see that π_S preserves orientation if and only if m is even. Since $\zeta_{-a} = -\zeta_a$, we similarly see that π_N preserves orientation if and only if m is odd. Hence, π_S and π_N do not form a positive atlas. To get one, change the sign of one coordinate in one of the projections. $\square$

(7.6) Orientation of inverse images. Let $f : M \to N$ be a differentiable $\mathcal{C}^r$ mapping of *oriented* manifolds, and let $a \in N \setminus \partial N$ be a regular value of both f and $f|_{\partial M}$. As we know, $f^{-1}(a)$ is a manifold (II.3.2, p. 62), and we can use the orientations of M and N to define an orientation on $f^{-1}(a)$. We do it as follows.

Let $x \in f^{-1}(a)$. By II.3.2, p. 62, $L = T_x f^{-1}(a)$ is the kernel of the derivative $d_x f$; hence this restricts to a linear isomorphism $E \to T_a N$ for any linear supplement $E \subset T_x M$ of L. Choose any such E with a basis $v_1, \dots, v_n$ that $d_x f$ maps onto a *positive* basis $w_1, \dots, w_n$ of $T_a N$. Then we define an orientation ξ_x on L by declaring positive any basis $u_1, \dots, u_{m-n}$ of L such that $v_1, \dots, v_n, u_1, \dots, u_{m-n}$ is a *positive* basis of $T_x M$.

We must see that this does not depend on the choice of data, so assume other v'_i, u'_j are given. Consider the matrix $A = \left(\begin{array}{c|c} A_1 & 0 \\ \hline * & A_2 \end{array}\right)$ of the base change from $\{v'_i, u'_j\}$ to $\{v_i, u_j\}$ and the matrix $C = (C_1 | 0)$ of $d_x f$ with respect to the bases $\{v_i, u_j\}$ and $\{w_i\}$. In view of the way orientations determine the choices, we have the following:

(i) The matrix C_1 is the base change from $\{d_x f(v_i)\}$ to $\{w_i\}$, so $0 < \det(C_1)$.

(ii) Since $CA = (C_1 A_1 | 0)$, the matrix $C_1 A_1$ is the base change from $\{d_x f(v'_i)\}$ to $\{w_i\}$; hence $0 < \det(C_1)\det(A_1)$, and by (i), $0 < \det(A_1)$.

(iii) The matrix A is the base change from $\{v'_i, u'_j\}$ to $\{v_i, u_j\}$; hence $0 < \det(A) = \det(A_1)\det(A_2)$, and by (ii), $0 < \det(A_2)$.

But A_2 is the base change from $\{u'_j\}$ to $\{u_j\}$, and consequently $\{u'_j\}$ defines the same orientation as $\{u_j\}$.

Since we know that the definition does not depend on choices, let us see why $\xi = \{\xi_x : x \in f^{-1}(a)\}$ is indeed a well-defined orientation on our inverse image. To do this, pick coordinates $(\mathtt{x}_1, \dots, \mathtt{x}_m)$ of M and $(\mathtt{y}_1, \dots, \mathtt{y}_n)$ of N as in the proof of II.3.2, p. 63, so that $(\mathtt{x}_1, \dots, \mathtt{x}_{m-n})$ are coordinates of $f^{-1}(a)$ and

$$\begin{cases} d_x f\left(\frac{\partial}{\partial \mathtt{x}_{m-n+i}}\Big|_x\right) = \frac{\partial}{\partial \mathtt{y}_i}\Big|_a & \text{for } i = 1, \dots, n, \\ \frac{\partial}{\partial \mathtt{x}_j}\Big|_x,\ j = 1, \dots, m-n, & \text{generate } T_x f^{-1}(a). \end{cases}$$

Thus we can use the vectors $v_i = \frac{\partial}{\partial \mathtt{x}_{m-n+i}}\big|_x$, $w_i = \frac{\partial}{\partial \mathtt{y}_i}\big|_a$, and $u_j = \frac{\partial}{\partial \mathtt{x}_j}\big|_x$. It is quite clear that depending on how $\mathtt{x}, \mathtt{y}$ respect orientations, either $(\mathtt{x}_1, \dots, \mathtt{x}_{m-n})$ or $(\mathtt{x}_1, \dots, -\mathtt{x}_{m-n})$ defines ξ consistently at all points of $f^{-1}(a)$ close to x. We are done.

(7.7) Orientation of boundaries. The definition of orientation applies of course to manifolds with boundary, as no boundaryless assumption has been made as far. If M has boundary ∂M, all tangent spaces $T_x M$ carry an orientation, despite whether or not x belongs to ∂M, In fact, it is easy to see that M is orientable if and only if the boundaryless manifold $M \setminus \partial M$ is; in other words, the orientation of M at boundary points is determined by the nearby interior points. A more interesting matter is that when M is oriented, its boundary can be oriented as a manifold itself. We describe this here.

Let M be a manifold with boundary ∂M, and fix an orientation ζ on M. Choose any point $a \in \partial M$. The tangent space $T_a(\partial M)$ is a hyperplane in $T_a M$, and we pick a coordinate system at a, $\mathtt{x} : U \to \mathbb{H}^m$ with $\mathtt{x}_1 \geq 0$. Then the linear isomorphism $d_a\mathtt{x} : u \mapsto t = (t_1, t_2, \dots, t_m)$ maps $T_a(\partial M) \subset T_a M$ onto $\{0\} \times \mathbb{R}^{m-1} \subset \mathbb{R}^m$. Thus we find two types of vectors that are tangent to M but not to ∂M: the *inward* vectors, with $t_1 > 0$, and the *outward* vectors, with $t_1 < 0$. As usual, via changes of coordinates, one sees that these conditions do not depend on the choice of the coordinates, and we have a consistent notion of outward tangency.

Curve germs provide a more geometric way to determine inward and outward vectors:

(1) If we have $\gamma : [0, \varepsilon) \to M$ with $\gamma(0) = a \in \partial M$ and $\gamma'(0) = u \in T_a M \setminus T_a(\partial M)$, then u is inward.

(2) If we have $\gamma : (1 - \varepsilon, 1] \to M$ with $\gamma(1) = a \in \partial M$ and $\gamma'(1) = u \in T_a M \setminus T_a(\partial M)$, then u is outward.

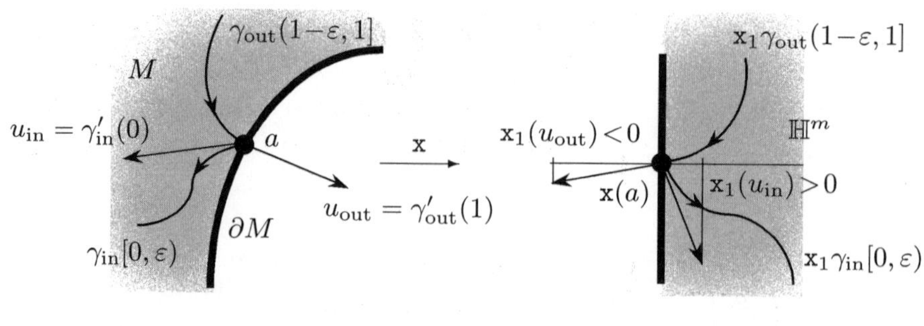

In (1), $\mathrm{x}_1 \circ \gamma : [0, \varepsilon) \to [0, \infty)$ must be increasing, and $t_1 = (\mathrm{x}_1 \circ \gamma)'(0)$ is positive; hence u is inward. In (2) we get an outward vector, because $\mathrm{x}_1 \circ \gamma : (1 - \varepsilon, 1] \to [0, \infty)$ is decreasing.

Finally, we define an orientation on $T_x(\partial M)$ by

$$\partial\zeta_x = [u_1, \dots, u_{m-1}] \quad \text{if} \quad \zeta_x = [v, u_1, \dots, u_{m-1}], \quad v \in T_x M \ \text{ outward.}$$

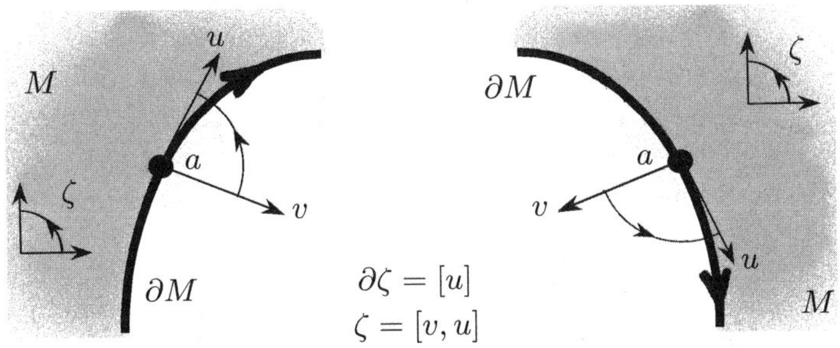

Again, some straightforward computations confirm that this indeed defines an orientation $\partial\zeta$ on ∂M.

Examples 7.8. (1) The orientation of $\mathbb{R}^m \equiv \mathbb{R}^m \times \{0\}$ as the boundary of $\mathbb{H}^{m+1} : x_1 \geq 0$ is the "wrong" one! Indeed, $\mathbb{H}^{m+1}$ carries the canonical orientation ζ^{m+1}, but since e_1 is inward tangent, we get $\partial\zeta^{m+1} = -\zeta^m$.

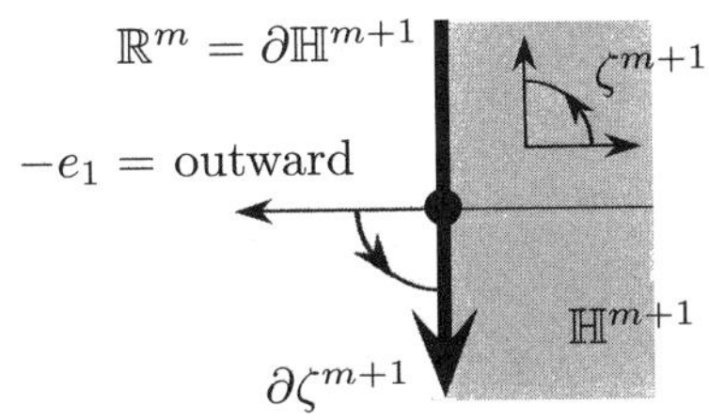

(2) Let us look at the unit sphere $\mathbb{S}^m$ as the boundary of the ball $\sum_i x_i^2 \leq 1$. Then the vector $\nu(x) = x$ is outward tangent to the ball, and we see that *the orientation of the sphere as boundary of the ball is the canonical orientation* defined in II.7.5(2), p. 88.

(3) A case of relevance for later discussion is that of a *cylinder*. Let M be an oriented boundaryless manifold and consider the cylinder $M' = [0, 1] \times M$ oriented as described in II.7.1, p. 85: if $u_1, \dots, u_m$ give the orientation ζ_x at $x \in M$, at $(t, x) \in M'$ take the orientation

$$\xi_{(t,x)} = [(1, 0), (0, u_1), \dots, (0, u_m)].$$

This ξ induces an orientation $\partial\xi$ on $\partial M'$, which is a disjoint union of two copies of M:

$$\begin{aligned}\partial M' &= M_0 \cup M_1,\\ \varphi_0 : M_0 &= \{0\}\times M \equiv M : (0,x) \equiv x,\\ \varphi_1 : M_1 &= \{1\}\times M \equiv M : (1,x) \equiv x.\end{aligned}$$

Then φ_1 preserves orientations and φ_0 reverses them. In other words, the orientation $\partial\xi$ induced on $\partial M'$ *is the right one on* M_1 *and the wrong one on* M_0. The picture below describes the situation.

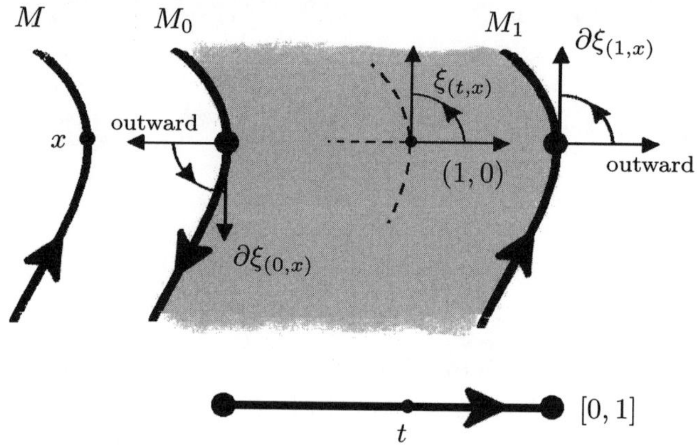

Indeed, it is enough to remark that $(1,0)$ is inward at $(0,x)$ and outward at $(1,x)$. Moreover, this is immediate, looking at the curve germs

$$\begin{cases}\gamma_0 : [0,1) \to M' : t \mapsto (t,x),\ \gamma_0(0) = (0,x),\ \gamma_0'(0) = (1,0),\\ \gamma_1 : (0,1] \to M' : t \mapsto (t,x),\ \gamma_1(1) = (1,x),\ \gamma_1'(1) = (1,0).\end{cases}$$ □

Exercises and problems

Number 1. Let M and N be diferentiable $\mathcal{C}^r$ manifolds of dimensions m and n, respectively, M boundaryless. Let ζ_M and ζ_N be orientations on M and N, respectively.

(1) Equip $M\times N$ and $N\times M$ with the orientations $\zeta_{M\times N}$ and $\zeta_{N\times M}$, and determine when the canonical diffeomorphism $\tau : M\times N \to N\times M : (x,y) \mapsto (y,x)$ preserves orientation.

(2) Note that $M\times\partial N = \partial(M\times N)$ and compare the orientations $\zeta_{M\times\partial N}$ and $\partial\zeta_{M\times N}$.

Number 2. Let $f : \mathbb{R}^m \to \mathbb{R}^m$ be a diffeomorphism. Show that there is a diffeotopy $F : [0,1]\times\mathbb{R}^m \to \mathbb{R}^m$ with $F_1 = f$ if and only if f preserves orientation.

Number 3. Show the following orientability criterion. A manifold M is orientable if and only if for any orientations ζ_U, ζ_V on two connected open sets $U, V \subset M$ and for any two points $x, y \in U\cap V$, $\zeta_{U,x} = \zeta_{V,x}$ if and only if $\zeta_{U,y} = \zeta_{V,y}$.

Number 4. Use the criterion of the preceding problem to show that:

(1) The Möbius band $M \subset \mathbb{R}^3$ parametrized by

$$x = \cos\theta + t\cos\theta\cos\tfrac{\theta}{2}, \quad y = \sin\theta + t\sin\theta\cos\tfrac{\theta}{2}, \quad z = t\sin\tfrac{\theta}{2}$$

is not orientable.

(2) The real projective plane is not orientable.

Number 5. Show that real projective spaces $\mathbb{RP}^m$ are orientable exactly for odd m. What about complex projective spaces?

Number 6. Let $M \subset \mathbb{R}^p$ be a boundaryless smooth manifold of codimension n with a smooth equation $f : \mathbb{R}^p \to \mathbb{R}^n$ (that is, $M = f^{-1}(0)$ and 0 is a regular value of f). Prove that M is orientable. Find another sufficient condition for orientability involving νM.

Number 7. Find the orientation of the cilinder $M \subset \mathbb{R}^3 : x^2 + y^2 = 1,\ 0 \leq z \leq 1$ such that the local diffeomorphism $\mathbb{R}^2 \to M : (t, x) \mapsto (\cos t, \sin t, z)$ preserves orientation. Then determine the orientation induced on each connected component C_0, C_1 of the boundary ∂M:

$$C_0 : x^2 + y^2 = 1,\ z = 0 \quad \text{and} \quad C_1 : x^2 + y^2 = 1,\ z = 1,$$

and study whether the mapping $C_0 \to C_1 : (x, y, 0) \mapsto (x, y, 1)$ preserves orientation or not.

Number 8. Let M be a boundaryless differentiable C^r manifold of dimension m. Prove the following:

(1) Let $a, b \in M$ be two points in a connected open set $A \subset M$ and let $\lambda : T_aM \to T_bM$ be a linear isomorphism, *preserving orientation if M is oriented.* Then there is a C^r diffeotopy joining a and b, which is the identity off a compact neighborhood $K \subset A$ of both a and b and such that $d_aF_1 = \lambda$.

(2) If M is oriented and connected and $\varphi, \psi : B \to \mathbb{M}$ are two parametrizations compatible with the given orientation, both defined on the unit open ball $B \subset \mathbb{R}^m$, then there is a C^r diffeotopy $F : [0, 1] \times M \to M$ such that $F_1 \circ \varphi = \psi$ on some smaller open ball $B' \subset B$.

Chapter III

The Brouwer-Kronecker degree

Here we present the beautiful *Brouwer-Kronecker degree theory* for arbitrary proper mappings from the differentiable viewpoint. To start with, §§1 and 3 contain the construction of the Brouwer-Kronecker degree. Also, we use it to prove the *Fundamental Theorem of Algebra.* In §2 we take a small diversion to describe the degree through integration by means of the *de Rham cohomology.* Next, we discuss in §4 the extension of the theory to arbitrary differentiable manifolds, which is based on the computation of the degree of a $\mathcal{C}^1$ mapping. In §5 we define the so-called *Hopf invariant,* for mappings between spheres of unequal dimension, which is based upon the notion of *link coefficient.* We devote §6 to a beautiful application of the theory: the *Jordan Separation Theorem.* We show that every closed differentiable hypersurface disconnects the Euclidean space, and from this we deduce that such a hypersurface has a global equation and is orientable. We conclude the chapter by proving in §7 the famous *Brouwer Fixed Point Theorem,* jointly with some interesting consequences, and the equally important theorem by Brouwer concerning the existence on spheres of vector fields without zeros.

1. The degree of a smooth mapping

Here we present the notion of degree for smooth data. The first step towards the definition of the Brouwer-Kronecker degree is as follows:

Proposition 1.1. *Let $f : M \to N$ be a proper smooth mapping of two oriented, boundaryless, smooth manifolds of dimension m. Then each regular value $a \in R_f \subset N$ of f has an open neighborhood $V \subset R_f$ such that for every $b \in V$, the inverse image $f^{-1}(b)$ is a finite set and the integer*

$$d(f, b) = \sum_{x \in f^{-1}(b)} \operatorname{sign}_x(f)$$

only depends on a.

Proof. First note that for any regular value b the compact set $f^{-1}(b)$ is discrete. This follows because f is a local diffeomorphism at each $x \in f^{-1}(b)$

by the Inverse Mapping Theorem. We must now show that $b \mapsto d(f, b)$ is locally constant in R_f (which is open because f is proper). To that end, we need an open neighborhood V of a such that $f^{-1}(V)$ is a disjoint union of finitely many connected open sets U_k such that each restriction $f|_{U_k}$ is a diffeomorphism onto V.

To obtain V, let $f^{-1}(a) = \{x_1, \dots, x_r\}$. By the Inverse Mapping Theorem, there are disjoint open neighborhoods U'_k of x_k, such that $f|_{U'_k}$ is a diffeomorphism onto some open neighborhood V' of a, which we can take common for all the k's. Then since f is closed, the set $f(M \setminus \bigcup_k U'_k)$ is closed. As that set does not contain a, we can find a connected open neighborhood $V \subset V'$ of a such that

$$V \cap f(M \setminus \bigcup_k U'_k) = \emptyset.$$

Hence, $f^{-1}(V) \subset \bigcup_k U'_k$, and we take $U_k = f^{-1}(V) \cap U'_k$, so that $f^{-1}(V) = \bigcup_k U_k$.

Note also that, U_k being connected, $\operatorname{sign}_x(f)$ is constant on U_k, say $\equiv \sigma_k$. Thus, for every $b \in V$ we have

$$f^{-1}(a) = \{x_1, \dots, x_r\}, \quad f^{-1}(b) = \{y_1, \dots, y_r\},$$

with $x_k, y_k \in U_k$. Then

$$d(f, a) = \sum_k \operatorname{sign}_{x_k}(f) = \sum_k \sigma_k = \sum_k \operatorname{sign}_{y_k}(f) = d(f, b).$$

□

Proposition 1.2 (Boundary Theorem). *Let X be an oriented smooth manifold of dimension $m+1$ with boundary $\partial X = Y$ and let N be an oriented, boundaryless, smooth manifold of dimension m. Let $H : X \to N$ be a proper smooth mapping and let a be a regular value of H and $H|_Y$. Then*

$$d(H|_Y, a) = 0.$$

Proof. Denote $f = H|_Y$, which is a proper smooth mapping, because Y is closed in X. By II.3.2, p. 62, the inverse image $C = H^{-1}(a)$ is a compact smooth curve with boundary

$$\partial C = C \cap Y = f^{-1}(a).$$

By the classification theorem for compact curves (II.3.3, p. 63), the finitely many connected components Γ_k of C are either circles (without boundary points) or arcs with two boundary points $p_k, q_k \in f^{-1}(a)$. Thus, the boundary points of ∂C come in pairs, and we have

$$d(f,a) = \sum_k \operatorname{sign}_{p_k}(f) + \sum_k \operatorname{sign}_{q_k}(f).$$

We claim that $\operatorname{sign}_{p_k}(f) = -\operatorname{sign}_{q_k}(f)$, which concludes the argument. Henceforth we omit the indices k in all notation. The situation is depicted in the figure below.

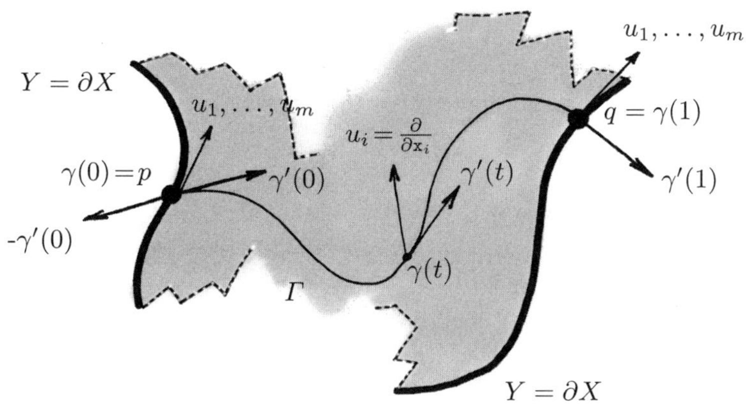

We consider in the component under consideration, Γ, the orientation it has as an inverse image (II.7.6, p. 89). Let $t \mapsto \gamma(t) \in \Gamma$ be a parametrization such that $\gamma(0) = p$, $\gamma(1) = q$. After renaming p, q as q, p and reparametrizing by $t \mapsto \gamma(1-t)$, we can suppose that the inverse image orientation in $T_{\gamma(t)}\Gamma$ is in fact given by the tangent vector $\gamma'(t)$. We next discuss orientations at the end points of Γ.

(i) *Orientations at p.* Let $u_1, \dots, u_m$ be a positive basis of T_pY. As Y is oriented as the boundary of X, $v, u_1, \dots, u_m$ is a positive basis of T_pX for any outward tangent vector v. But $\gamma'(0)$ is inward (II.7.7, p. 90); hence we can take $v = -\gamma'(0)$. Thus

$$-\gamma'(0), u_1, \dots, u_m$$

is a positive basis of T_pX. On the other hand, we know that $\gamma'(0)$ gives the orientation of Γ as an inverse image; hence

$$u_1, \dots, u_m, \gamma'(0)$$

is a positive basis of T_pX if and only if d_pf maps $u_1, \dots, u_m$ onto a positive basis of T_aN, that is, if and only if $\mathrm{sign}_p(f) = +1$. But the determinant of the change from the first base to the second is $(-1)^{m+1}$, and so

$$\mathrm{sign}_p(f) = +1 \textit{ if and only if } m \textit{ is odd.}$$

(ii) *Orientations at* q. The argument is the same, but at q the tangent vector $\gamma'(1)$ is outward; hence we do not need $-\gamma'(1)$, and the sign of the base change is $(-1)^m$. Thus:

$$\mathrm{sign}_q(f) = +1 \textit{ if and only if } m \textit{ is even.}$$

From these two equivalences, we see that the signs $\mathrm{sign}_p(f)$ and $\mathrm{sign}_q(f)$ are opposite, as desired.

This concludes the proof of the proposition. □

Next we consider smooth homotopies:

Proposition 1.3 (Homotopy invariance). *Let $H : [0,1] \times M \to N$ be a proper smooth homotopy of oriented, boundaryless, smooth manifolds, and let a be a regular value of both H_0 and H_1. Then*

$$d(H_0, a) = d(H_1, a).$$

Proof. By the Sard-Brown Theorem, II.3.4, p. 63, we can choose regular values b of H arbitrarily close to a, and then

$$d(H_0, a) = d(H_0, b), \quad d(H_1, a) = d(H_1, b)$$

by III.1.1, p. 95. We apply the preceding proposition with $X = [0,1] \times M$, so that $Y = \{0\} \times M \cup \{1\} \times M$, and get $d(H|_Y, b) = 0$. But, since Y is a union of two disjoint components,

$$(*) \qquad d(H|_Y, b) = d(H|_{\{0\}\times M}, b) + d(H|_{\{1\}\times M}, b).$$

We can identify

$$\begin{aligned} \{0\} \times M \equiv M \quad &\text{and} \quad H|_{\{0\}\times M} \equiv H_0, \\ \{1\} \times M \equiv M \quad &\text{and} \quad H|_{\{1\}\times M} \equiv H_1 \end{aligned}$$

and then see what the degrees are. But we must care for the orientations induced on Y as a boundary. Indeed, as we know (II.7.8, p. 91), in a cylinder like X the orientations induced in $\{0\} \times M$ and $\{1\} \times M$ are reversed; hence if, say, the equivalence $\{0\} \times M \equiv M$ preserves orientations, then $\{1\} \times M \equiv M$ does not, and we have

$$d(H|_{\{0\}\times M}, b) = d(H_0, b) \quad \text{and} \quad d(H|_{\{1\}\times M}, b) = -d(H_1, b).$$

Substituting in (*), we see that $d(H_0, b) = d(H_1, b)$. Clearly, if $\{0\}\times M \equiv M$ reverses orientations, the conclusion is the same. □

Proposition and Definition 1.4. *Let $f : M \to N$ be a proper smooth mapping of two oriented, boundaryless, smooth manifolds of dimension m; furthermore, N is connected. Then the integer $d(f, a)$ does not depend on the choice of the regular value $a \in R_f$. It is called the* degree of f *and is denoted by* $\deg(f)$.

Proof. Let b be another regular value of f. Since N is connected, by II.6.5, p. 82, there is a smooth diffeotopy F_t of N with $F_1(b) = a$. Then $F_t \circ f$ is a proper smooth homotopy of f and $F_1 \circ f$. Since a is a regular value of these two mappings, from III.1.3, p. 98, we get

$$d(f, a) = d(F_1 \circ f, a).$$

But F_1 is a diffeomorphism that preserves orientation everywhere (II.7.3, p. 86); hence by the chain rule:

$$d(F_1 \circ f, a) = \sum_{x\in(F_1\circ f)^{-1}(a)} \operatorname{sign}_x(F_1 \circ f) = \sum_{x\in f^{-1}(b)} \operatorname{sign}_x(f) = d(f, b).$$

□

Remarks 1.5. (1) It is clear that the above definition depends on the chosen orientations. The essential thing here is that if the orientation of M is reversed, then the degree changes sign.

(2) It is also clear from the definition that every non-surjective smooth mapping has degree zero. □

Examples 1.6. Let $\mathbb{S}^m \subset \mathbb{R}^{m+1}$ denote the standard unit sphere.

(1) Of course the identity has degree 1. Now let us look at the antipodal map $f : x \mapsto -x$. It is a diffeomorphism. Hence its degree is ± 1 according to the behavior of orientations, which can be checked at any point. Pick the north pole $a = (0, \dots, 0, 1)$. Then $f(a) = -a = (0, \dots, 0, -1)$ is the south pole, and the two tangent hyperplanes $T_a\mathbb{S}^m$ and $T_{-a}\mathbb{S}^m$ coincide as linear spaces, but they carry opposite orientations. Now, d_af is the symmetry: $u \mapsto -u$, which maps any basis $\{u_1, \dots, u_m\}$ to $\{-u_1, \dots, -u_m\}$. The determinant of the latter with respect to the former is $(-1)^m$, that is:

$$\begin{cases} +1, \text{ and so the bases define equal orientations, if } m \text{ is even,} \\ -1, \text{ and so the bases define opposite orientations, if } m \text{ is odd.} \end{cases}$$

Hence, a positive basis of $T_a\mathbb{S}^m$ is mapped to

$$\begin{cases} \text{a negative basis of } T_{-a}\mathbb{S}^m \text{ if } m \text{ is even,} \\ \text{a positive basis of } T_{-a}\mathbb{S}^m \text{ if } m \text{ is odd,} \end{cases}$$

and we conclude that d_af

$$\begin{cases} \text{reverses orientation if } m \text{ is even,} \\ \text{preserves orientation if } m \text{ is odd.} \end{cases}$$

All in all, $\deg(f) = -1$ if m is even, and $\deg(f) = +1$ if m is odd.

(2) To produce a mapping with degree -1 in any dimension, we turn to symmetries, like $h : x \mapsto (x_1, \dots, x_m, -x_{m+1})$. This h maps again the north to the south pole, but here d_ah is the identity, because the tangent vectors have their last component zero. Consequently, a discussion as above shows that $d_ah : T_a\mathbb{S}^m \to T_{-a}\mathbb{S}^m$ reverses orientations, and $\deg(h) = -1$.

(3) We can identify the unit circle $\mathbb{S}^1 \subset \mathbb{R}^2$ with the group of complex numbers of module 1. Then *the mapping $\mathbb{S}^1 \to \mathbb{S}^1 : z \to z^d$ (complex multiplication) has degree d.*

If $d = 0$, this mapping is constant and its degree is indeed 0. Hence, we assume $d \neq 0$. In polar coordinates, we can represent the mapping by

$$z = \cos\theta + i\sin\theta \mapsto z^d = \cos(d\cdot\theta) + i\sin(d\cdot\theta).$$

Clearly, there are no critical values, and each value has $|d|$ preimages. To check orientations, just look at the representation

$$\mathbb{S}^1 \to \mathbb{S}^1 : (\cos\theta, \sin\theta) \mapsto (\cos(d\cdot\theta), \sin(d\cdot\theta)),$$

which preserves orientation for $d > 0$ and reverses it for $d < 0$. In fact, we have just formalized the idea of winding $\mathbb{S}^1$ around itself. We will generalize this later (V.1.1, p. 183). This example shows quite clearly why degree is called degree! □

The degree of a composite mapping is easy to compute:

Proposition 1.7. *Let $f : M \to N$ and $g : N \to P$ be two proper smooth mappings of oriented, boundaryless, smooth manifolds, with N and P connected. Then*

$$\deg(g \circ f) = \deg(g) \cdot \deg(f).$$

Proof. Let $a \in P$ be a regular value of $g \circ f$. Then every $y \in g^{-1}(a)$ is a regular value of f. Indeed, let $x \in f^{-1}(y)$. As a is a regular value of $g \circ f$, the derivative $d_x(g \circ f)$ is an isomorphism. But by the Chain Rule,

$$d_x(g \circ f) = d_y g \circ d_x f$$

and we conclude that $d_x f$ is injective, hence an isomorphism, because all spaces have dimension m. Consequently, we have

$$\deg(f) = \sum_{x \in f^{-1}(y)} \operatorname{sign}_x(f) \quad \text{for every } y \in g^{-1}(a).$$

Thus

$$\begin{aligned}\deg(g \circ f) &= \sum_{x \in (g \circ f)^{-1}(a)} \operatorname{sign}_x(g \circ f) = \sum_{y \in g^{-1}(a)} \sum_{x \in f^{-1}(y)} \operatorname{sign}_y(g) \operatorname{sign}_x(f) \\ &= \sum_{y \in g^{-1}(a)} \operatorname{sign}_y(g) \sum_{x \in f^{-1}(y)} \operatorname{sign}_x(f) = \sum_{y \in g^{-1}(a)} \operatorname{sign}_y(g) \deg(f) \\ &= \deg(g) \cdot \deg(f).\end{aligned}$$

□

Now we prove two fundamental properties of deg:

Proposition 1.8 (Boundary Theorem). *Let X be an oriented smooth manifold of dimension $m+1$ with boundary $\partial X = Y$, and let N be a connected, oriented, boundaryless, smooth manifold of dimension m. Let $H : X \to N$ be a proper smooth mapping. Then*

$$\deg(H|_Y) = 0.$$

Proof. The smooth mapping $H|_Y$ is proper, because Y is closed in X. Then, by the Sard-Brown Theorem, we can pick a point $a \in N$ that is a regular value of both H and $H|Y$. Hence

$$\deg(H|_Y) = d(H|_Y, a),$$

and the result follows readily from III.1.2, p. 96. □

Proposition 1.9. *Let $H : [0,1] \times M \to N$ be a proper smooth homotopy of oriented, boundaryless, smooth manifolds, N connected. Then*

$$\deg(H_0) = \deg(H_1).$$

Proof. By the Sard-Brown Theorem, the two proper smooth mappings H_0 and H_1 have some common regular value $a \in N$, and then

$$\deg(H_0) = d(H_0, a), \quad \deg(H_1) = d(H_1, a).$$

From this and III.1.3, p. 98, we get $\deg(H_0) = \deg(H_1)$. □

We end this section with a proof of the Fundamental Theorem of Algebra, which in the first chapter was presented as the origin of degree theory. The proof that follows brings in the preceding notions and makes clear the interest of developing the theory for proper mappings. Later we will give a second proof of this result, closer to Gauss's ideas, appealing to the Euclidean degree (IV.2.7, p. 150)

Proposition 1.10 (Fundamental Theorem of Algebra). Every algebraic equation $z^p + c_1 z^{p-1} + \cdots + c_p = 0$ with complex coefficients has some complex solution.

Proof. The mapping

$$P : \mathbb{R}^2 \equiv \mathbb{C} \to \mathbb{C} : (x, y) \equiv x + iy = z \mapsto z^p + c_1 z^{p-1} + \cdots + c_p$$

is smooth and proper (because $\lim_{|z|\to\infty} |P(z)| = \infty$). By III.1.5(2), p. 99, it is enough to show that $\deg(P) \neq 0$. To do that, consider the homotopy

$$P_t(z) = z^p + tc_1 z^{p-1} + \cdots + tc_p, \quad 0 \leq t \leq 1,$$

which is proper and smooth. Hence $\deg(P) = \deg(P_0)$. But the degree of the mapping $P_0 : z \mapsto z^p$ is exactly p. Indeed, one checks immediately that (i) any $a \neq 0$ is a regular value of P_0 with p different roots c in $\mathbb{C}$ and (ii) at each of them P_0 preserves orientation: $d_c P_0$ is multiplication by the complex number pc^{p-1}. □

Exercises and problems

Number 1. For each $i = 1, \ldots, r$, consider a proper smooth mapping $f_i : M_i \to N_i$ with $\dim(M_i) = \dim(N_i)$, N_i connected. Prove that $\deg(f_1 \times \cdots \times f_r) = \deg(f_1) \cdots \deg(f_r)$. Deduce that on the torus $T^m = \mathbb{S}^1 \times \cdots \times \mathbb{S}^1$ there are smooth mappings $T^m \to T^m$ of arbitrary degree.

Number 2. Let $p, q, m_1, \ldots, m_r$ be even positive integers, and let $n = p + q$, $m = \sum_i m_i$.

(1) Show that the following smooth mapping has degree 2:

$$f_2 : \mathbb{S}^p \times \mathbb{S}^q \to \mathbb{S}^n : (x_1, \ldots, x_{p+1}, y_1, \ldots, y_{q+1}) \mapsto (x_1 y_1, \ldots, x_{p+1} y_1, y_2, \ldots, y_{q+1}).$$

(2) Define by induction a mapping $f_r : \mathbb{S}^{m_1} \times \cdots \times \mathbb{S}^{m_r} \to \mathbb{S}^m$ of degree 2^{r-1}, hence not null-homotopic.

Number 3. Check that the following quadratic mapping is well defined, and compute its degree:

$$h : \mathbb{S}^m \to \mathbb{S}^m : (x_1, \dots, x_{m+1}) \mapsto (2x_1^2 - 1, 2x_1x_2, \dots, 2x_1x_{m+1}).$$

Number 4. Let $m = 2n + 1$ be an odd positive integer.

(1) Compute the degree of the canonical surjection

$$\pi : \mathbb{S}^m \to \mathbb{RP}^m : (x_0, \dots, x_m) \mapsto (x_0 : \dots : x_m).$$

(2) Show that the mapping h of the preceding problem factorizes through π, and obtain a smooth mapping $f : \mathbb{RP}^m \to \mathbb{S}^m$ with $\deg(f) = 1$.

(3) Take a different view of f via stereographic projections.

Number 5. Fix an orientation on $\mathbb{C}^m \equiv \mathbb{R}^{2m}$. Let $U \subset \mathbb{C}^m$ be an open set, and let $f : U \to \mathbb{C}^m$ be a proper holomorphic function. Show that $\deg(f)$ is the cardinal of the generic fiber of f: the inverse image $f^{-1}(a)$ of every regular value a of f has exactly $\deg(f)$ points. In particular, $\deg(f) \geq 0$.

Number 6. It can be shown that the degree of a smooth mapping $\mathbb{CP}^2 \to \mathbb{CP}^2$ must be a perfect square. Construct all of them.

Number 7. Fix an orientation on the complex projective space $\mathbb{CP}^m$ to compute the degree of the mapping

$$\mathbb{CP}^m \to \mathbb{CP}^m : z = (z_0 : \dots : z_m) \mapsto \bar{z} = (\bar{z}_0 : \dots : \bar{z}_m),$$

where $\bar{z}_i$ stands for the complex conjugate of $z_i \in \mathbb{C}$.

Number 8. Given a proper smooth mapping $f : M \to N$ of smooth manifolds with boundary, N connected, such that $f(\partial M) \subset \partial N$, define

$$\deg(f) = \sum_{x \in f^{-1}(a)} \operatorname{sign}_x(f),$$

where a is a regular value of f off ∂N. Develop the corresponding theory, using homotopies H_t such that $H_t(\partial M) \subset \partial N$. Prove also that $\deg(f) = \deg(f|_P)$, where $P \subset \partial M$ is the inverse image $f^{-1}(Q) \cap \partial M$ of any chosen connected component Q of ∂N.

2. The de Rham definition

In this section we describe the Brouwer-Kronecker degree of a smooth mapping in terms of differential forms and integration. We will not fully depict all details behind the constructions, only those directly linked to the notion of degree.

(2.1) Forms and de Rham cohomology. Let M be a connected, oriented, boundaryless, smooth manifold of dimension m. We denote by

$\Gamma_c^k(M)$ the linear space of all smooth differential forms of degree k, k-forms for short, with compact support. For $k = 0$, $\Gamma_c^0(M)$ consists of all smooth functions $f : M \to \mathbb{R}$ with compact support. As is well known, the derivatives of such a function f give its total differential $\mathrm{d}f : x \mapsto d_x f$, and this induces a linear operator $\mathrm{d} : \Gamma_c^0(M) \to \Gamma_c^1(M)$. Then, this operator extends to $\mathrm{d} : \Gamma_c^k(M) \to \Gamma_c^{k+1}(M)$ for all k, so that $\mathrm{d} \circ \mathrm{d} = 0$ and

$$\mathrm{d}(\alpha \wedge \beta) = \mathrm{d}\alpha \wedge \beta + (-1)^r \alpha \wedge \mathrm{d}\beta$$

for every r-form α and s-form β; d is called *exterior differentiation.*

A k-form ω with compact support is called *closed* when its exterior differential is zero, i.e., $\mathrm{d}\omega = 0$, and *exact* when $\omega = \mathrm{d}\alpha$ for some $(k-1)$-form with compact support α. The closed k-forms with compact support are a linear subspace of $\Gamma_c^k(M)$ and the exact k-forms with compact support another linear subspace. As $\mathrm{d} \circ \mathrm{d} = 0$, exact implies closed, and we can consider the quotient linear space

$$H_c^k(M, \mathbb{R}) = \frac{\{\text{closed } k\text{-forms with compact support}\}}{\{\text{ exact } k\text{-forms with compact support}\}}.$$

This is the so-called k-th de Rham cohomology group of M with compact support. It is a deep theorem that this cohomology coincides with the singular cohomology $H_c^k(M, \mathbb{Z})$ after extension of coefficients ($H_c^k(M, \mathbb{R}) = \mathbb{R} \otimes_{\mathbb{Z}} H_c^k(M, \mathbb{Z})$), but we will not discuss this matter here. Of course, if M is compact, everything has compact support and we get the ordinary cohomology groups, but in the non-compact case, it is important to consider compact supports. For instance, $H_c^m(\mathbb{R}^m, \mathbb{R}) = \mathbb{R}$, but disregarding compact supports in all definitions above, we get $H^m(\mathbb{R}^m, \mathbb{R}) = 0$.

Here we will concentrate on the maximum degree m and look solely at the m-th cohomology group $H_c^m(M, \mathbb{R})$. This group can be beautifully described by integration. Indeed, since M is boundaryless, Stokes' Theorem says simply that

$$\int_M \mathrm{d}\alpha = 0,$$

which means that the linear mapping

$$\int_M : H_c^m(M, \mathbb{R}) \to \mathbb{R} : \omega \mapsto \int_M \omega$$

is well defined. But in fact, it can be shown that it is an isomorphism. By some standard reduction using partitions of unity and local coordinate systems, this amounts to proving the following assertion:

Let $h : \mathbb{R}^m \to \mathbb{R}$ be a smooth function with compact support, such that $\int_{\mathbb{R}^n} h = 0$. Then

$$h = \sum_{i=1}^{m} \frac{\partial h_i}{\partial x_i}$$

for some smooth functions $h_1, \dots, h_m$ with compact support.

In other words, h is the *divergence* of some vector field with compact support. This is proved in parametric form, by induction on m, using elementary integration.

(2.2) Homomorphisms on cohomology. Let M and N be two connected, oriented, boundaryless, smooth manifolds of dimension m. Every proper smooth mapping $f : M \to N$ induces linear mappings

$$f^* : \Gamma_c^k(N) \to \Gamma_c^k(M) : \omega \mapsto f^*\omega.$$

These *pull-back* mappings are compatible with all operations with forms, including exterior differentiation, so that they induce homomorphisms on the cohomology groups, which we still denote f^*. Again, we concentrate on the maximum degree group. Then, we have a commutative diagram of linear mappings

$$\begin{array}{ccc} H_c^m(N,\mathbb{R}) & \xrightarrow{f^*} & H_c^m(M,\mathbb{R}) \\ \int_N \downarrow \cong & & \cong \downarrow \int_M \\ \mathbb{R} & \xrightarrow{\lambda} & \mathbb{R} \end{array}$$

Thus we are ready to prove the main result concerning degree and cohomology.

Theorem 2.3. *Let $f : M \to N$ be a proper smooth mapping of connected, oriented, boundaryless, smooth manifolds of dimension m. Then there is an integer d such that*

$$\int_M f^*\omega = d \cdot \int_N \omega$$

for every m-form ω with compact support on N. Moreover, for any regular value $a \in N$ of f we have

$$d = \sum_{x \in f^{-1}(a)} \operatorname{sign}_x(f) \qquad (0 \textit{ if } f^{-1}(a) = \emptyset).$$

Proof. The mapping λ in the preceding commutative diagram is a linear mapping $\mathbb{R} \to \mathbb{R}$; hence it must be multiplication by some real number δ, and commutativity reads:

$$\int_M f^*\omega = \delta \cdot \int_N \omega$$

for every m-form ω with compact support on N. Then, by II.3.4, p. 63, we can pick a regular value a of f, and the theorem follows at once if we show that

$$\delta = \sum_{x \in f^{-1}(a)} \operatorname{sign}_x(f)$$

(note that the right-hand side is indeed an integer). In other words, it is enough to see that

$$(*) \qquad \int_M f^*\omega = \sum_{x \in f^{-1}(a)} \operatorname{sign}_x(f) \int_N \omega$$

for a suitable ω that we construct now.

First of all, since a is a regular value, its inverse image $f^{-1}(a)$ is finite, suppose non-empty, say $f^{-1}(a) = \{x_1, \dots, x_r\}$. Actually, as in the proof of III.1.1, p. 95, we find an open neigborhood V of a in N and r disjoint open neighborhoods $U_1, \dots, U_r$ of $x_1, \dots, x_r$ in M, such that the restrictions $f_k = f|_{U_k} : U_k \to V$ are diffeomorphisms and $f^{-1}(V) = U_1 \cup \dots \cup U_r$. Finally, we shrink V to a domain of coordinates diffeomorphic to $\mathbb{R}^m$.

Next, using any diffeomorphism $V \to \mathbb{R}^m$, we pull back to V the m-form

$$\varphi(x) dx_1 \wedge \dots \wedge dx_m \,,$$

where $\varphi : \mathbb{R}^m \to [0, 1]$ is a smooth bump function such that

$$\varphi(x) = \begin{cases} 1 & \text{for } \|x\| \le 1, \\ 0 & \text{for } \|x\| \ge 2. \end{cases}$$

This construction guarantees that the pull back can be extended by zero to the whole manifold N and that it has compact support. We denote that extension by ω, and we will check the equality $(*)$ for it.

On the one hand, since the support of ω is contained in V, we have

$$\int_N \omega = \int_V \omega = \pm \int_{\mathbb{R}^m} \varphi(x) dx_1 \dots dx_m \neq 0$$

(change of variables for the difeomorphism $V \to \mathbb{R}^m$). On the other hand, the support of the form $f^*\omega$ is contained in $f^{-1}(V) = \bigcup U_k$, and since the U_k's are disjoint, we have

$$\int_M f^*\omega = \sum_k \int_{U_k} f^*\omega.$$

Here we can compute each summand using the corresponding diffeomorphism $f|_{U_k} : U_k \to V$, and again by the change of variables for integrals, we get

$$\int_{U_k} f^*\omega = \sigma_k \int_V \omega,$$

where $\sigma_k = \pm 1$ according to whether or not $f|_{U_k}$ preserves or reverses orientation (which holds on the whole U_k because this open set is connected). In other words, since $x_k \in U_k$,

$$\sigma_k = \operatorname{sign}_{x_k}(f),$$

and we conclude that

$$\int_M f^*\omega = \sum_k \sigma_k \int_V \omega = \sum_k \operatorname{sign}_{x_k}(f) \int_N \omega\,,$$

as desired. Note that this determines the integer $\sum_k \operatorname{sign}_{x_k}(f)$, because $\int_N \omega \neq 0$.

The reader will easily supply the simplified argument that settles the case $f^{-1}(a) = \emptyset$. □

Thus, the de Rham approach gives the following most general form of the change of variables for integrals:

Corollary 2.4. *Let $f : M \to N$ be a proper smooth mapping of connected, oriented, boundaryless, smooth manifolds of dimension m. Then*

$$\int_M f^*\omega = \deg(f) \cdot \int_N \omega$$

for every m-form ω on N with compact support. In particular, the integral is zero if f is not surjective

Of course, this statement contains the classical change of variables of calculus, when f is a diffeomorphism.

On the other hand, the invariance of homotopy follows quite easily in this context:

Corollary 2.5. *Let M, N be two connected, oriented, boundaryless, smooth manifolds of dimension m, and let $H : [0,1] \times M \to N$ be a proper smooth homotopy. Then*

$$\deg(H_0) = \deg(H_1).$$

Proof. Let ω be an m-form with compact support on N, and consider $H^*\omega$. By the properties of exterior diferentiation we have $\mathrm{d}H^*\omega = H^*\mathrm{d}\omega$. But $\mathrm{d}\omega = 0$, because $m = \dim(N)$, so that

$$0 = \int_{[0,1]\times M} H^*\mathrm{d}\omega = \int_{[0,1]\times M} \mathrm{d}H^*\omega.$$

Now we compute the latter integral using Stokes' Theorem. Since the boundary of $[0,1]\times M$ is a disjoint union of two copies M_0 and M_1 of M, with wrong and right orientations, respectively (II.7.8(3), p. 91), we have

$$0 = \int_{[0,1]\times M} \mathrm{d}H^*\omega = \int_M H_1^*\omega - \int_M H_0^*\omega.$$

Thus,

$$\deg(H_0)\int_N \omega = \int_M H_0^*\omega = \int_M H_1^*\omega = \deg(H_1)\int_N \omega\,;$$

hence $\deg(H_0) = \deg(H_1)$. □

Example 2.6. One of the deepest theorems in mathematics, the *Gauss-Bonnet Theorem,* can be formulated as the computation of a degree. Let us describe this.

Fix a connected, closed, boundaryless smooth hypersurface $M \subset \mathbb{R}^{m+1}$ of dimension m. Then M is orientable, disconnects $\mathbb{R}^{m+1}$, and has a global equation $f = 0$ (all of this is the Jordan Separation Theorem; see III.6.4, p. 129). What is relevant here is that

$$\nu = \mathrm{grad}(f)/\|\,\mathrm{grad}(f)\| : M \to \mathbb{S}^m \subset \mathbb{R}^{m+1}$$

is a unitary global normal field (II.7.5(1), p. 88). Replacing f by $-f$ if needed, ν is compatible with the orientation chosen on M (II.7.4, p. 87), and it describes the volume element of M: $\Omega_M = \det(\nu,\cdot)$. This ν is the so-called *Gauss mapping.* Then, at each point $x \in M$ we have the derivative

$$d_x\nu : T_xM \to T_{\nu(x)}\mathbb{S}^m = \text{ hyperplane perpendicular to } \nu(x) \; = T_xM.$$

This is the *Weingarten endomorphism,* and it is *self-adjoint.* Indeed, a straightforward computation shows that for every $u, v \in \mathbb{R}^{m+1}$ perpendicular to $\nu(x)$, that is, for every $u, v \in T_xM$,

$$\langle v, d_x\nu(u)\rangle = \frac{1}{\|\,\mathrm{grad}(f)\|}\langle v, d_x\big(\mathrm{grad}(f)\big)(u)\rangle$$

and $d_x\big(\operatorname{grad}(f)\big)$ is a self-adjoint endomorphism of $\mathbb{R}^{m+1}$, its symmetric matrix being $\left(\frac{\partial^2 f}{\partial x_i \partial x_j}(x)\right)$.

One can accept that this endomorphism $d_x\nu$ of the tangent space T_xM measures the way $M \subset \mathbb{R}^m$ twists at x, and its determinant $K(x) = \det(d_x\nu)$ is the *Gauss curvature of M at x*. As $d_x\nu$ is self-adjoint, its eigenvalues are all real and the endomorphism diagonalizes. Those real eigenvalues are the *principal curvatures of M at x*, and their product is precisely $K(x)$.

Next, suppose M is compact. Then the number

$$\kappa = \int_M K\Omega_M$$

is well defined, and quite naturally, it is called the *integral curvature* of M. Thus we find the integral considered by Dyck in 1888 (I.3, p. 31).

To explore this integral curvature further, consider the volume element $\Omega_{\mathbb{S}^m}$ of the sphere. A straightforward computation shows that $K\Omega_M = \nu^*\Omega_{\mathbb{S}^m}$, and we have

$$\kappa = \int_M K\Omega_M = \int_M \nu^*\Omega_{\mathbb{S}^m} = \deg(\nu)\int_{\mathbb{S}^m} \Omega_{\mathbb{S}^m} = \operatorname{vol}_m(\mathbb{S}^m)\deg(\nu).$$

Now the Gauss-Bonnet Theorem says that if m is even,

$$\kappa = \frac{1}{2}\operatorname{vol}(\mathbb{S}^m)\chi(M),$$

where $\chi(M)$ is the Euler characteristic of M. We conclude that the Gauss-Bonnet Theorem just says:

The degree of the Gauss map of an even-dimensional compact hypersurface is one half of its Euler characteristic.

This statement can be proved by means of the *Poincaré-Hopf Index Theorem*. We will discuss this in V.7. □

Exercises and problems

Number 1. Compact supports are essential. Show that any smooth form $\alpha = f dx_1 \wedge \cdots \wedge dx_m$ in $\mathbb{R}^m$ has some primitive, but this may well not have compact support.

Number 2. Let $\varphi : \mathbb{R} \to \mathbb{S}^1$ be the local diffeomorphism $t \mapsto (\cos t, \sin t)$. Let α be a 1-form on $\mathbb{S}^1$. Show that:

(1) There is a periodic smooth function $g : \mathbb{R} \to \mathbb{R}$ such that $\varphi^*\alpha = gdt$, and $f(t) = \int_0^t g$ is a smooth function such that $f(t+2\pi) = f(t) + \int_{\mathbb{S}^1} \alpha$.

(2) If $\int_{\mathbb{S}^1} \alpha = 0$, f induces a smooth function $h : \mathbb{S}^1 \to \mathbb{R}$ with $\alpha = \mathrm{d}h$.

Deduce from the above that the linear map $\int_{\mathbb{S}^1} : H^1(\mathbb{S}^1, \mathbb{R}) \to \mathbb{R}$ is indeed injective.

Number 3. Consider the torus $T^m = \mathbb{S}^1 \times \cdots \times \mathbb{S}^1$. Prove that $H^k(T^m, \mathbb{R}) \neq 0$ for $0 \leq k \leq m$.

Number 4. Let $T = \mathbb{S}^1 \times \mathbb{S}^1$.

(1) From each factor in $T = \mathbb{S}^1 \times \mathbb{S}^1$ obtain a linear subspace of $H^1(T, \mathbb{R})$ isomorphic to $H^1(\mathbb{S}^1, \mathbb{R})$.

(2) Use the usual periodic local diffeomorphism $\varphi : \mathbb{R}^2 \to T$ to represent the 1-forms α on T by forms $f\mathrm{d}x + g\mathrm{d}y$ on $\mathbb{R}^2$, with f and g periodic, and prove that the linear mapping

$$H^1(T, \mathbb{R}) \to \mathbb{R}^2 : \alpha \mapsto \left(\int_0^{2\pi} f(s,0)ds, \int_0^{2\pi} g(0,t)dt \right)$$

is injective.

Conclude that $\dim_{\mathbb{R}}(H^1(T, \mathbb{R})) = 2$, and so $H^1(\mathbb{S}^1 \times \mathbb{S}^1, \mathbb{R}) = \mathbb{R}^2$.

Number 5. Consider the smooth surface $M \subset \mathbb{R}^3$ given by $x^2 + y^2 + z^4 = 1$, and consider the smooth mapping $f : M \to \mathbb{S}^2 : (x, y, z) \mapsto (x, y, z^2)$. Compute the integral $\int_M f^*\omega$ for the form $\omega = x^2\mathrm{d}y \wedge \mathrm{d}z - y^2\mathrm{d}x \wedge \mathrm{d}z + z^2\mathrm{d}x \wedge \mathrm{d}y$ on $\mathbb{S}^2$. Is the computation really necessary?

Number 6. Consider in $\mathbb{R}^3$ the two diffeomorphic surfaces $x^2 + y^2 = z$ and $x^4 + y^4 = z$. Check that their Gauss curvatures are, respectively,

$$K = \frac{4}{(4x^2 + 4y^2 + 1)^2} \quad \text{and} \quad K = \frac{144x^2y^2}{(16x^6 + 16y^6 + 1)^2}.$$

Study the extrema of both, as well as their lines of constant curvature.

Number 7. Let $M \subset \mathbb{R}^3$ be the torus parametrized by

$$x = (R + r\cos u)\cos v, \quad y = (R + r\cos u)\sin v, \quad z = r\sin u.$$

Then:

(1) Show that the Gauss mapping of M is

$$\nu(x, y, z) = (-\cos u \cos v, -\cos u \sin v, -\sin u).$$

(2) Obtain the matrix of the Weingarten endomorphism with respect to the basis $\partial/\partial u$, $\partial/\partial v$ of $T_{(x,y,z)}M$.

(3) Deduce that $K(x, y, z) = \frac{1}{r^2}\left(1 - \frac{R}{\sqrt{x^2+y^2}}\right)$.

(4) Compute the integral curvature of M.

3. The degree of a continuous mapping

So far, we have defined the Brouwer-Kronecker degree for proper smooth mappings only. To have the notion for arbitrary proper mappings, it is enough to recall that every proper mapping is homotopic to a proper smooth mapping by a proper homotopy:

Proposition and Definition 3.1. *Let $f : M \to N$ be a proper mapping of two oriented, boundaryless, smooth manifolds of dimension m; furthermore, N is connected. Then all proper smooth mappings $g : M \to N$ properly homotopic to f have the same degree, and we define the* degree of f *by* $\deg(f) = \deg(g)$.

Proof. If $g, g' : M \to N$ are properly homotopic to f, then they are properly homotopic themselves, and by II.5.4, p. 79, there is a proper smooth homotopy H_t such that $H_0 = g$ and $H_1 = g'$. But then, $\deg(g) = \deg(g')$ by III.1.9, p. 101. That indeed there are smooth mappings g homotopic to f follows from II.5.2, p. 76, and II.5.3, p. 78. □

As for smooth mappings, we have the product formula for composite continuous mappings:

Proposition 3.2. *Let $f : M \to N$ and $g : N \to P$ be two proper mappings of oriented, boundaryless, smooth manifolds, with N and P connected. Then*

$$\deg(g \circ f) = \deg(g) \cdot \deg(f).$$

Proof. Let f' and g' be proper smooth mappings properly homotopic to f and g, respectively. Then $g' \circ f'$ is a proper smooth mapping properly homotopic to $g \circ f$. Thus:

$$\deg(g \circ f) = \deg(g' \circ f') = \deg(g') \cdot \deg(f') = \deg(g) \cdot \deg(f). \qquad \square$$

We also have the following:

Proposition 3.3 (Boundary Theorem). *Let X be an oriented, smooth manifold of dimension $m + 1$ with boundary $\partial X = Y$, and let N be a connected, oriented, boundaryless, smooth manifold of dimension m. Let $H : X \to N$ be a proper mapping. Then*

$$\deg(H|_Y) = 0.$$

Proof. Let H' be a proper smooth mapping properly homotopic to H. Then $H'|_Y$ is a proper smooth mapping properly homotopic to $H|_Y$, so that $\deg(H|_Y) = \deg(H'|_Y)$, and the latter degree is zero by III.1.8, p. 101. □

Similarly, we deduce invariance by homotopy:

Proposition 3.4. *Let* $H : [0,1]\times M \to N$ *be a proper homotopy of oriented, boundaryless, smooth manifolds,* N *connected. Then*

$$\deg(H_0) = \deg(H_1).$$

Proof. Let $H' : [0,1] \times M \to N$ be a proper smooth mapping properly homotopic to H, so that H'_0 and H'_1 are proper smooth mappings properly homotopic to H_0 and H_1, respectively. We have

$$\deg(H_0) = \deg(H'_0) \quad \text{and} \quad \deg(H_1) = \deg(H'_1),$$

and by III.1.9, p. 101, $\deg(H'_0) = \deg(H'_1)$. □

Thus we have an invariant attached to every proper mapping between two manifolds M and N, the Brouwer-Kronecker degree, which only depends on the homotopy class of the mapping.

Example 3.5. Consider the degree -1 mapping $f : \mathbb{S}^1 \to \mathbb{S}^1 : z \mapsto 1/z$ (this is III.1.6(3), p. 100, for $d = -1$). Then define on the torus $T = \mathbb{S}^1 \times \mathbb{S}^1$ the mapping $F = f \times f : T \to T$. This map has degree 1, but it is not homotopic to the identity.

Indeed, to compute $\deg(F)$, look at the derivative of F at the fixed point $a = (1,1)$:

$$d_aF(u,v) = (d_1f(u), d_1f(v)) = (-u, -v).$$

This preserves orientation; hence $\deg(F) = 1$. Next, suppose there is a homotopy H_t with $H_0 = F$, $H_1 = \mathrm{Id}_T$. Denote by $j : \mathbb{S}^1 \to T$ the injection $z \mapsto (z,1)$ and by $p : T \to \mathbb{S}^1$ the projection $(z,z') \mapsto z$. Then $h_t = p \circ H_t \circ j$ is a homotopy with $h_0 = f$ and $h_1 = \mathrm{Id}_{\mathbb{S}^1}$. But such a homotopy cannot exist, because $\deg(f) = -1$ and $\deg(\mathrm{Id}_{\mathbb{S}^1}) = 1$.

Thus we see that, in general, degree does not fully classify homotopy classes. We will see that for spheres it does classify: this is the content of the famous Hopf Theorems (V.2.1, p. 191, and V.3.1, p. 196). □

Exercises and problems

Number 1. Prove that a proper mapping $f : \mathbb{R} \to \mathbb{R}$ has degree $0, +1$, or -1. Show that the three cases occur. Compute the degree of a real polynomial mapping $f(x) = x^d + a_1x^{d-1} + \cdots + c_d$. What does this say about the solutions of the equation $f(x) = 0$?

Number 2. Construct two proper mappings $f, g : \mathbb{R}^2 \to \mathbb{R}^2$ which are homotopic but have different degree (of course, they will not be properly homotopic).

Number 3. Prove a product formula for the degree of continuous mappings, like that of Problem Number 1 of III.1.

Number 4. Construct a proper mapping $f : \mathbb{R}^2 \to \mathbb{R}^2$ of degree 1 such that $f^{-1}(\{y = 0, x \neq 0\})$ is the graph of the topological sinus $y = \sin(1/x)$, $x \neq 0$. Extend f to $\mathbb{S}^2$ via stereographic projection from the north pole, and describe the inverse image by that extension of the meridians corresponding to the coordinate axes.

Number 5. Fix integers $m, n \geq 0$. Every point z in $\mathbb{S}^{m+n+1} \subset \mathbb{R}^{m+n+2} = \mathbb{R}^{m+1} \times \mathbb{R}^{n+1}$ can be written uniquely as

$$z = \cos(t)x + \sin(t)y, \quad \text{where } x \in \mathbb{S}^m,\ y \in \mathbb{S}^n,\ 0 \leq t \leq \pi/2,$$

except that x (resp., y) is not determined for $t = \pi/2$ (resp., $t = 0$). Given two continuous mappings $f : \mathbb{S}^m \to \mathbb{S}^m$ and $f : \mathbb{S}^n \to \mathbb{S}^n$, their *join* is defined by

$$f \star g : \mathbb{S}^{m+n+1} \to \mathbb{S}^{m+n+1} : z \mapsto \cos(t)f(x) + \sin(t)g(y).$$

Prove that $f \star g$ is continuous, and its degree is $\deg(f)\deg(g)$.

Number 6. Consider the exponential mapping

$$\exp : \mathbb{R} \to \mathbb{S}^1 : x \mapsto \exp(x) = (\cos(2\pi x), \sin(2\pi x)).$$

Now, let $f : \mathbb{S}^1 \to \mathbb{S}^1$ be a continuous mapping of degree d, and let $\bar{h} : [0,1] \to \mathbb{R}$ be a continuous lifting of $h = f \circ \exp|_{[0,1]}$ (that is, $h = \exp \circ \bar{h}$). Show that $d = \bar{h}(1) - \bar{h}(0)$.

Number 7. Let $f, g : \mathbb{S}^m \to \mathbb{S}^m$ be continuous mappings, $m \geq 2$. Check that the mapping $h = -f + 2\langle f, g\rangle g : \mathbb{S}^m \to \mathbb{S}^m$ is well defined and then compute its degree as follows:

(1) Reduce to the case that $f \equiv (1, 0, \ldots, 0)$ on the upper hemisphere and $g \equiv (1, 0, \ldots, 0)$ on the lower one.

(2) Note that h coincides on the upper hemisphere with $(2g_1^2 - 1, 2g_1g_2, \ldots, 2g_1g_{m+1})$ and compute the degree of the latter in terms of $\deg(g)$.

(3) Note that h coincides on the lower hemisphere with $(f_1, -f_2, \ldots, -f_{m+1})$, and compute the degree of the latter in terms of $\deg(f)$.

(4) Conclude that $\deg(h) = (-1)^m \deg(f) + (1 - (-1)^m)\deg(g)$.

Number 8. Let $\mu : \mathbb{S}^m \times \mathbb{S}^m \to \mathbb{S}^m$ be a continuous mapping.

(1) Show that for every $y \in \mathbb{S}^m$ all mappings $\mu(\cdot, y) : \mathbb{S}^m \to \mathbb{S}^m$ have the same degree, which we denote $\deg_1(\mu)$.

(2) Define a degree $\deg_2(\mu)$ similarly.

Now let $g, f : \mathbb{S}^m \to \mathbb{S}^m$ be two continuous mappings, and consider the mapping $\mu(f, g) : \mathbb{S}^m \to \mathbb{S}^m$. Show that

$$\deg(\mu(f, g)) = \deg(f)\deg_1(\mu) + \deg(g)\deg_2(\mu).$$

How does this generalize the previous problem?

4. The degree of a differentiable mapping

So far, the degree of a mapping f is computed through some smooth approximation by the formula in III.1.4, p. 99. However, it is natural to expect that in case the mapping f is itself differentiable, the same formula works directly for f, without any interposed smooth approximation. This is indeed true but, somehow surprisingly, a quite delicate technical matter. Indeed, a first naive attempt would be to revise all constructions in III.1 replacing smooth by differentiable $\mathcal{C}^r$ mappings. Then, for homotopy invariance, we must apply the Sard-Brown Theorem to a $\mathcal{C}^r$ mapping $[0,1] \times M \to N$: this requires the difficult version for finite class mappings, and even more, it needs class $r > \dim\big([0,1] \times M\big) - \dim(N) = 1$. Thus we see the essential difficulty that differentiable $\mathcal{C}^1$ mappings are excluded by this approach. In fact, as we will see, this is the only obstruction to a $\mathcal{C}^1$ degree theory as neat as the smooth one we have previously presented. This is the part of the theory that relies on some results that we do not prove, and it is intentionally confined to this section.

After this preamble, we get into the main result:

Theorem 4.1. *Let $f : M \to N$ be a proper $\mathcal{C}^1$ mapping of two oriented, boundaryless, smooth manifolds of dimension m; furthermore, N is connected. Let $a \in N$ be a regular value of f. Then $f^{-1}(a)$ is a finite set and*

$$\deg(f) = \sum_{x \in f^{-1}(a)} \operatorname{sign}_x(f).$$

Proof. We will use a close smooth approximation $g : M \to N$ as given by II.5.2, p. 76, with some additional conditions. Since a is a regular value and f is proper, $f^{-1}(a)$ consists of finitely many points, say $c_1, \dots, c_r$. Then we pick r local coordinates that preserve orientations on disjoint domains $U_k \equiv \mathbb{R}^m$ at each point c_k and another coordinate domain $V \equiv \mathbb{R}^m$ at a such that $c_k \equiv 0$, $a \equiv 0$, and $f(U_k) \subset V$. Denote by $f_k : \mathbb{R}^m \to \mathbb{R}^m$ the localizations of f at each U_k. We claim the following

Let $\zeta = f_k$, and let $\det\Big(\frac{\partial \zeta_i}{\partial x_j}(0)\Big) \neq 0$. There is a closed ball $B = \{\|x\| \leq \rho\}$ such that for $\varepsilon > 0$ small enough, if a smooth mapping $\xi : B \to \mathbb{R}^m$ verifies

$$\|\xi(x) - \zeta(x)\| < \varepsilon, \quad \Big|\frac{\partial \xi_i}{\partial x_j}(x) - \frac{\partial \zeta_i}{\partial x_j}(x)\Big| < \varepsilon, \quad \textit{for all } i, j,$$

then:

(1) $0 = \xi(b)$ *for a unique* b, *and* $\|b\| < \rho$,

(2) $\det\left(\frac{\partial \zeta_i}{\partial x_j}(0)\right) \cdot \det\left(\frac{\partial \xi_i}{\partial x_j}(b)\right) > 0$ *(that is,* $\operatorname{sign}_0(\zeta) = \operatorname{sign}_b(\xi)$*).*

Assume this for the moment. Pick a smooth approximation g of f such that, for every $x \in M$,

$$\|g(x) - f(x)\| < \operatorname{dist}\big(f(M \setminus W), a\big), \operatorname{dist}\big(f(T), N \setminus V\big),$$

where W is the union of the sets $W_k \equiv \{\|x\| < \rho\}$ and T is that of the sets $T_k \equiv \{\|x\| \leq \rho\}$ (by construction these distances are not zero). These bounds give:

(i) $a \notin g(M \setminus W)$. Indeed, if $a = g(x)$ with $x \notin W$, we have

$$\operatorname{dist}\big(f(M \setminus W), a\big) \leq \|f(x) - g(x)\|,$$

which is impossible.

(ii) $g(T_k) \subset V$. For, if $g(x) \notin V$ with $x \in T_k$, we get

$$\operatorname{dist}\big(f(T), N \setminus V\big) \leq \|f(x) - g(x)\|,$$

a contradiction.

Then, the approximation g has well-defined localizations $g_k : B \to \mathbb{R}^m$, and the crucial point here is that g can be chosen so that each $\xi = g_k$ *verifies the bounds on the derivatives required by the claim.* This is the part we assume without proof.

We also have that $g^{-1}(a) \subset W$; hence

$$g^{-1}(a) \equiv \bigcup_k g_k^{-1}(0),$$

and by the conclusions of the claim, there are exactly r points $b_k \in g_k^{-1}(0)$, none is critical, and

$$\operatorname{sign}_{b_k}(g_k) = \operatorname{sign}_0(f_k) = \operatorname{sign}_{c_k}(f).$$

We conclude that

$$\deg(f) = \deg(g) = \sum_k \operatorname{sign}_{b_k}(g_k) = \sum_k \operatorname{sign}_{c_k}(f).$$

The proof is thus complete, except for the following:

Proof of the Claim. This is a review of the the proof of the Inverse Mapping Theorem. Consider the continuous mapping

$$\bar{\zeta} : \mathbb{R}^m \times \cdots \times \mathbb{R}^m \to \mathbb{R} : \bar{x} = (\bar{x}_1, \ldots, \bar{x}_m) \mapsto \det\Big(\frac{\partial \zeta_i}{\partial x_j}(\bar{x}_i)\Big).$$

Since $\bar{\zeta}(\bar{0}) = \alpha \neq 0$, there is an $\eta > 0$ such that

$$(*) \qquad \Big|\alpha - \det\Big(\frac{\partial \zeta_i}{\partial x_j}(0) + t_{ij}\Big)\Big| < \tfrac{1}{2}|\alpha| \qquad \text{for } |t_{ij}| < \eta.$$

Then choose $\rho > 0$ such that

$$\Big|\frac{\partial \zeta_i}{\partial x_j}(\bar{x}_i) - \frac{\partial \zeta_i}{\partial x_j}(\bar{0})\Big| < \tfrac{1}{2}\eta \quad (\text{hence } |\bar{\zeta}(\bar{x}) - \alpha| < \tfrac{1}{2}|\alpha|),$$

for $\|\bar{x}_1\|, \ldots, \|\bar{x}_m\| < \rho$.

Next, denote by $\delta > 0$ the minimum of $\|\zeta(x)\|$ on the sphere $\|x\| = \rho$ (this minimum is not zero, because by construction 0 is the unique zero of ζ). Then, choose $\varepsilon > 0$ smaller than $\frac{1}{2}\eta$ and $\frac{1}{2}\delta$. We will check that this ε does the job.

Define $\bar{\xi}$ mimicking $\bar{\zeta}$. Hence

$$\bar{\xi}(\bar{x}) = \det\Big(\frac{\partial \zeta_i}{\partial x_j}(\bar{0}) + t_{ij}\Big), \quad \text{where} \quad t_{ij} = \frac{\partial \xi_i}{\partial x_j}(\bar{x}_i) - \frac{\partial \zeta_i}{\partial x_j}(\bar{0}),$$

and on $\|\bar{x}_i\| < \rho$ we have

$$|t_{ij}| \leq \Big|\frac{\partial \xi_i}{\partial x_j}(\bar{x}_i) - \frac{\partial \zeta_i}{\partial x_j}(\bar{x}_i)\Big| + \Big|\frac{\partial \zeta_i}{\partial x_j}(\bar{x}_i) - \frac{\partial \zeta_i}{\partial x_j}(\bar{0})\Big| < \varepsilon + \tfrac{1}{2}\eta \leq \eta,$$

so that by $(*)$, $|\alpha - \bar{\xi}(\bar{x})| < \frac{1}{2}|\alpha|$, which clearly implies $\alpha \cdot \bar{\xi}(\bar{x}) > 0$; hence

$$\alpha \cdot \det\Big(\frac{\partial \xi_i}{\partial x_j}(x)\Big) > 0 \ \text{ for } \|x\| < \rho.$$

This already guarantees (2) in the statement of the Claim, as soon as b is given.

To find it, note that the b we sought is a minimum of the function $\mu(x) = \|\xi(x)\|^2$ in the compact set B. But on its boundary $\|x\| = \rho$ we have

$$\|\xi(x)\| \geq \|\zeta(x)\| - \|\xi(x) - \zeta(x)\| > \delta - \varepsilon > \varepsilon > \|\xi(0) - \zeta(0)\| = \|\xi(0)\|,$$

which means that the minimum does not belong to the boundary. Hence, b is a critical point of the smooth mapping μ on the open ball $\|x\| < \rho$, so that its gradient is zero at that point. But a simple computation gives

$$\operatorname{grad}_b(\mu) = 2\xi(b)\Big(\frac{\partial \xi_i}{\partial x_j}(b)\Big),$$

and since the determinant of the latter matrix is not zero, we conclude $\xi(b) = 0$ as desired.

Finally, we turn to the uniqueness of b. Suppose that $\xi(x) = 0$ with $\|x\| < \rho$. We apply the Mean Value Theorem to each component ξ_i on the segment $[b, x]$ and find points $\bar{x}_i \in [b, x]$, hence $\|\bar{x}_i\| < \rho$, such that

$$0 = \xi_i(x) - \xi_i(b) = \sum_j \frac{\partial \xi_i}{\partial x_j}(\bar{x}_i)(x_j - b_j).$$

This is a linear system in the unknowns $z_j = x_j - b_j$, and its determinant is $\bar{\xi}(\bar{x}) \neq 0$ (see above). Hence the only solution is $z_j = 0$, that is, $x = b$. □

(4.2) Degree theory for differentiable manifolds. Once we know that computations are the same for differentiable as for smooth mappings, the theory goes through for differentiable manifolds. As explained before (II.2.7, p. 57), the key result needed is smoothness: *every differentiable $\mathcal{C}^r$ manifold is $\mathcal{C}^r$ diffeomorphic to a smooth manifold.* Such a smooth manifold is constructed by approximation, in a more delicate sense than the one developed in II.5.2, p. 76. Actually, it is the kind of approximation also used without proof for III.4.1, p. 114: *approximation of derivatives.* As mentioned already, here is where we put the limits on our presentation.

Now, we sketch the construction of the Brouwer-Kronecker degree in the differentiable setting. Let $f : M \to N$ be a proper $\mathcal{C}^1$ mapping of oriented, boundaryless, $\mathcal{C}^1$ manifolds of dimension m, N connected. Then there are $\mathcal{C}^1$ diffeomorphisms $\varphi : M \to M'$ and $\psi : N \to N'$ onto smooth manifolds M' and N'; we choose in these *smooth models* the orientations compatible with φ and ψ and define

$$\deg(f) = \deg(\psi \circ f \circ \varphi^{-1}).$$

Of course, this $\deg(f)$ does not depend on the diffeomorphisms φ and ψ. To prove it, pick any regular value $a \in N$ of f. Then, from the preceding theorem, III.4.1, p. 114, it easily follows that

$$\deg(f) = \sum_{x \in f^{-1}(a)} \operatorname{sign}_x(f),$$

which clearly does not depend on φ or ψ.

After this, one extends the theory to $\mathcal{C}^1$ manifolds using smooth models: proper mappings are properly homotopic to proper $\mathcal{C}^1$ mappings whose degree we can compute by the sum of signs formula, the Boundary Theorem holds, and it follows that properly homotopic mappings have the same degree, and the degree of a composite mapping is the product of the degrees of the factors. All arguments follow without surprise.

Exercises and problems

Number 1. Let $\zeta : \mathbb{R}^m \to \mathbb{R}^m$ be a differentiable mapping with $\zeta(0) = 0$ such that $d_0\zeta$ is a linear isomorphism. Show that for $\rho > 0$ small enough

$$\big\|d_0\zeta^{-1} \circ \big(d_z\zeta - d_0\zeta\big)\big\| < \tfrac{1}{4} \quad \text{for } \|z\| < \rho.$$

Pick a positive real number $\delta < \|d_0\zeta^{-1}\|^{-1}$. Prove that for $\eta > 0$ small enough any smooth mapping $\xi : \mathbb{R}^m \to \mathbb{R}^m$ such that

$$\|\xi(z) - \zeta(z)\| < \eta \quad \text{and} \quad \|d_z\xi - d_z\zeta\| < \eta$$

verifies the two conditions

$$\delta < \|d_0\xi^{-1}\|^{-1} \quad \text{and} \quad \big\|d_0\xi^{-1} \circ \big(d_z\xi - d_0\xi\big)\big\| < \tfrac{3}{4} \quad \text{for } \|z\| < \rho.$$

Use these bounds and the Finite Increment Theorem to deduce that ξ has a unique zero c with $\|c\| < \rho$ and the signs at c of the Jacobian determinants of ξ and ζ coincide.

Number 2. Show that the mapping $f : \mathbb{S}^1 \to \mathbb{S}^1$ defined by

$$f(x, y) = \big(\cos\big(2\pi\sqrt{|y|^3}\,\big), \sin\big(2\pi\sqrt{|y|^3}\,\big)\big)$$

is $\mathcal{C}^1$ but not $\mathcal{C}^2$, and compute its degree.

Number 3. Let $f : M \to N$ be a proper mapping of two oriented, boundaryless, smooth manifolds of dimension m, N connected. Suppose that there is an $a \in N$ such that f is smooth on some neighborhood U of $f^{-1}(a)$ and a is a regular value of $f|_U$. Prove that there is a close smooth approximation $g : M \to N$ of f that coincides with f on U (up to shrinking) and such that $g^{-1}(a) \subset U$. Deduce that

$$\deg(f) = \sum_{x \in f^{-1}(a)} \operatorname{sign}_x(f).$$

What is the difference between this and III.4.1, p. 114?

Number 4. Compute the degrees of the following two mappings $\mathbb{S}^1 \to \mathbb{S}^1$:

$$(1)\ g(x, y) = \begin{cases} (x^2 - y^2, -2xy) & \text{for } x \geq 0, \\ (x^2 - y^2, 2xy) & \text{for } x \leq 0. \end{cases}$$

$$(2)\ h(x, y) = \begin{cases} (x, -y) & \text{for } x \geq 0, \\ ((x^2 - 3y^2)x, (3x^2 - y^2)y) & \text{for } x \leq 0. \end{cases}$$

Number 5. Complete the details of $\mathcal{C}^1$ mapping degree theory as suggested in III.4.2, p. 117.

Number 6. Let $M \subset \mathbb{R}^2$ denote the graph of the function $h : \mathbb{R} \to \mathbb{R}$ given by

$$h(x) = \begin{cases} 0 & \text{for } x \leq 0, \\ x^{3/2} & \text{for } x \geq 0. \end{cases}$$

Prove that M is a $\mathcal{C}^1$ curve, but not $\mathcal{C}^2$. Fix an orientation on M and find proper mappings $f : \mathbb{R} \to M$ of all possible degrees.

5. The Hopf invariant

Although the notion of degree is mainly intended for discussing mappings of manifolds of the same dimension, it can be profitably used to treat some mappings between spheres of different dimensions. The pioneering ideas in this respect are due to Hopf (see I.3, p. 30).

We fix the following notation. Consider the north and south poles, $p = (0, \dots, 0, 1)$ and $-p = (0, \dots, 0, -1)$, of the unit sphere $\mathbb{S}^{2m-1} \subset \mathbb{R}^{2m}$, $m \geq 2$, and let $\phi : \mathbb{S}^{2m-1} \setminus \{-p\} \to T_p\mathbb{S}^{2m-1} \equiv \mathbb{R}^{2m-1}$ be the stereographic projection from the south pole. Let $\Omega \subset \mathbb{S}^{2m-1} \times \mathbb{S}^{2m-1}$ be the open set defined by the inequalities $x \neq y$, $x \neq -p$, $y \neq -p$, and consider the smooth mapping

$$\Phi : \Omega \to \mathbb{S}^{2m-2} : (x, y) \mapsto \frac{\phi(y) - \phi(x)}{\|\phi(y) - \phi(x)\|}.$$

We have the following:

Proposition and Definition 5.1. *Let $f : \mathbb{S}^{2m-1} \to \mathbb{S}^m$ be a smooth mapping.*

(1) *If f is not surjective, we let $H(f) = 0$.*

(2) *If f is surjective, consider any two distinct regular values $a, b \neq f(-p)$ of f. Then $f^{-1}(a) \times f^{-1}(b) \subset \Omega$ is a compact, oriented, boundaryless, smooth manifold of dimension $2m-2$, and Φ restricts to a smooth mapping*

$$\phi_{a,b} : f^{-1}(a) \times f^{-1}(b) \to \mathbb{S}^{2m-2},$$

whose degree does not depend on the choice of a and b. We let

$$H(f) = \deg(\phi_{a,b}).$$

The integer $H(f)$ is called the Hopf invariant *of f.*

Proof. Suppose f surjective. By II.3.2, p. 62, and II.7.6, p. 89, the inverse image of every regular value is an oriented manifold of dimension $m-1$; hence the product $f^{-1}(a) \times f^{-1}(b)$ is an oriented manifold of dimension $2m-2$. On the other hand, since the compact sets $f^{-1}(a)$ and $f^{-1}(b)$ are disjoint and do not contain the south pole $-p$, their product is contained in Ω. All in all, the degree $\deg(\phi_{a,b})$ is well defined.

Next, let us see why that degree does not depend on a (the proof for b would be the same). Let $a' \neq f(-p)$ be another regular value of f, and let F_t be a diffeotopy of $\mathbb{S}^m$ that fixes b and $f(-p)$ and is the identity off a neighborhood W of a, a' such that $F_1(a') = a$. We are to apply a boundary argument twice to conclude that $\deg(\phi_{a,b}) = \deg(\phi_{a',b})$.

Suppose first that a' is close to a. Then we take $W \subset R_f$, which implies that a is a regular value of the smooth mapping $H(t,x) = F_t(f(x))$. Indeed, if $H_t(x) = a$, we have $F_t(f(x)) = a$, and $f(x)$ must be in W, so that x is a regular point of f. Hence $d_x f$ is surjective, and since F_t is a diffeomorphism, the derivative

$$d_x H_t = d_{f(x)} F_t \circ d_x f$$

is surjective. With this settled, $H^{-1}(a)$ is a compact smooth manifold with

$$-p \notin H_t^{-1}(a) \quad \text{and} \quad H_t^{-1}(a) \cap f^{-1}(b) = \emptyset \quad \text{for all } t.$$

For the first condition, notice that if $a = H_t(-p) = F_t(f(-p))$, then $f(-p) \in W$, a contradiction. For the second, if $a = H_t(x) = F_t(f(x))$, then $f(x) \in W$ and $b \notin W$.

Thus the smooth mapping

$$\overline{\Phi} : X = H^{-1}(a) \times f^{-1}(b) \to \mathbb{S}^{2m-2} : (t,x,y) \mapsto \frac{\phi(y) - \phi(x)}{\|\phi(y) - \phi(x)\|}$$

is well defined, and by the Boundary Theorem (III.1.8, p. 101)

$$0 = \deg(\overline{\Phi}|_{\partial X}), \quad \text{where} \quad \partial X = \partial H^{-1}(a) \times f^{-1}(b).$$

Note that the boundary of $H^{-1}(a)$ consists of the two pieces $\{0\} \times f^{-1}(a)$ and $\{1\} \times f^{-1}(a')$, one with the right and the other with the wrong orientation. Thus ∂X consists of

$$\{0\} \times f^{-1}(a) \times f^{-1}(b) \quad \text{and} \quad \{1\} \times f^{-1}(a') \times f^{-1}(b),$$

one with the right and the other with the wrong orientation. Clearly, $\overline{\Phi}$ restricts to $\phi_{a,b}$ on the first component and to $\phi_{a',b}$ on the second, and by the remark concerning orientations we conclude

$$0 = \deg(\overline{\Phi}|_{\partial X}) = \deg(\phi_{a,b}) - \deg(\phi_{a',b}).$$

Once this case is proved, assume a' is arbitrary. Then choose a diffeotopy F_t as above, but now a need not be a regular value of $H_t = F_t \circ f$. Then, by the Sard-Brown Theorem, there is a regular value a_1 of H close to a so that $F_1^{-1}(a_1) = a_1'$ is close to a'. By the case already proved, we have

$$\deg(\phi_{a,b}) = \deg(\phi_{a_1,b}) \quad \text{and} \quad \deg(\phi_{a',b}) = \deg(\phi_{a_1',b}).$$

On the other hand, the argument above with $H^{-1}(a_1)$ instead of $H^{-1}(a)$ shows that

$$\deg(\phi_{a_1,b}) = \deg(\phi_{a_1',b}),$$

and, as desired, we deduce

$$\deg(\phi_{a,b}) = \deg(\phi_{a',b}).$$

□

Remarks 5.2. (1) The degree of the mapping $\phi_{a,b}$ is in fact the *link coefficient* of the two manifolds $f^{-1}(a)$ and $f^{-1}(b)$ as was introduced by Brouwer (see I.2, p. 28). We use the stereographic projection to set the two manifolds in a Euclidean space. We will abuse notation and write

$$\ell(f^{-1}(a), f^{-1}(b)) = \deg(\phi_{a,b}),$$

although this disregards the sign on the left-hand side; *for our use here this is only a notation.*

(2) After the above definition of the Hopf invariant, it is only natural to ask whether for non-surjective mappings we can also compute it through link coefficients. The answer is that we can. Namely, if $f : \mathbb{S}^{2m-1} \to \mathbb{S}^m$ *is not surjective and* $a, b \neq f(-p)$ *are two regular values such that* $f^{-1}(a) \neq \emptyset$ *and* $f^{-1}(b) \neq \emptyset$, *then*

$$\ell(f^{-1}(a), f^{-1}(b)) = 0.$$

To see this, just repeat the proof of III.5.1, p. 119, by choosing a second regular value $a' \notin f(\mathbb{S}^{2m-1})$. Then $f^{-1}(a') = \emptyset$ and

$$\partial H^{-1}(a) = \{0\} \times f^{-1}(a),$$

so that

$$\partial X = \partial H^{-1}(a) \times f^{-1}(b) = \{0\} \times f^{-1}(a) \times f^{-1}(b).$$

Consequently, $\overline{\Phi}|_{\partial X} \equiv \phi_{a,b}$, and from $\deg(\overline{\Phi}|_{\partial X}) = 0$ we deduce

$$\ell(f^{-1}(a), f^{-1}(b)) = \deg(\phi_{a,b}) = \deg(\overline{\Phi}|_{\partial X}) = 0.$$

(3) We have fixed the center of projection at the south pole and need no more freedom for our purposes, but this choice is actually irrelevant. The

argument to show this is quite similar to those above, using a rotation to move the center elsewhere. We will not give any details here.

(4) In odd dimension, the Hopf invariant vanishes: *if m is odd, $H(f) = 0$ for every smooth mapping $f : \mathbb{S}^{2m-1} \to \mathbb{S}^m$.*

Indeed, we have the commutative diagram

$$\begin{array}{ccc} f^{-1}(a) \times f^{-1}(b) & \xrightarrow{\phi_{a,b}} & \mathbb{S}^{2m-2} \\ {\scriptstyle\tau}\downarrow\approx & & \approx\downarrow{\scriptstyle\sigma} \\ f^{-1}(b) \times f^{-1}(a) & \xrightarrow{\phi_{b,a}} & \mathbb{S}^{2m-2} \end{array}$$

where τ is the permutation $(x, y) \mapsto (y, x)$ and σ is the antipodal mapping $z \mapsto -z$. But an easy computation shows that $\deg(\tau) = (-1)^{(m-1)^2} = +1$ (m is odd) and $\deg(\sigma) = -1$ (III.1.6(1), p. 99), so that

$$H(f) = \deg(\phi_{a,b}) = \deg(\sigma)\deg(\phi_{b,a})\deg(\tau) = -\deg(\phi_{b,a}) = -H(f),$$

and hence $H(f) = 0$. □

Next we see that we have constructed a homotopy invariant:

Proposition 5.3 (Homotopy invariance). *Let $f, g : \mathbb{S}^{2m-1} \to \mathbb{S}^m$ be two homotopic smooth mappings. Then, $H(f) = H(g)$.*

Proof. Since $[0, 1] \times \mathbb{S}^{2m-1}$ is compact, the homotopy (which we can take to be smooth) is uniformly continuous; hence there is an $\varepsilon > 0$ such that

$$\|H_t(x) - H_s(x)\| < \tfrac{1}{2} \quad \text{for all } x \in \mathbb{S}^{2m-1} \text{ and } |t - s| < \varepsilon.$$

The result follows immediately if we see that for $0 < s - t < \varepsilon$, H_t and H_s have the same Hopf invariant. In other words, we can merely assume

$$\|H_t(x) - H_s(x)\| < \tfrac{1}{2} \quad \text{for all } x,\ t,\ s.$$

Now, the result is trivial if $H(f) = H(g) = 0$; hence we suppose that one of the mappings is surjective, say f. We will use a boundary argument (as in the proof of III.5.1, p. 119) to deduce first

$$\ell(f^{-1}(a), f^{-1}(b)) = \ell(g^{-1}(a), f^{-1}(b))$$

and then

$$\ell(g^{-1}(a), f^{-1}(b)) = \ell(g^{-1}(a), g^{-1}(b)).$$

Notice that what matters here is only that the smooth mappings

$$\overline{\Phi} : X = H^{-1}(a) \times f^{-1}(b) \to \mathbb{S}^{2m-2} : (t, x, y) \mapsto \tfrac{\phi(y)-\phi(x)}{\|\phi(y)-\phi(x)\|} \quad \text{and}$$

$$\overline{\Psi} : Y = g^{-1}(a) \times H^{-1}(b) \to \mathbb{S}^{2m-2} : (x, t, y) \mapsto \tfrac{\phi(y)-\phi(x)}{\|\phi(y)-\phi(x)\|}$$

are well defined. But by the bound that we have assumed, this follows immediately if the regular values verify $\|a - b\| \geq \frac{1}{2}$ and are taken off $H([0,1] \times \{-p\})$ (and the latter is possible by the Easy Sard Theorem (II.3.4, p. 63), because $m \geq 2$).

Apart from this, we only recall what was stated in III.5.2(1), p. 121, for non-surjective mappings. As mentioned there, if g is not surjective, we can take the regular value a off $g(\mathbb{S}^{2m-1})$, so that $\partial H^{-1}(a) = \{0\} \times f^{-1}(a)$; hence in the first step we get

$$\ell(f^{-1}(a), f^{-1}(b)) = \deg(\overline{\Phi}|_{\partial X}) = 0 = H(g),$$

and this case is finished here. □

As for degree, the preceding homotopy invariance result gives grounds for defining the Hopf invariant of arbitrary continuous mappings.

Exercises and problems

Number 1. Let $f : \mathbb{S}^{2m-1} \to \mathbb{S}^m$ be a smooth mapping ($m \geq 2$). We assume both spheres oriented. Let ω be a form on $\mathbb{S}^m$ of maximum degree m, such that $\int_{\mathbb{S}^m} \omega = 1$. Then $f^*\omega$ is a closed form in $\mathbb{S}^{2m-1}$, hence exact (because $H^m(\mathbb{S}^{2m-1}, \mathbb{R}) = 0$), and there is a form α of degree $m-1$ such that $d\alpha = f^*\omega$. Consequently, we have on $\mathbb{S}^{2m-1}$ the $2m-1$ form $\alpha \wedge d\alpha$ and can compute the integral

$$\varrho(f) = \int_{\mathbb{S}^{2m-1}} \alpha \wedge d\alpha.$$

Show that $\varrho(f)$ is the same for any choice of ω and α under the conditions above. How does it depend on the orientations?

This is in fact another definition of the Hopf invariant, but we cannot prove this now and consequently we use a different notation.

Number 2. Show that the invariant ϱ defined in the preceding problem is a homotopy invariant and thus it can be defined for arbitrary continuous mappings.

Number 3. Compute the invariant $\varrho(f)$ of a continuous mapping $f : \mathbb{S}^{2m-1} \to \mathbb{S}^m$ when m is odd.

Number 4. Compute $H(f)$ and $\varrho(f)$ for the mapping $f : \mathbb{S}^3 \to \mathbb{S}^2$ defined by

$$f(x_1, x_2, x_3, x_4) = \big(2x_1\sqrt{1 - x_1^2 - x_2^2},\, 2x_2\sqrt{1 - x_1^2 - x_2^2},\, 1 - 2x_1^2 - 2x_2^2\big).$$

6. The Jordan Separation Theorem

We devote this section to one very important application of degree theory: a proof of the Jordan Separation Theorem for closed differentiable hypersurfaces of the Euclidean space. In fact, this is the proof involving the Cauchy index that triggered Hadamard's, and then Brouwer's, research on the notion of degree (I.2, p. 14). In fact, this proof does not require the notion of degree in full, but only a simpler version which we quickly describe next.

(6.1) Mod 2 degree theory. All constructions and definitions in the preceding sections can be carried over disregarding orientations, which means that *manifolds need not be orientable.* The resut is a mod 2 invariant defined as follows:

Let $f : M \to N$ be a proper mapping of two boundaryless differentiable manifolds of dimension m; furthermore, N is connected. Let $g : M \to N$ be a proper differentiable mapping close to f and let $a \in N$ be a regular value of g. Then the parity

$$\deg_2(f) = \#g^{-1}(a) \mod 2$$

depends only on the homotopy class of f and is called the degree mod 2 of f.

We leave the reader the task of reviewing all proofs involved and only notice for the key result, III.1.2, p. 96, that the curves Γ used there have either no boundary points or two boundary points; hence their union C has an *even* number of boundary points.

Using this mod 2 degree, we can now prove the following:

Theorem 6.2 (Jordan Separation Theorem). Let $M \subset \mathbb{R}^{m+1}$ be a compact, boundaryless, differentiable hypersurface. Then M disconnects $\mathbb{R}^{m+1}$.

Proof. Pick any point $p \notin M$ and define the differentiable mapping

$$f_p : M \to \mathbb{S}^m : x \mapsto \frac{x-p}{\|x-p\|},$$

which is proper because M is compact. The mod 2 degree of this mapping is called the *mod 2 winding number of M around p* and is denoted by $w_2(M,p)$ (compare IV.4.1, p. 156). It is easy to see that the winding number is constant on each connected component of $\mathbb{R}^{m+1} \setminus M$.

Indeed, if p and q are in the same component, that component, which is an open set in $\mathbb{R}^{m+1}$, contains a continuous (even piecewise linear) path $\gamma = \gamma(t)$ with $\gamma(0) = p$ and $\gamma(1) = q$. Thus we can define a homotopy by

$$H : [0,1] \times M \to \mathbb{S}^m : x \mapsto \frac{x - \gamma(t)}{\|x - \gamma(t)\|}.$$

Hence, by homotopy invariance for mod 2 degree,

$$w_2(M,p) = \deg_2(f_p) = \deg_2(H_0) = \deg_2(H_1) = \deg_2(f_q) = w_2(M,q).$$

Thus, to show that M disconnects, it is enough to find two points p and q with different mod 2 winding numbers.

To that end, pick any point $a \in M$ and line ℓ through a whose direction is perpendicular to T_aM. Then pick $p \in \ell \setminus M$. We see that a is a regular point of f_p by direct computation. For $v \in T_pM$ we have

$$d_af_p(v) = \frac{\|a-p\|^2 v - \langle a-p, v\rangle(a-p)}{\|a-p\|^3} = \frac{v}{\|a-p\|},$$

because $\langle a-p, v\rangle = 0$ by the choice of ℓ. Thus, d_af_p is a linear isomorphism, as desired.

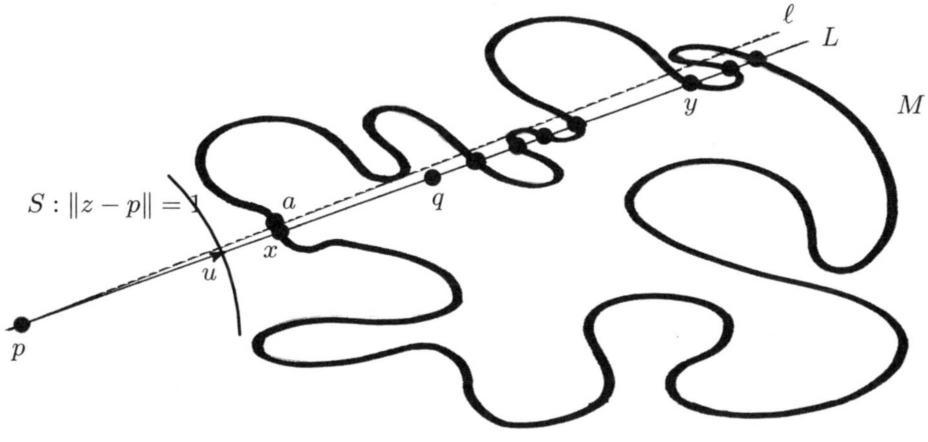

Consequently, by the Inverse Mapping Theorem, f_p is a local diffeomorphism at a. This implies that $f_p(M)$ has non-empty interior in $\mathbb{S}^m$, and by the Sard-Brown Theorem, f_p has some regular value $u \in f_p(M)$. We denote by L the half-line $x = p + \lambda u$, $\lambda > 0$. Then, if p is far enough from our compact hypersurface M, we have $f_p^{-1}(u) = M \cap L$, and

$$w_2(M,p) = \deg_2(f_p) = \#(M \cap L) \mod 2.$$

Now let x be the first point we find in $M \cap L$ when starting from p, and pick a second point $q \in L \setminus M$ before the next (we can do this, because $M \cap L = f_p^{-1}(u) \neq \emptyset$). In this situation,

$$f_q^{-1}(u) = M \cap L \setminus \{x\},$$

and u is a regular value of f_q.

Indeed, the first assertion is clear, and for the second note that

$$y - q = \mu(y - p) \quad \text{with } \mu > 0 \text{ for every } y \in f_q^{-1}(u).$$

Then, for $v \in T_y M$ we have

$$\begin{aligned} d_y f_q(v) &= \frac{\|y-q\|^2 v - \langle y-q, v\rangle (y-q)}{\|y-q\|^3} \\ &= \frac{\|y-p\|^2 v - \langle y-p, v\rangle (y-p)}{\mu\|y-p\|^3} = \tfrac{1}{\mu} d_y f_p(v), \end{aligned}$$

which shows that $d_y f_q = \frac{1}{\mu} d_y f_p$ is an isomorphism.

Thus,

$$w_2(M, q) = \deg_2(f_q) = \#(M \cap L \setminus \{x\}) \mod 2.$$

Obviously this implies $w_2(M,p) \neq w_2(M,q)$, and the proof is finished. □

As it stands, the above proof cannot be used for non-compact hypersurfaces: for them, f_p is not proper any more. However, the idea behind it can be put to work in that case too. In fact, we have the following:

Theorem 6.3 (Jordan Separation Theorem). Let $M \subset \mathbb{R}^{m+1}$ be a closed, boundaryless, differentiable hypersurface. Then M disconnects $\mathbb{R}^{m+1}$.

Proof. We will use the following notation. Let $D \subset \mathbb{R}^{m+1}$ be an open ball with boundary a sphere $S = \overline{D} \setminus D$, and let $p \in \mathbb{R}^{m+1} \setminus \overline{D}$. We denote by π_p the conic projection with center p into the sphere S. This mapping is defined on a compact cone K_p with vertex p and its image is the spherical region $K_p \cap S$. The boundary of K_p is the tangent cone to the sphere with vertex p, and the interior T_p of K_p in $\mathbb{R}^{m+1}$ is the open cone generated by p and the interior N_p of $K_p \cap S$ in S; note that $N_p = \pi_p(T_p)$. We need the following simple property of π_p:

$$d_x \pi_p(u) = 0 \textit{ if and only if } u \textit{ is parallel to } x - p.$$

Indeed, consider the curve germ $\gamma(t) = x + tu$. Then

$$\pi_p \circ \gamma(t) = p + \lambda(t)(x + tu - p)$$

for some smooth $\lambda(t)$, and

$$d_x\pi_p(u) = (\pi_p \circ \gamma)'(0) = \lambda'(0)(x - p) + \lambda(0)u = v.$$

Since $\pi_p(x) = p + \lambda(0)(x - p) \in S$ and $p \notin S$, we have $\lambda(0) \neq 0$, and v is vanishes if and only if u is parallel to $x - p$.

We can reformulate the above remark by saying that $d_x\pi_p$ *is injective on any linear hyperplane* L *transversal to* $x - p$.

Now, pick a point $e \in M$ and a line ℓ through e not tangent to M (that is, $\ell \not\subset e + T_eM$). This choice guarantees that e is isolated in $\ell \cap M$ (II.2.12, p. 60): there is a segment $[\alpha, c] \subset \ell$ containing $e(\neq \alpha, c)$, such that $[\alpha, c] \cap M = \{e\}$. In particular, c is not in the closed set M, and there is an open ball D centered at c with $M \cap \overline{D} = \emptyset$.

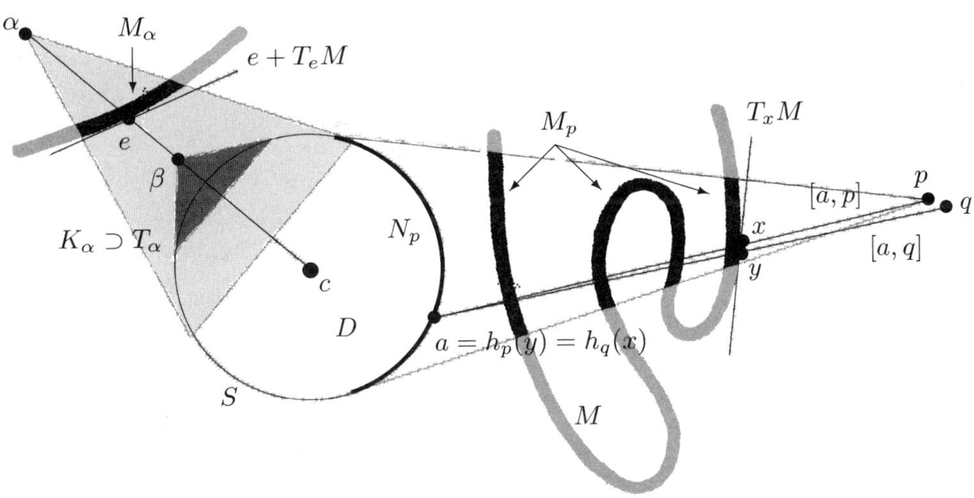

Note that for D small enough, π_α restricts to a diffeomorphism $h_\alpha : M_\alpha = M \cap T_\alpha \to N_\alpha$. Indeed, for a first choice of D, the preceding remarks on the derivatives of conic projections tell that d_eh_α is a linear isomorphism. Consequently, h_α is a local diffeomorphism at the point e, and the claim follows by shrinking D.

Once this setting is ready, consider an arbitrary point $p \notin M \cup \overline{D}$ and the corresponding π_p, K_p, T_p, N_p. If there is some point $a \in N_p$ such that $[a, p] \cap M = \emptyset$, we write

$$w_2(p) = 0.$$

Otherwise, $M_p = T_p \cap M \neq \emptyset$ is an m-manifold open in M and π_p induces a differentiable mapping $h_p = \pi_p|_{M_p} : M_p = T_p \cap M \to N_p$. This mapping is proper (the restriction of π_p to the compact set $K_p \cap M$ is proper), and we write

$$w_2(p) = \deg_2(h_p).$$

To compute this $w_2(p)$, we use regular values. Now, $a \in N_p$ is a regular value of h_p if and only if $h_p^{-1}(a)$ is finite and for each x in that inverse image, the vector $a - p$ is not tangent to M at x. Again this follows from what was earlier remarked concerning the derivatives of conic projections. In short we say that $[a, p]$ meets M *transversally,* and then

$$w_2(p) = \#([a, p] \cap M) \mod 2.$$

Note that there is some overlapping here. Indeed, M_p may be non-empty and h_p will be defined, even if $[a, p] \cap M = \emptyset$ for some $a \in N_p$. But then a is a regular value of h_p, so that $\deg_2(h_p) = 0$. Of course, the computation through any other regular value will give the same degree 0.

Next, note that our careful choice of data guarantees $w_2(\alpha) = 1$. On the other hand, take any β in the segment $\big(e, \pi_\alpha(e)\big)$. Clearly, $M_\beta = \emptyset$, and $w_2(\beta) = 0$. Thus, the invariant w_2 is not constant. We claim that it is locally constant.

To prove this, fix p. If $[a, p] \cap M = \emptyset$ for some $a \in N_p$, as $[a, p]$ is compact and M is closed, we see immediately that $[a, q] \cap M = \emptyset$ for q very close to p. Hence suppose we have the mapping h_p well defined, with a regular value $a \in N_p$ so that $[a, p]$ meets M transversally. We want the same for q close enough to p and also that

$$\#([a, q] \cap M) = \#([a, p] \cap M),$$

to conclude $w_2(p) = w_2(q)$. For this we argue as follows. For every $x \in [a, p] \cap M$, let $\gamma_{a,x}$ be the conic projection with center a onto $x + T_x M$: the same remarks on derivatives of conic projections apply to γ_a to deduce that its restriction to a neighborhood V of x in M is a diffeomorphism from V onto an open set of $x + T_x M$. This implies that for q close to p, the segment $[a, q]$ meets M exactly as $[a, p]$ does, and we are done.

Summing up, w_2 is locally constant, but not constant, on the open set $\mathbb{R}^{m+1} \setminus M \cup \overline{D}$: thus this set cannot be connected. Finally, if the open set $\mathbb{R}^{m+1} \setminus M$ were connected, no *closed* ball would disconnect it, and we conclude as desired that the complement of the hypersurface M is not connected. □

From the fact that hypersurfaces disconnect, much more information can be deduced. We gather all of it into a single statement:

Proposition 6.4. *Let $M \subset \mathbb{R}^{m+1}$ be a connected, closed, boundaryless, differentiable hypersurface. Then M disconnects $\mathbb{R}^{m+1}$ into two connected components D and E, whose closures $\overline{D}$ and $\overline{E}$ are closed manifolds with boundary M. Furthermore, M has a global equation f such that*

$$M = \{f = 0\}, \quad D = \{f > 0\}, \quad \textit{and} \quad E = \{f < 0\}.$$

In particular, M is orientable and has a unitary global normal vector field

$$\nu = \operatorname{grad}(f)/\|\operatorname{grad}(f)\|.$$

If M is compact, one of the two connected components is bounded (the interior *of M) and the other is not (the* exterior *of M).*

Proof. As we know (II.2.3, p. 55), M has local equations; hence every point in M has an open neighborhood U in $\mathbb{R}^{m+1}$ such that $U \cap M = \{f = 0\}$ for some differentiable function $f : U \to \mathbb{R}$ with 0 a regular value. In fact, by II.2.3, p. 55, we can assume that there is a diffeomorphism $\varphi : V \to U$ from an open ball V centered at $0 \in \mathbb{R}^{m+1}$, such that

$$\varphi^{-1}(U \cap M) = \{x \in V : x_{m+1} = 0\}.$$

In this situation:

(a) *If a differentiable function $h : U \to \mathbb{R}$ vanishes on $U \cap M$, then h/f is a well-defined differentiable function on U.*

(b) *If $h : U \to \mathbb{R}$ is another equation of M in U, then $\alpha = h/f$ has no zeros in U.*

(c) *The set $U \setminus M$ has two connected components, $U^+ = \{f > 0\}$ and $U^- = \{f < 0\}$, so that $M \cap U \subset \overline{U^+} \cap \overline{U^-}$.*

For (a), we note that $h \circ \varphi(x_1, \dots, x_m, 0) = 0$; hence

$$h \circ \varphi(x) = \int_0^1 \frac{\partial}{\partial t}\left(h \circ \varphi(x_1, \dots, x_m, t x_{m+1})\right) dt = h^*(x) x_{m+1},$$

with

$$h^*(x) = \int_0^1 \frac{\partial}{\partial x_{m+1}}\left(h \circ \varphi(x_1, \dots, x_m, t x_{m+1})\right) dt.$$

Similarly, $f \circ \varphi(x) = f^*(x)x_{m+1}$, but in this case we can say more: f^* never vanishes. Indeed, we have

$$d_x(f \circ \varphi) = x_{m+1} d_x f^* + f^*(x)\pi,$$

where $\pi : \mathbb{R}^{m+1} \to \mathbb{R}$ is the last coordinate projection. If $f^*(x) = 0$, then $f(\varphi(x)) = 0$; hence $\varphi(x) \in M$ and $x_{m+1} = 0$. Thus

$$d_x(f \circ \varphi) = x_{m+1} d_x f^* + f^*(x)\pi = 0,$$

and $\varphi(x) \in U \cap M$ would be a critical point of f, a contradiction. Consequently

$$h \circ \varphi(x) = h^*(x)x_{m+1} = \frac{h^*(x)}{f^*(x)}(f \circ \varphi)(x),$$

which shows that $(h/f) \circ \varphi = h^*/f^*$ is differentiable on V.

For (b), note that if h is an equation itself, the same argument above shows that h^* has no zeros, and neither does h^*/f^*.

Finally, assertion (c) is evident for the equation $h = \pi \circ \varphi^{-1}$ and follows for $f = \alpha h$, because α never vanishes, hence has constant sign on the connected set U.

After these three local properties, we obtain a covering of M by open sets $U_i \subset \mathbb{R}^{m+1}$ with equations f_i as above, to which we add a covering of $\mathbb{R}^{m+1} \setminus M$ by open connected sets with trivial equations $\equiv 1$. Recall that by the Jordan Separation Theorem, III.6.3, p. 126, we know that M disconnects $\mathbb{R}^{m+1}$. In fact,

(d) $\mathbb{R}^{m+1} \setminus M$ *has exactly two connected components* D *and* E, *and*

$$M = \overline{D} \setminus D = \overline{E} \setminus E.$$

Indeed, we claim first that $\overline{W} \setminus W = M$ for every connected component W of $\mathbb{R}^{m+1} \setminus M$. Note that any other component W' of $\mathbb{R}^{m+1} \setminus M$ is open in $\mathbb{R}^{m+1}$ and does not meet W; hence it does not meet $\overline{W}$ either. Thus, $\overline{W} \setminus W \subset M$. On the other hand, the open set W cannot be closed in $\mathbb{R}^{m+1}$; hence there is some point $x \in \overline{W} \setminus W \subset M$. Let $y \in M$ be arbitrary. As M is connected, there is a chain $U_{i_0}, \dots, U_{i_r}$ with $x \in U_{i_0}$, $y \in U_{i_r}$, and $U_{i_k} \cap U_{i_{k+1}} \cap M \neq \emptyset$, $U_{i_k} \setminus M = U_{i_k}^+ \cup U_{i_k}^-$, for every k:

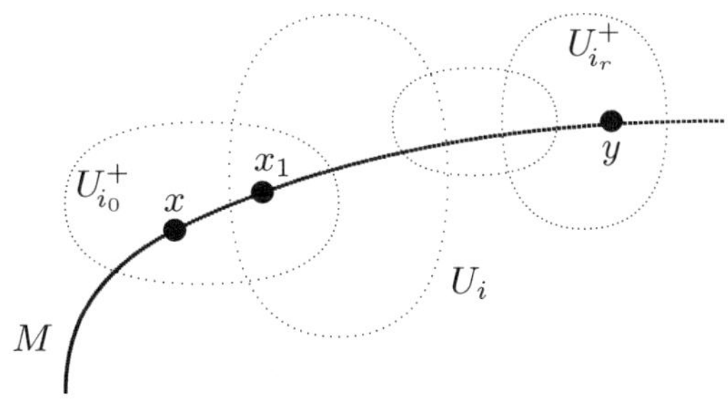

From $U_{i_0} \setminus M = U^+_{i_0} \cup U^-_{i_0}$ and $x \in \overline{W}$, we deduce $U^+_{i_0} \subset W$ or $U^-_{i_0} \subset W$. It follows that $M \cap U_{i_0} \subset \overline{U^+_{i_0}} \cap \overline{U^-_{i_0}} \subset \overline{W}$, and looking at a point $x_1 \in M \cap U_{i_0} \cap U_{i_1}$, we see $x_1 \in U_{i_1} \cap \overline{W}$; hence $U^+_{i_1} \subset W$ or $U^-_{i_1} \subset W$. Thus $M \cap U_{i_1} \subset \overline{W}$.

Repeating the argument, we conclude that $y \in M \cap U_{i_r} \subset \overline{W}$. This completes the proof of our claim.

Now, let $D = W$ be a component as in the argument above, and let E be a second component (recall that M does disconnect $\mathbb{R}^{m+1}$). Arguing as above, we find $U^+_{i_0} \subset E$ or $U^-_{i_0} \subset E$. As D and E are disjoint, there are two possibilities:

$$U^+_{i_0} \subset D,\ U^-_{i_0} \subset E \quad \text{or} \quad U^+_{i_0} \subset E,\ U^-_{i_0} \subset D,$$

which gives no extra room for anymore components!

Once we have (d), we choose signs ε_i as follows:

$$\text{Case } U_i \cap M = \emptyset: \quad \varepsilon_i = \begin{cases} +1 & \text{if } U_i \subset D, \\ -1 & \text{if } U_i \not\subset D \quad (\text{that is, } U_i \cap D = \emptyset). \end{cases}$$

$$\text{Case } U_i \cap M \neq \emptyset: \quad \varepsilon_i = \begin{cases} +1 & \text{if } U^+_i \subset D, \\ -1 & \text{if } U^+_i \not\subset D \quad (\text{that is, } U^-_i \subset D). \end{cases}$$

With these signs fixed, set $g_i = \varepsilon_i f_i$, so that

$$\{g_i > 0\} \cap U_i = D \cap U_i, \quad \{g_i < 0\} \cap U_i = E \cap U_i\,.$$

Consequently, the function g_k/g_i, which is defined on $U_k \cap U_i$ and has no zeros by (a) and (b), is > 0.

Finally, let $\{\theta_i\}$ be a differentiable partition of unity for the U_i's. The formula

$$(*) \qquad f(x) = g_i(x) \exp\left[\sum_k \theta_k(x) \log\left(\frac{g_k(x)}{g_i(x)}\right)\right] \quad \text{for } x \in U_i$$

gives the equation $f:\mathbb{R}^{m+1}\to\mathbb{R}$ that we sought.

First note that the sum is locally finite, and for $x\in U_i$ and $\theta_k(x)\neq 0$, it is $x\in U_k\cap U_i$, and in that intersection $g_k/g_i>0$; hence the logarithm exists. Moreover, in every non-empty intersection $U_i\cap U_j$ we have

$$\frac{g_i\exp\big[\sum_k\theta_k\log(g_k/g_i)\big]}{g_j\exp\big[\sum_k\theta_k\log(g_k/g_j)\big]}=\frac{g_i}{g_j}\exp\Big[\sum_k\theta_k\Big(\log\frac{g_k}{g_i}-\log\frac{g_k}{g_j}\Big)\Big]$$
$$=\frac{g_i}{g_j}\exp\Big[\Big(\sum_k\theta_k\Big)\log\Big(\frac{g_k}{g_i}\Big/\frac{g_k}{g_j}\Big)\Big]=\frac{g_i}{g_j}\frac{g_j}{g_i}=1.$$

By all of this, $(*)$ is a consistent definition.

Now, on U_i we have $f=e_if_i$, where $e_i=\varepsilon_i\exp[\cdots]$ has no zero. Thus,

$$\{f=0\}\cap U_i=\{f_i=0\}\cap U_i=U_i\cap M,$$

and for every $x\in U_i\cap M$, derivation gives

$$d_xf=f_i(x)d_xe_i+e_i(x)d_xf_i=e_i(x)d_xf_i\neq 0.$$

Summing up, f is a global equation for M. Furthermore, as $\exp>0$, we have $D=\{f>0\}$ and $E=\{f<0\}$, and by (d),

$$\overline{\{f>0\}}\setminus\{f>0\}=M=\{f=0\},$$

that is, $\overline{\{f>0\}}=\{f\geq 0\}$. This is a manifold with boundary $\{f=0\}=M$ (II.3.2, p. 62). The same works for $\overline{\{f<0\}}$.

The other assertions in the statement come from II.7.5(1), p. 88. □

Exercises and problems

Number 1. Complete the details for a mod 2 degree theory as suggested in III.6.1. Note that for oriented manifolds $\deg_2=\deg \bmod 2$.

Number 2. Let $\bullet$ denote a (bilinear) multiplication in $\mathbb{R}^{m+1}$ such that $x\bullet y=0$ only if $x=0$ or $y=0$. Fix any nonzero $u\in\mathbb{R}^{m+1}$ and define the mapping

$$f:\mathbb{S}^m\to\mathbb{S}^m: x\mapsto f(x)=\frac{u\bullet x}{\|u\bullet x\|}.$$

Show that f is a homeomorphism whose homotopy class does not depend on u. Deduce that f is homotopic to $-f$ and that m is odd.

Number 3. Let $\bullet$ denote a multiplication in $\mathbb{R}^{m+1}$ as above that furthermore is commutative. Consider

$$f:\mathbb{RP}^m\to\mathbb{S}^m: x\mapsto f(x)=\frac{x\bullet x}{\|x\bullet x\|}.$$

Show that:

(1) f is a well-defined injective continuous mapping.

(2) $\deg_2(f) = 1$; hence f is surjective.

Deduce that $\mathbb{RP}^m$ is homeomorphic to $\mathbb{S}^m$. As is well known, this is the case only for $m = 1$; hence for $m \geq 2$ such a multiplication does not exist.

Number 4. Let $M \subset \mathbb{R}^{m+1}$ be a connected, compact, boundaryless, differentiable hypersurface. Denote by D (resp., E) the bounded (resp., unbounded) connected component of $\mathbb{R}^{m+1} \setminus M$. For every point $p \notin M$ set

$$f_p : M \to \mathbb{S}^m : x \mapsto \frac{x - p}{\|x - p\|}.$$

Prove that $\deg(f_p)$ is 0 if $p \in E$ and ± 1 for $p \in D$, the sign depending solely on the orientation of M.

Number 5. Let $N \subset \mathbb{R}^{m+1}$ be a closed smooth manifold of dimension n, $N \neq \mathbb{R}^{m+1}$. Show that:

(1) If $n < m$, the complement $\mathbb{R}^{m+1} \setminus N$ is connected.

(2) If $n = m + 1$, N has a boundary $M = \partial N \neq \emptyset$, and if M is connected, so is $\mathbb{R}^{m+1} \setminus N$. What if M is not connected?

Number 6. Let $M \subset \mathbb{R}^{m+1}$ be a closed, boundaryless, differentiable hypersurface. Suppose that M has r connected components. How many does its complement $\mathbb{R}^{m+1} \setminus M$ have?

Number 7. Let $M \subset \mathbb{R}^{m+1}$ be a connected, boundaryless, smooth hypersurface. Let $U \subset \mathbb{R}^{m+1}$ be a tubular neighborhood associated to νM, and let $\rho : U \to M$ be the corresponding retraction. Let $\nu : M \to \mathbb{R}^{m+1}$ be a unitary global normal vector field. Prove that there is a smooth function $\delta : U \to \mathbb{R}$ such that

$$\delta(x)\nu(\rho(x)) = x - \rho(x).$$

Check that δ is an equation of M in U and that $\operatorname{grad}(\delta) = \nu$.

7. The Brouwer Theorems

We will prove here two famous theorems due to Brouwer. We start with the following immediate consequence of the Boundary Theorem:

Proposition 7.1. *Let M be an oriented differentiable manifold with boundary $\partial M = N$. Then N is not a proper retraction of M; that is, there is no proper mapping $\rho : M \to N$ whose restriction to N is the identity.*

Proof. Suppose there is such a retraction $\rho : M \to N$. Let N' be a connected component of N, and let M' be the connected component of M that contains N'. Since $N' = \rho(N') \subset \rho(M') \subset N$ and $\rho(M')$ is connected, we conclude

$$N' = \rho(M') = M' \cap N = \partial M',$$

because M' is open in M. Furthermore, $\rho|_{M'}$ is proper, because M' is closed in M. Now apply the Boundary Theorem, III.1.8, p. 101, with $X = M'$, $Y = \partial M'$, and $H = \rho|_{M'}$. We get $0 = \deg(H|_Y) = \deg(\rho|_{N'}) = \deg(\mathrm{Id}_{N'})$, which is impossible. □

The condition that the retraction is proper is essential: for instance, look at the (non-proper) continuous retraction $\mathbb{S}^1 \times [0, +\infty) \to \mathbb{S}^1 : (x,t) \mapsto x$.

From the preceding result we deduce (see I.2, p. 20 and p. 28):

Proposition 7.2 (Brouwer Fixed Point Theorem). *Let $M \subset \mathbb{R}^{m+1}$ be homeomorphic to a closed ball. Then every continuous mapping $f : M \to M$ has some fixed point.*

Proof. Clearly, it is enough to prove the statement for $M = \{x \in \mathbb{R}^{m+1} : \|x\| \leq 1\}$, which is a compact manifold with boundary $\partial M = \mathbb{S}^m$. Suppose there is a continuous mapping $f : M \to M$ without fixed points. Then define a mapping $\rho : M \to \mathbb{S}^m$ as shown in the picture

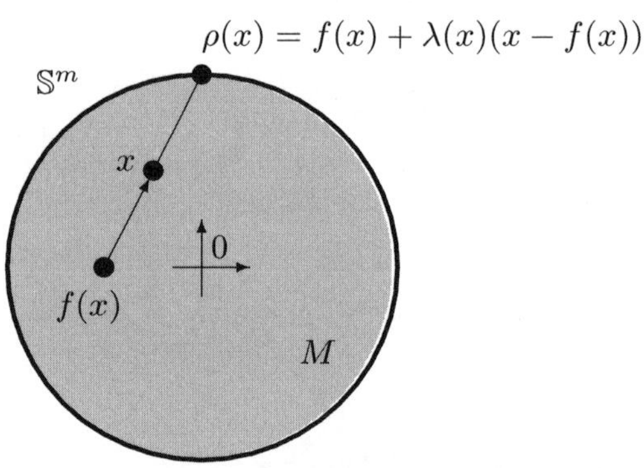

with $\lambda(x) > 0$ such that $\|\rho(x)\| = 1$. This mapping is proper and fixes every point in $\mathbb{S}^m$. Thus $\mathbb{S}^m$ is a proper retraction of M, contrary to the preceding proposition. This contradiction means that f must have some fixed point. □

This theorem will be refined in the next chapter (IV.2.9, p. 152) using the Euclidean degree. For the moment, here there is a nice application of it:

Proposition 7.3 (Perron-Fröbenius Theorem). *If a regular real matrix has non-negative entries, then it has some positive eigenvalue, corresponding to an eigenvector whose coordinates are all ≥ 0.*

Proof. Suppose A is an $(m+1)\times(m+1)$ matrix. Since A is regular, we can define the following continuous mapping:

$$h : \mathbb{S}^m \to \mathbb{S}^m : x \mapsto \frac{A\cdot x}{\|A\cdot x\|}.$$

Now the hypothesis on the entries of A guarantees that h keeps invariant the quadrant

$$Q = \{x \in \mathbb{S}^m : x_1 \geq 0, \ldots, x_{m+1} \geq 0\}.$$

But this quadrant is homeomorphic to a closed ball, and by the Brouwer Fixed Point Theorem, it must contain some fixed point x of h. Consequently

$$A\cdot x = \lambda x, \quad \text{where} \quad \lambda = \|A\cdot x\| > 0.$$

Thus, λ is the eigenvalue and x is the eigenvector we sought. □

We finish the section with the important fact, also proved by Brouwer, that only on spheres of odd dimension can we find tangent vector fields without zeros (I.2, p. 20 and p. 28):

Proposition 7.4 (Hedge-hog Theorem). *A sphere $\mathbb{S}^m \subset \mathbb{R}^{m+1}$ has a continuous tangent vector field without zeros if and only if m is odd.*

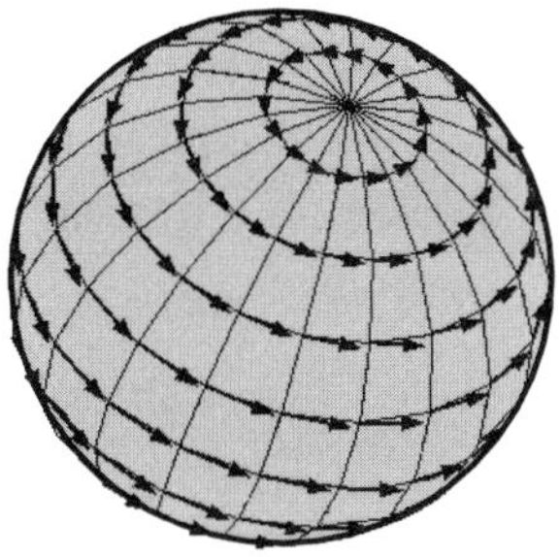

Tangent vector field on $\mathbb{S}^2$ vanishing at both poles
$\xi = (-y, x, 0)$.

Proof. If m is odd, then $m+1 = 2k$, and we have the smooth tangent vector field

$$\xi_x = (-x_2, x_1, \ldots, -x_{2k}, x_{2k-1}) \in T_x\mathbb{S}^m.$$

Next, suppose that $x \mapsto \xi_x \in T_x\mathbb{S}^m$ is a continuous tangent vector field on $\mathbb{S}^m$ without zeros. After dividing by its norm, we have $\|\xi_x\| = 1$ for all $x \in \mathbb{S}^m$; in other words, we have a continuous mapping $\mathbb{S}^m \to \mathbb{S}^m$ such that $\langle x, \xi_x\rangle = 0$ for all $x \in \mathbb{S}^m$. By this condition,

$$H_t(x) = \cos(\pi t)x + \sin(\pi t)\xi_x$$

is a well-defined homotopy, and $\deg(H_0) = \deg(H_1)$. Clearly, H_0 is the identity and H_1 the antipodal diffeomorphism; hence we conclude that the latter has the same degree as the former, that is, $+1$. But in III.1.6(1), p. 99, we saw this is possible only for m odd. This finishes the proof. □

Exercises and problems

Number 1. Let M be an oriented differentiable manifold with boundary $\partial M = N$. Prove that there is no proper mapping $\rho : M \to N$ whose restriction to N is homotopic to the identity.

Number 2. Let D denote an open ball in $\mathbb{R}^p$, $\overline{D}$ its closure, and S its boundary. Produce the following examples:

(1) a continuous mapping $D \to D$ with no fixed point,

(2) a continuous mapping $\overline{D} \to \overline{D}$ with no fixed point in D,

(3) a continuous mapping $\overline{D} \times S \to \overline{D} \times S$ with no fixed point. Here, where does the proof of III.7.2, p. 134, fail?

Number 3. Let $K \subset \mathbb{R}^{m+1}$ be a convex compact set. Prove that every continuous mapping $f : K \to K$ has some fixed point. Can *convex* be replaced by *connected*?

Number 4. Let $D \subset \mathbb{R}^{m+1}$ be the unit open ball, and let $f : \overline{D} \to \mathbb{R}^{m+1}$ be a continuous mapping such that $\langle x, f(x)\rangle < 1$ for all $x \in \mathbb{S}^m$. Show that f has some fixed point $a \in D$.

Number 5. Let $D \subset \mathbb{R}^{m+1}$ be the unit open ball, and let $f : \overline{D} \to \mathbb{R}^{m+1}$ be a continuous mapping such that $\langle x, f(x)\rangle < 0$ for all $x \in \mathbb{S}^m$. Show that f has some zero $a \in D$.

Chapter IV

Degree theory in Euclidean spaces

In this chapter we develop *Euclidean degree theory*. This could be done in full generality for proper mappings on closures of arbitrary open sets. However, here we have chosen the simplest way, enough for the applications, and restricted ourselves to bounded open sets. The reader will find in §§1–3 that everything can be done with a little calculus (and some very ingenious arguments); on the other hand, it is quite straightforward to extend the construction to the general unbounded case, a profitable exercise we recommend to the reader. Along the way we include Gauss's proof of the *Fundamental Theorem of Algebra* and a refinement of the *Brouwer Fixed Point Theorem.* Next, in §4, we come back to the origins and define the *winding number,* which is the agent behind the scenes for the equivalence of the Brouwer-Kronecker and Euclidean degrees. In §5, we deduce the famous *Borsuk-Ulam Theorem* and an important consequence: the *Invariance of Domain Theorem.* After this, §6 contains one fundamental computation: the *Multiplication Formula* for the Euclidean degree of a composite mapping. Finally, that formula is used in §7 to deduce a most general purely topological version of the *Jordan Separation Theorem* and from it the *Invariance of Domain Theorem* once again.

1. The degree of a smooth mapping

From now on, D stands for a bounded open set in $\mathbb{R}^{m+1}$, and $X = \overline{D} \setminus D$ stands for its (topological) boundary. Our purpose here is to define the degree of a continuous mapping $\overline{D} \to \mathbb{R}^{m+1}$. This we will do by the usual method: first we consider smooth mappings, and then we approximate by them any given continuous mapping. Here, the reader will recognize some familiar arguments, which are repeated to make the construction fully self-contained.

Note that, $\overline{D}$ being compact, a continuous mapping $f : \overline{D} \to \mathbb{R}^{m+1}$ is proper (hence closed), and in particular $f(\overline{D})$ and $f(X)$ are closed in $\mathbb{R}^{m+1}$. Furthermore, the *norm*

$$\|f\| = \max\{\|f(x)\| : x \in \overline{D}\}$$

is well defined. As usual, this norm measures the approximation to functions on $\overline{D}$.

As mentioned above, we will consider the case when f is smooth. In particular we will be dealing with the set $R_{f|_D} \subset \mathbb{R}^{m+1}$ of regular values of $f|_D$. Recall that $a \in R_{f|_D}$ when the derivative $d_x f$ is bijective at every point $x \in D$ with $f(x) = a$ (trivially true if $a \notin f(D)$). The Sard-Brown Theorem, II.3.4, p. 63, says that $R_{f|_D}$ is a residual set, which allows us to find regular values everywhere. On the other hand, although the set $C_{f|_D}$ of all critical points of $f|_D$ (the points at which the Jacobian determinant vanishes) is closed in D, the set $R_{f|_D} = \mathbb{R}^{m+1} \setminus f(C_{f|_D})$ need not be open in $\mathbb{R}^{m+1}$. However we can say something useful: $C_{f|_D} \cup X$ is closed in $\overline{D}$; hence $f(C_{f|_D} \cup X)$ is closed in $\mathbb{R}^{m+1}$ (f is proper), and

$$R_{f|_D} \setminus f(X) = \mathbb{R}^{m+1} \setminus f(C_{f|_D} \cup X)$$

is open in $\mathbb{R}^{m+1}$.

These remarks will often be implicit in our arguments. Now, down to the matter of discussion, we introduce the degree of a smooth mapping:

Proposition and Definition 1.1. *Let $f : \overline{D} \to \mathbb{R}^{m+1}$ be a smooth mapping, and let $a \in \mathbb{R}^{m+1} \setminus f(X)$ be a regular value of $f|_D$. Then $f^{-1}(a)$ is a finite set (possibly empty), and we define the* degree of f *by*

$$d(f, D, a) = \sum_{x \in f^{-1}(a)} \operatorname{sign}_x(f)$$

(of course, the sum is 0 in case $f^{-1}(a) = \emptyset$), where

$$\operatorname{sign}_x(f) = \operatorname{sign}\det\left(\frac{\partial f_i}{\partial x_j}(x)\right) = \pm 1$$

is the sign of f at x.

Proof. Note that $f^{-1}(a) \subset D$, and the derivative $d_x f$ is a linear isomorphism at every $x \in D$ with $f(x) = a$. Thus, by the Inverse Mapping Theorem, all points of $f^{-1}(a)$ are isolated, and $f^{-1}(a)$ is discrete. But $\overline{D}$ is compact, and $f^{-1}(a)$ does not meet $X = \overline{D} \setminus D$; hence $f^{-1}(a)$ is finite. $\square$

We stress that $\operatorname{sign}_x(f)$ tells whether the derivative $d_x f : \mathbb{R}^{m+1} \to \mathbb{R}^{m+1}$ preserves orientation (sign $= +1$) or reverses it (sign $= -1$) as seen when we discussed orientations in the general setting of manifolds (II.7.2). Note also that from this definition it is clear that:

(1) If $f|_D$ is a diffeomorphism, then $d(f, D, a) = \pm 1$.

(2) If $\tau : \mathbb{R}^{m+1} \to \mathbb{R}^{m+1}$ is a translation, then $d(f, D, a) = d(\tau \circ f, D, \tau(a))$.

Now, we prove a first property of this degree.

Proposition 1.2 (Degree is locally constant)**.** There exists an open neighborhood of a, $W \subset R_{f|_D} \setminus f(X)$, such that $d(f, D, a) = d(f, D, b)$ for all $b \in W$.

Proof. As remarked above, $\overline{D}$ is closed in $\mathbb{R}^{m+1}$, and $\mathbb{R}^{m+1} \setminus f(\overline{D})$ is open in $\mathbb{R}^{m+1}$ and is contained in $R_{f|_D} \setminus f(X)$. Now, if $f^{-1}(a) = \emptyset$, then $a \in \mathbb{R}^{m+1} \setminus f(\overline{D})$, and there is an open neighborhood $W \subset R_{f|_D} \setminus f(X)$ of a that does not meet $f(\overline{D})$. Consequently, $d(f, D, b) = 0$ for all $b \in W$.

After this, we can suppose $f^{-1}(a) = \{x_1, \dots, x_r\}$. Then, by the Inverse Mapping Theorem, we find a connected open neigborhood $V \subset \mathbb{R}^{m+1} \setminus f(X)$ of a and r disjoint open neighborhoods $U_1, \dots, U_r$ of $x_1 \dots, x_r$, respectively, such that by restriction we have r diffeomorphisms

$$f_k = f|_{U_k} : U_k \to V$$

whose sign is constant (because U_k is connected) and is denoted by σ_k. We claim that

$$W = V \setminus f(\overline{D} \setminus \bigcup_k U_k)$$

is the open neighborhood we seek. Indeed, we have:

1. W is open, because $\overline{D} \setminus \bigcup_k U_k$ is a compact set; hence its image is closed.

2. $a \in W$, because all its preimages x_k are in $\bigcup_k U_k$.

3. Every $b \in W$ has exactly r preimages $y_k \in U_k$. Indeed, b has at least these r preimages, and if it had others, say $y \notin \bigcup_k U_k$, then $y \in \overline{D} \setminus \bigcup_k U_k$, so that $f(y) \in f(\overline{D} \setminus \bigcup_k U_k)$ and $b = f(y) \notin W$.

Thus, for every $b \in W$ we deduce as claimed

$$d(f, D, b) = \sum_k \operatorname{sign}_{y_k}(f) = \sum_k \sigma_k = \sum_k \operatorname{sign}_{x_k}(f) = d(f, D, a).$$ □

Next, we turn to homotopy. We start with a somewhat technical statement:

Proposition 1.3 (Homotopy invariance, 1st). *Let* $H : [0,1] \times \overline{D} \to \mathbb{R}^{m+1}$ *be a smooth mapping, and consider a point* $a \in \mathbb{R}^{m+1} \setminus H([0,1] \times X)$ *that is a regular value of the three restrictions* $H_0|_D, H_1|_D$*, and* $H|_{[0,1]\times D}$*. Then*

$$d(H_0, D, a) = d(H_1, D, a).$$

Proof. Denote as usual $H_t(x) = H(t,x)$ and let

$$H(t,x) = (\varphi_1(t,x), \dots, \varphi_{m+1}(t,x)).$$

Since a is a regular value of $H|_{[0,1]\times D}$ and $C = H^{-1}(a)$ does not meet X, this inverse image

$$C : \begin{cases} \varphi_1(t,x) & = & a_1, \\ & \vdots & \\ \varphi_{m+1}(t,x) & = & a_{m+1} \end{cases}$$

is, by II.3.2, p. 62, a smooth compact curve possibly with boundary

$$\partial C = \big(C \cap (\{0\} \times D)\big) \cup \big(C \cap (\{1\} \times D)\big) \equiv H_0^{-1}(a) \cup H_1^{-1}(a).$$

We denote by $\Gamma_1, \dots, \Gamma_r$ the connected components of C. By the classification theorem for smooth curves, each $\Gamma = \Gamma_k$ is either a compact interval or a circle. In any case, II.3.3, p. 63, gives a smooth parametrization of Γ:

$$\gamma : s \mapsto (t(s), x_1(s), \dots, x_{m+1}(s)) = (t(s), x(s)), \quad 0 \le s \le 1.$$

In particular, $\frac{d\gamma}{ds}$ never vanishes. Moreover, from the identity $\varphi_i(\gamma(s)) \equiv a_i$ it follows that

$$0 \equiv \frac{d}{ds}(\varphi_i \circ \gamma)(s) = \frac{\partial \varphi_i}{\partial t}\frac{dt}{ds} + \sum_j \frac{\partial \varphi_i}{\partial x_j}\frac{dx_j}{ds}. \tag{1}$$

Now consider the determinant

$$\Delta(s) = \left| \begin{array}{c|ccc} \frac{dt}{ds} & \frac{dx_1}{ds} & \cdots & \frac{dx_{m+1}}{ds} \\ \hline \frac{\partial \varphi_1}{\partial t} & \frac{\partial \varphi_1}{\partial x_1} & \cdots & \frac{\partial \varphi_1}{\partial x_{m+1}} \\ \vdots & \vdots & & \vdots \\ \frac{\partial \varphi_{m+1}}{\partial t} & \frac{\partial \varphi_{m+1}}{\partial x_1} & \cdots & \frac{\partial \varphi_{m+1}}{\partial x_{m+1}} \end{array} \right|.$$

We claim that $\Delta(s)$ never vanishes. Indeed, the last $m+1$ rows are independent because a is a regular value of H, and by (1) the first row is perpendicular to the others, hence independent. Thus we conclude that $\Delta(s)$ has no zero; hence it has constant sign:

(2) *$\Delta(s)$ has the same sign at $s = 0$ and $s = 1$.*

Next, we look again at the equations in (1), which tell how to combine the columns in our determinant to get

$$(3)\qquad \frac{dt}{ds}\cdot\Delta = \left|\begin{array}{c|ccc} \|\frac{d\gamma}{ds}\|^2 & \frac{dx_1}{ds} & \cdots & \frac{dx_{m+1}}{ds} \\ \hline 0 & \frac{\partial\varphi_1}{\partial x_1} & \cdots & \frac{\partial\varphi_1}{\partial x_{m+1}} \\ \vdots & \vdots & & \vdots \\ 0 & \frac{\partial\varphi_{m+1}}{\partial x_1} & \cdots & \frac{\partial\varphi_{m+1}}{\partial x_{m+1}} \end{array}\right| = \left\|\frac{d\gamma}{ds}\right\|^2 \det\left(\frac{\partial\varphi_i}{\partial x_j}(\gamma(s))\right).$$

After this preparation, we compute degrees:

$$(4)\qquad \begin{cases} d(H_0, D, a) = \sum_k \sum_{(0,y)\in\Gamma_k} \operatorname{sign}_y(H_0), \\ d(H_1, D, a) = \sum_k \sum_{(1,z)\in\Gamma_k} \operatorname{sign}_z(H_1). \end{cases}$$

For that computation, we are only interested in the components $\Gamma = \Gamma_k$ that meet the levels $t = 0, 1$. Those components are the bordered components among the Γ_k's, that is, the components diffeomorphic to a compact interval; hence they have two distinct boundary points, $\gamma(0) = p$ and $\gamma(1) = q$, both at those t-levels. The following picture shows the various possibilities:

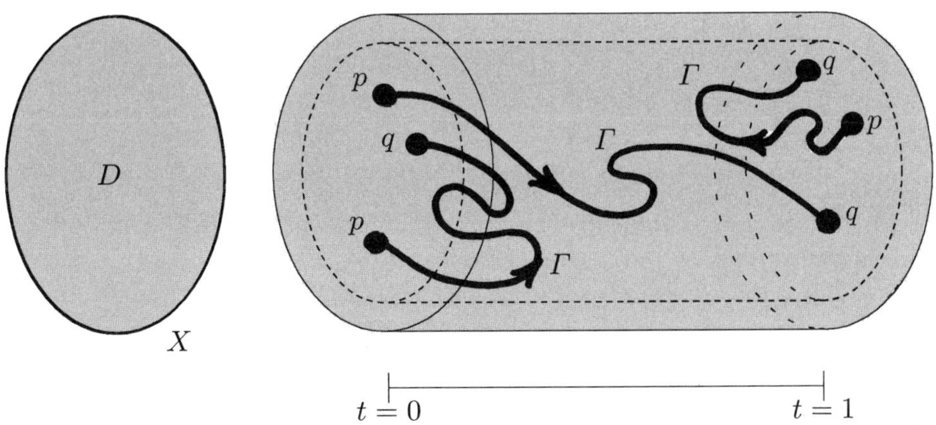

Clearly, the required equality of degrees $d(H_0, D, a) = d(H_1, D, a)$ will come from the following facts:

1. If p and q are at the same t-level, they contribute to the same degree in (4) with opposite signs; hence they cancel each other.

2. If p and q are at different t-levels, they contribute to different degrees in (4) with the same signs in both.

This is easily proved. Consider the first component $t : [0,1] \to [0,1]$ of the parametrization $\gamma = (t, x)$. Since $\big(\frac{\partial \varphi_i}{\partial x_j}(\gamma(s))\big)$ is the Jacobian matrix of $H_{t(s)}$ and a is a regular value of H_0 and H_1, the right-hand side in (3) is not zero for $t(s) = 0$ and 1; hence $\frac{dt}{ds}(0) \neq 0$ and $\frac{dt}{ds}(1) \neq 0$. We have:

1(i) If p and q are at the same level $t = 0$, that is, $t(0) = t(1) = 0$, then t is increasing at 0 and decreasing at 1, so $\frac{dt}{ds}(0) > 0$ and $\frac{dt}{ds}(1) < 0$. From (2) and (3) we deduce that the signs of H_0 at $x(0)$ and $x(1)$ are opposite.

1(ii) If p and q are at the same level $t = 1$, that is, $t(0) = t(1) = 1$, then t is decreasing at 0 and increasing at 1, so $\frac{dt}{ds}(0) < 0$ and $\frac{dt}{ds}(1) > 0$. By (2) and (3) again, the signs of H_1 at $x(0)$ and $x(1)$ are opposite.

2(i) If p and q are at levels $t = 0$ and $t = 1$, that is, $t(0) = 0$ and $t(1) = 1$, then t is increasing at 0 and 1, so $\frac{dt}{ds}(0) > 0$ and $\frac{dt}{ds}(1) > 0$. By (2) and (3) as before, we see that the signs of H_0 at $x(0)$ and H_1 at $x(1)$ are the same.

2(ii) If p and q are at levels $t = 1$ and $t = 0$, that is, $t(0) = 1$ and $t(1) = 0$, then t is decreasing at 0 and 1, so $\frac{dt}{ds}(0) < 0$ and $\frac{dt}{ds}(1) < 0$. Thus, from (2) and (3) we deduce that the signs of H_0 at $x(1)$ and H_1 at $x(0)$ are the same.

This completes the proof. □

The previous statement can be simplified a bit:

Proposition 1.4 (Homotopy invariance, 2nd). *Let $H : [0,1] \times \overline{D} \to \mathbb{R}^{m+1}$ be a smooth mapping, and consider a point $a \in \mathbb{R}^{m+1} \setminus H([0,1] \times X)$ that is a regular value of the two restrictions $H_0|_D$ and $H_1|_D$. Then*

$$d(H_0, D, a) = d(H_1, D, a).$$

Proof. This follows from the preceding results. By IV.1.2, p. 139, there is an open neighborhood W of a that does not meet the compact set $H([0,1]\times X)$ and consists solely of regular values of $H_0|_D$ and $H_1|_D$ and is such that

$$d(H_0, D, a) = d(H_0, D, b) \quad \text{and} \quad d(H_1, D, a) = d(H_1, D, b)$$

for all $b \in W$. Now, by the Sard-Brown Theorem, W contains some regular value b of the mapping $H|_{[0,1]\times D}$, and applying IV.1.3, p. 140, we get $d(H_0, D, b) = d(H_1, D, b)$, which jointly with the preceding equalities proves the statement. $\square$

Once this is established, we can improve on IV.1.2, p. 139, and define degree *for non-regular values:*

Proposition and Definition 1.5. *Let $f : \overline{D} \to \mathbb{R}^{m+1}$ be a smooth mapping, and let $a \in \mathbb{R}^{m+1} \setminus f(X)$. Let Ω be the connected component of $\mathbb{R}^{m+1} \setminus f(X)$ that contains a. Then $d(f, D, b)$ is constant for all $b \in \Omega \cap R(f|_D)$, and we define the* degree of f *by*

$$d(f, D, a) = d(f, D, b)$$

for any such b (which exists by the Sard-Brown Theorem).

Proof. Let $b, b' \in \Omega \cap R(f|_D)$. Since Ω is an open connected set in a Euclidean space, there is a smooth arc $\gamma : [0,1] \to \Omega$ with $\gamma(0) = b, \gamma(1) = b'$, and we define

$$H : [0,1] \times \overline{D} \to \mathbb{R}^{m+1} : (t, x) \mapsto f(x) - \gamma(t).$$

We have:

(1) $0 \notin H([0,1] \times X)$*:* if $f(x) = \gamma(t)$, for some $x \in X$, then $f(x) \in f(X) \cap \Omega$, which is a contradiction.

(2) 0 *is a regular value of* $H_0|_D$*:* since $H_0 = f - b$, the derivatives $d_x H_0$ and $d_x f$ coincide, and we know that b is a regular value of f.

(3) 0 *is a regular value of* $H_1|_D$*:* same argument for $H_1 = f - b'$.

Thus, by IV.1.4, p. 142,

$$d(H_0, D, 0) = d(H_1, D, 0).$$

Finally, since both H_0 and H_1 are translations of f, we obtain

$$d(H_0, D, 0) = d(f, D, b) \quad \text{and} \quad d(H_1, D, 0) = d(f, D, b').$$

All together, we get what we want. $\square$

For this extended definition, IV.1.4, p. 142, remains true:

Proposition 1.6 (Homotopy invariance, 3rd). *Let $H : [0,1] \times \overline{D} \to \mathbb{R}^{m+1}$ be a smooth mapping, and consider a point $a \in \mathbb{R}^{m+1} \setminus H([0,1] \times X)$. Then*

$$d(H_0, D, a) = d(H_1, D, a).$$

Proof. Let Ω be the connected component of $\mathbb{R}^{m+1} \setminus H([0,1] \times X)$ that contains a. By the Sard-Brown Theorem, Ω contains some regular value b of both $H_0|_D$ and $H_1|_D$, and then

$$d(H_0, D, a) = d(H_0, D, b) \quad \text{and} \quad d(H_1, D, a) = d(H_1, D, b)$$

(IV.1.5, p. 143). But by IV.1.4, p. 142, $d(H_0, D, b) = d(H_1, D, b)$, and we are done. □

Exercises and problems

Number 1. Consider the *quadratic transform*

$$f : \mathbb{R}^n \to \mathbb{R}^n : (x_1, \dots, x_n) \mapsto (x_1, x_1x_2, \dots, x_1x_n)$$

and the open set $D \subset \mathbb{R}^n$ defined by $\alpha_1 < x_1 < \beta_1, \dots, \alpha_n < x_n < \beta_n$. Describe the function $a \mapsto d(f, D, a)$.

Number 2. ([Picard 1891]; I.1, p. 12) Let $D \subset \mathbb{R}^{m+1}$ be a bounded open set, let $X = \overline{D} \setminus D$, and let$f_1, \dots, f_{m+1} : \overline{D} \to \mathbb{R}$ be smooth functions. Suppose that $0 \notin f(X)$ is a regular value of $f|_D$, where $f = (f_1, \dots, f_{m+1}) : \overline{D} \to \mathbb{R}^{m+1}$. Consider the bounded open set $E = D \times (-1,1) \subset \mathbb{R}^{m+1} \times \mathbb{R} = \mathbb{R}^{m+2}$ and the smooth function

$$f_{m+2} : \overline{E} \to \mathbb{R} : (x,t) \mapsto t \cdot \det\Big(\frac{\partial f_i}{\partial x_j}(x)\Big).$$

Finally, consider

$$g : \overline{E} \to \mathbb{R}^{m+2} : (x,t) \mapsto (f_1(x), \dots, f_{m+1}(x), f_{m+2}(x,t)).$$

Prove that $d(g, E, 0)$ is well defined and that it coincides with the number of solutions in D of the system

$$f_1(x) = \cdots = f_{m+1}(x) = 0.$$

Number 3. ([Heinz 1959]; I.4, p. 38) Let $D \subset \mathbb{R}^{m+1}$ be a bounded open set, let $X = \overline{D} \setminus D$, and let $f : \overline{D} \to \mathbb{R}^{m+1}$ be a smooth mapping. Let $a \in \mathbb{R}^{m+1} \setminus f(X)$ be a regular value of f. Let $\varphi_\varepsilon : \mathbb{R}^{m+1} \to \mathbb{R}$ be a smooth mapping whose closed support K_ε is contained in the open ball of radius $\varepsilon > 0$ centered at the origin and such that $\int_{\mathbb{R}^{m+1}} \varphi_\varepsilon = 1$ (called a *convolution kernel*). Prove the Heinz Integral Formula:

$$d(f, D, a) = \int_D \varphi_\varepsilon(f(x) - a) \det\big(\tfrac{\partial f_i}{\partial x_j}(x)\big) dx$$

for ε small enough.

Number 4. Explain and prove the following product formula for smooth mappings:

$$d(f \times g, D \times E, (a, b)) = d(f, D, a) \cdot d(g, E, b).$$

Number 5. Let $D \subset \mathbb{R}^2$ denote the open disc of radius 2, and consider the smooth mapping $f = (f_1, f_2) : \overline{D} \to \mathbb{R}^2$ given by

$$\begin{cases} f_1(x, y) = (-x^4 + y)e^{xy} + x^2y - x^2 + y + 1, \\ f_2(x, y) = x^3y - 4xy. \end{cases}$$

Study the variation of $d(f, D, (a, 0))$ for $a \in \mathbb{R}$, when defined.

Number 6. Let $X \subset \mathbb{R}^2$ be the curve union of the following pieces:

(i) the graph of the topological sinus $y = \sin(1/x)$, $0 < x \leq \sqrt{3}$,

(ii) the segment $x = 0$, $-1 \leq y \leq 1$,

(iii) the arc of the circle $x^2 + (y + 1)^2 = 4$ off the first quadrant, and

(iv) the segment $x = \sqrt{3}$, $0 \leq y \leq \sin(1/\sqrt{3})$.

Show that X is not homeomorphic to $\mathbb{S}^1$, but nonetheless it bounds an open set. Then prove that the mappings $f, g : X \to \mathbb{R}^2 \setminus \{(0, 0)\}$ given by

$$f(x, y) = (x^3 + y^3 + 1, x - y) \quad \text{and} \quad g(x, y) = (x^3 + y^3 + 1, x + y)$$

are not homotopic.

Number 7. Let $D \subset \mathbb{R}^{m+1}$ be a bounded open set and let X be its boundary. Suppose we are given two smooth functions $f, g : \overline{D} \to \mathbb{R}^{m+1}$, a homotopy $H : [0, 1] \times \overline{D} \to \mathbb{R}^{m+1}$ with $H_0 = f, H_1 = g$, and a point $a \in \mathbb{R}^{m+1} \setminus H([0, 1] \times X)$ that is a regular value of f and g. Prove that the fibers $f^{-1}(a)$ and $g^{-1}(a)$ have the same parity. Is it always true that they have the same number of points?

2. The degree of a continuous mapping

After the discussion of smooth mappings in the preceding section, we are ready to define degree for arbitrary continuous mappings:

Proposition and Definition 2.1. *Let $f : \overline{D} \to \mathbb{R}^{m+1}$ be a continuous mapping and let $a \in \mathbb{R}^{m+1} \setminus f(X)$. Then there exists a smooth mapping $g : \overline{D} \to \mathbb{R}^{m+1}$ such that $\|f - g\| < \operatorname{dist}(a, f(X))$. For all such g's the degree $d(g, D, a)$ is defined ($a \in \mathbb{R}^{m+1} \setminus g(X)$) and is the same, and we define the* degree of f *by*

$$d(f, D, a) = d(g, D, a).$$

Furthermore, g can be chosen such that a is a regular value of $g|_D$, and then

$$d(f, D, a) = \sum_{x \in g^{-1}(a)} \operatorname{sign}_x(g).$$

Proof. First of all, by the Weierstrass Approximation Theorem, we find a polynomial (hence smooth) mapping P such that

$$\|P - f\| = \max\left\{\|P(x) - f(x)\| : x \in \overline{D}\right\} < \tfrac{1}{2}\operatorname{dist}(a, f(X))$$

(recall that $a \notin f(X)$ and $f(X)$ is closed). Next, by the Sard-Brown Theorem, there is some regular value b of $P|_D$ such that

$$\|a - b\| < \tfrac{1}{2}\operatorname{dist}(a, f(X)).$$

One easily checks that $g = P + a - b$ is the function we sought. To see that $d(g, D, a) = d(g', D, a)$ for two mappings g, g' that are close to f, consider the smooth homotopy

$$H : [0,1] \times \overline{D} \to \mathbb{R}^{m+1} : (x,t) \mapsto (1-t)g(x) + tg'(x).$$

Clearly, $H_0 = g$ and $H_1 = g'$; hence to apply IV.1.6, p. 144, we only must check that $a \notin H([0,1] \times X)$. But suppose there is an $x \in X$ such that $a = H_t(x)$. Then

$$\begin{aligned}
\|a - f(x)\| &= \big\|H_t(x) - \big((1-t)f(x) + tf(x)\big)\big\| \\
&\le (1-t)\|g(x) - f(x)\| + t\|g'(x) - f(x)\| \\
&< (1-t)\operatorname{dist}(a, f(X)) + t\operatorname{dist}(a, f(X)) = \operatorname{dist}(a, f(X)),
\end{aligned}$$

a contradiction. □

Remark 2.2. As is to be expected, the approximations g used above to define the degree of f are *homotopic to f off a on X*. Indeed, the homotopy

$$H_t(x) = tf(x) + (1-t)g(x)$$

does not touch a. For, if there is an $x \in X$ with $H_t(x) = a$, then a lies in the segment between $f(x)$ and $g(x)$; hence

$$\|f(x) - g(x)\| \ge \|f(x) - a\| \ge \operatorname{dist}(a, f(X)),$$

contrary to the choice of g. □

From the definition, we get the following:

Proposition 2.3. *Let $f : \overline{D} \to \mathbb{R}^{m+1}$ be a continuous mapping. Then the degree $d(f, D, \cdot)$ is constant on every connected component of $\mathbb{R}^{m+1} \setminus f(X)$.*

Proof. Let $a \in \mathbb{R}^{m+1} \setminus f(X)$, and, according to IV.2.1, p. 145, pick a smooth mapping $g : \overline{D} \to \mathbb{R}^{m+1}$ such that $\|f-g\| < \operatorname{dist}(a, f(X))$. We can compute the degree $d(f, D, a)$ through g, but in fact, we can also compute $d(f, D, b)$ for any other $b \in \mathbb{R}^{m+1} \setminus f(X)$ close enough to a: namely whenever

$$\|a - b\| < \eta = \operatorname{dist}(a, f(X)) - \|f - g\|.$$

Indeed, this implies

$$\operatorname{dist}(b, f(X)) \geq \operatorname{dist}(a, f(X)) - \|a - b\| > \operatorname{dist}(a, f(X)) - \eta = \|f - g\|,$$

and since $\eta \leq \operatorname{dist}(a, g(X))$ (check it!), the two points a, b are in the same connected component of $\mathbb{R}^{m+1} \setminus g(X)$. Then, by IV.1.5, p. 143, we see that

$$d(f, D, a) = d(g, D, a) = d(g, D, b) = d(f, D, b).$$

This means that the function $a \mapsto d(f, D, a)$ is locally constant on $\mathbb{R}^{m+1} \setminus f(X)$, and, since its values are integers, it must be constant on every connected component of $\mathbb{R}^{m+1} \setminus f(X)$. □

Of course, we can improve on the homotopy invariance once the definition is most general:

Proposition 2.4 (Homotopy invariance). *Let $H : [0, 1] \times \overline{D} \to \mathbb{R}^{m+1}$ be a continuous mapping, and let $\gamma : [0, 1] \to \mathbb{R}^{m+1}$ be a continuous path such that $\gamma(t) \notin H_t(X)$ for $0 \leq t \leq 1$. Then $d(H_t, D, \gamma(t))$ does not depend on t.*

Proof. Fix $0 \leq t_0 \leq 1$. The function $(t, x) \mapsto \|\gamma(t) - H_t(x)\|$ is continuous and > 0 on the compact set $[0, 1] \times X$ and hence has a minimum $\varepsilon > 0$. On the other hand, H is uniformly continuous on the compact set $[0, 1] \times \overline{D}$; hence there is an $\eta > 0$ such that

$$\|H_t(x) - H_{t'}(x')\| < \tfrac{1}{2}\varepsilon \quad \text{if} \quad |t - t'|, \ \|x - x'\| < \eta.$$

Then, by the Weierstrass Approximation Theorem, there is a polynomial mapping g such that

$$\|g(x) - H_{t_0}(x)\| < \tfrac{1}{2}\varepsilon \qquad \text{for} \quad x \in \overline{D}.$$

We deduce that

$$\|g(x) - H_t(x)\| \leq \|g(x) - H_{t_0}(x)\| + \|H_{t_0}(x) - H_t(x)\| < \varepsilon$$

for $x \in \overline{D}$, $|t - t_0| < \eta$. Thus,

$$\|g(x) - H_t(x)\| < \operatorname{dist}(\gamma(t), H_t(X)) \qquad \text{for} \quad x \in \overline{D},\ |t - t_0| < \eta.$$

In particular, for $x \in X$, $t = t_0$, we get

$$\|g(x) - H_{t_0}(x)\| < \operatorname{dist}(\gamma(t_0), H_{t_0}(X));$$

hence $g(x) \neq \gamma(t_0)$. As $g(X)$ is compact, we deduce that

$$\operatorname{dist}(\gamma(t_0), g(X)) > 0.$$

Thus, for a possibly smaller η, we can assume that $|t - t_0| < \eta$ implies

$$\|\gamma(t) - \gamma(t_0)\| < \operatorname{dist}(\gamma(t_0), g(X)).$$

Now we are ready to compute the degree. Suppose $|t - t_0| < \eta$. First, in view of the construction, we can apply IV.2.1, p. 145, to get

$$d(H_t, D, \gamma(t)) = d(g, D, \gamma(t)).$$

Then, by translation

$$\cdots = d(g - \gamma(t), D, 0).$$

Next,

$$\cdots = d(g - \gamma(t_0), D, 0),$$

because

$$\begin{aligned} \|(g - \gamma(t)) - (g - \gamma(t_0))\| &= \|\gamma(t) - \gamma(t_0)\| \\ &< \operatorname{dist}(\gamma(t_0), g(X)) = \operatorname{dist}\big(0, (g - \gamma(t_0))(X)\big) \end{aligned}$$

and IV.2.1, p. 145, applies again. We continue the computation, and by translation

$$\cdots = d(g, D, \gamma(t_0)).$$

Finally, by a third use of IV.2.1, p. 145, we obtain

$$\cdots = d(H_{t_0}, D, \gamma(t_0)).$$

Hence, $d(H_t, D, \gamma(t))$ is locally constant and consequently constant in the connected interval [0,1]. □

We have thus defined the degree and proved its homotopy invariance in full generality. As an easy consequence we deduce the other properties that characterize degree axiomatically:

Corollary 2.5 (Nagumo, Führer, Deimling Axiomatizations).

(1) Normality: *If f is the inclusion* $\mathrm{Id}_{\overline{D}} : \overline{D} \to \overline{D} \subset \mathbb{R}^{m+1}$, *then*

$$d(f, D, a) = \begin{cases} 1 & \text{if } a \in D, \\ 0 & \text{if } a \notin \overline{D}. \end{cases}$$

(2) Existence of solutions: *Given a continuous mapping $f : \overline{D} \to \mathbb{R}^{m+1}$ and a point $a \notin f(X)$, if $d(f, D, a) \neq 0$, the equation $f(x) = a$ has some solution in D.*

(3) Additivity: *Given two disjoint open sets $D_1, D_2 \subset D$, a continuous mapping $f : \overline{D} \to \mathbb{R}^{m+1}$, and a point $a \notin f(\overline{D} \setminus D_1 \cup D_2)$, then*

$$d(f, D, a) = d(f, D_1, a) + d(f, D_2, a).$$

Proof. For (1) use IV.1.1, p. 138.

For (2) we argue by way of contradiction. Suppose $f^{-1}(a) = \emptyset$. Then

$$\mathrm{dist}(a, f(X)) \geq \mathrm{dist}(a, f(\overline{D})) = \varepsilon > 0,$$

and by the Weierstrass Approximation Theorem there is a polynomial mapping $g = g(x)$ such that $\|g - f\| < \frac{1}{2}\varepsilon$; in particular, $a \notin g(X)$. Then, $d(f, D, a) = d(g, D, a)$. For $x \in \overline{D}$ we have

$$\|a - g(x)\| \geq \|a - f(x)\| - \|g(x) - f(x)\| > \varepsilon - \tfrac{1}{2}\varepsilon = \tfrac{1}{2}\varepsilon.$$

This means that $\mathrm{dist}(a, g(\overline{D})) > \frac{1}{2}\varepsilon$. Now, we pick, in the connected component of $\mathbb{R}^{m+1} \setminus g(X)$ that contains a, a regular value b of $g|_D$ such that $\|a - b\| < \frac{1}{2}\varepsilon$. Then $g^{-1}(b) = \emptyset$ and $d(g, D, a) = d(g, D, b) = 0$, a contradiction.

For (3) notice first that $\overline{D} \setminus D_1 \cup D_2$ contains the three boundaries

$$\overline{D} \setminus D, \quad \overline{D_1} \setminus D_1, \quad \overline{D_2} \setminus D_2.$$

Now, $\mathrm{dist}(a, f(\overline{D} \setminus D_1 \cup D_2)) = \varepsilon > 0$, and we can choose a polynomial mapping $g = g(x)$ with $\|g - f\| < \varepsilon$. Then $a \notin g(\overline{D} \setminus D_1 \cup D_2)$, and it follows that

$$d(f, D, a) = d(g, D, a) \quad \text{and} \quad \begin{cases} d(f, D_1, a) = d(g, D_1, a), \\ d(f, D_2, a) = d(g, D_2, a). \end{cases}$$

Finally, let Ω be the connected component of $\mathbb{R}^{m+1} \setminus g(\overline{D} \setminus D_1 \cup D_2)$ that contains a, and pick in Ω a regular value b of $g|_D$. Then

$$d(g, D, a) = d(g, D, b) \quad \text{and} \quad \begin{cases} d(g, D_1, a) = d(g, D_1, b), \\ d(g, D_2, a) = d(g, D_2, b). \end{cases}$$

All in all, assertion (3) follows from IV.2.1, p. 145, because

$$g^{-1}(b) = g|_{\overline{D}_1}^{-1}(b) \cup g|_{\overline{D}_2}^{-1}(b). \qquad \square$$

We complete the construction of the Euclidean degree with another very important fact:

Proposition 2.6 (Boundary Theorem). *Given two continuous mappings $f, g : \overline{D} \to \mathbb{R}^{m+1}$ such that $f|_X = g|_X$ and given a point $a \notin f(X) = g(X)$, then*

$$d(f, D, a) = d(g, D, a).$$

Proof. Apply IV.2.4, p. 147, to the homotopy $H_t(x) = tf(x) + (1-t)g(x)$ and the path $\gamma \equiv a$. Note that $H_t = f = g$ on X. $\square$

Example 2.7 (Gauss). We come back to the solution of complex algebraic equations, already discussed earlier (III.1.10, p. 102). There we used the Brouwer-Kronecker degree for proper mappings, but we can instead apply the more elementary Euclidean degree, along Gauss's lines (see I.1, p. 2).

Let $P(z) = z^p + c_1 z^{p-1} + \cdots + c_p$ be a polynomial with complex coefficients. If $\zeta \in \mathbb{C}$ is a root of P with $|\zeta| \geq 1$, then

$$|\zeta|^p \leq |c_1||\zeta|^{p-1} + \cdots + |c_p|;$$

hence

$$|\zeta| \leq \frac{|c_1||\zeta|^{p-1} + \cdots + |c_p|}{|\zeta|^{p-1}} \leq |c_1| + \cdots + |c_p|.$$

Thus we set $\rho = 1 + |c_1| + \cdots + |c_p|$, and consider in $\mathbb{R}^2 \equiv \mathbb{C}$ the open disc $D = \{|z| < \rho\}$, with boundary $X = \{|z| = \rho\}$. Then $f = P|_{\overline{D}}$ is a continuous mapping with $0 \notin f(X)$, and we are to compute the degree $d(P, D, 0)$. To that end consider the homotopy

$$P_t(z) = z^p + tc_1 z^{p-1} + \cdots + tc_p, \quad 0 \leq t \leq 1.$$

Using the same bound on the coefficients, we see that P_t has no root in X; hence (IV.2.4, p. 147)

$$d(f, D, 0) = d(P_1, D, 0) = d(P_0, D, 0).$$

Now the mapping $P_0 : z \mapsto z^p$ has degree p, for, any small enough $a \neq 0$ is a regular value of P_0 with p different roots c in D. Moreover, at each c, P_0 preserves orientation: $d_c P_0$ is multiplication times the complex number pc^{p-1}.

We have thus seen that the degree of a complex polynomial is exactly its ordinary degree as a polynomial. Quite predictable, but very important, is the fact that degree is not zero; hence $f = P|_D$ is surjective onto the connected component Ω of $\mathbb{C} \setminus f(X)$ that contains 0 (IV.2.5(2), p. 149). In particular, P has some root in D: this is again the *Fundamental Theorem of Algebra*.

Furthermore, we can even see that degree counts the number of roots of P. Indeed, the derivative of $f = P|_D$ at any given point z is the linear mapping defined by complex multiplication by $P'(z)$; such a multiplication preserves orientations. Thus, at every regular value $a \in \Omega$ of f we get

$$p = d(f, D, a) = \sum_{z \in f^{-1}(a)} \operatorname{sign}_z(f) = \# f^{-1}(a);$$

that is, the equation $P(z) = a$ has p roots. If a is not a regular value, the roots must be counted with multiplicities; details are left to the reader. □

We close this section with a variation on the Brouwer Fixed Point Theorem as was stated in III.7.2, p. 134. First we formulate a somewhat technical version:

Proposition 2.8. *Consider the unit open ball $D^{m+1} = \{x \in \mathbb{R}^{m+1} : \|x\| < 1\}$ and the unit sphere $\mathbb{S}^m \subset \mathbb{R}^{m+1}$.*

(1) Let $f : \overline{D}^{m+1} \to \mathbb{R}^{m+1}$ be a continuous mapping such that $f(x) \neq 0$ for all $x \in \overline{D}^{m+1}$. Then there is a $z \in \mathbb{S}^m$ such that $f(z) = \lambda z$ for some $\lambda > 0$.

(2) Let $g : \overline{D}^{m+1} \to \mathbb{R}^{m+1}$ be a continuous mapping such that $g(x) \neq x$ for all $x \in \overline{D}^{m+1}$. Then there is a $z \in \mathbb{S}^m$ such that $g(z) = \lambda z$ for some $\lambda > 1$.

Proof. (1) Suppose by way of contradiction that $f(x) \neq \lambda x$ for all $x \in \mathbb{S}^n$ and $\lambda > 0$. Then we can apply IV.2.4, p. 147, to the homotopy

$$H_t(x) = (1 - t) f(x) - tx,$$

so that

$$\begin{aligned} d(f, D^{m+1}, 0) &= d(H_0, D^{m+1}, 0) = d(H_1, D^{m+1}, 0) \\ &= d(-\operatorname{Id}_{\overline{D}^{m+1}}, D^{m+1}, 0) = (-1)^{m+1} \neq 0. \end{aligned}$$

Hence, by IV.2.5(2), p. 149, there must be some $x \in D^{m+1}$ with $f(x) = 0$, a contradiction.

(2) Apply (1) to the mapping $f(x) = g(x) - x$. □

Now the mentioned variation on III.7.2, p. 134, is immediate:

Corollary 2.9. *Let $D \subset \mathbb{R}^{m+1}$ be an open ball with boundary S. Then every continuous mapping $f : \overline{D} \to \mathbb{R}^{m+1}$ such that $f(S) \subset \overline{D}$ has some fixed point.*

Proof. Let D^{m+1} be the unit open ball, and consider the mapping

$$g : \overline{D}^{m+1} \to \mathbb{R}^{m+1} : x \mapsto \frac{f(rx + c) - c}{r},$$

where c is the center of D and r is its radius. By the hypothesis on f, $g(\mathbb{S}^m) \subset \overline{D}$, so that for all $x \in \mathbb{S}^m$ we have $\|g(x)\| \leq 1$; hence $g(x) \neq \lambda x$ if $\lambda > 1$. Consequently, by IV.2.8(2) above, g must have some fixed point x, and $rx + c$ is a fixed point of f. □

Remark 2.10. We have thus weakened the condition $f(\overline{D}) \subset \overline{D}$ to $f(S) \subset \overline{D}$, and this is indeed weaker. For instance, look at

$$g(x) = \begin{cases} 3x & \text{for } \|x\| \leq \frac{1}{2}, \\ (5 - 4\|x\|)x & \text{for } \|x\| \geq \frac{1}{2}. \end{cases}$$

This mapping fixes $\mathbb{S}^m$ but sends all points in $\frac{1}{3} < \|x\| < 1$ off $\|x\| \leq 1$. □

Exercises and problems

Number 1. Extend Euclidean degree to unbounded domains starting with the following. Let $\Omega \subset \mathbb{R}^{m+1}$ be open, possibly not bounded. Then consider continuous mappings $f : \overline{\Omega} \to \mathbb{R}^{m+1}$ verifying

$$\sup_{x \in \overline{\Omega}} \|x - f(x)\| < +\infty.$$

Prove that for every $a \notin f(\partial\overline{\Omega})$, $f^{-1}(a)$ is bounded and for all bounded open sets $D \subset \Omega$ containing $f^{-1}(a)$, the degrees $d(f, D, a)$ are the same. This common value is the degree $d(f, \Omega, a)$.

Number 2. ([Poincaré 1883]; I.1, p. 11) Let $a_1, \ldots, a_n$ be positive real numbers, and for each $i = 1, \ldots, n$, let $f_i : [-a_1, a_1] \times \cdots \times [-a_n, a_n] \to \mathbb{R}$ be a continuous function positive on $x_i = a_i$ and negative on $x_i = -a_i$. Prove that the system

$$f_1(x) = \cdots = f_n(x) = 0$$

has some solution in $(-a_1, a_1) \times \cdots \times (-a_n, a_n)$.

Number 3. Let $D \subset \mathbb{R}^{m+1}$ be a bounded open set, let $X = \overline{D} \backslash D$, and let $f : \overline{D} \to \mathbb{R}^{m+1}$ be a continuous function such that $f(D)$ is contained in a hyperplane. Then $d(f, D, a) = 0$ for all $a \in \mathbb{R}^{m+1} - f(X)$.

Number 4. Let $a < b$ be two real numbers and let D be a bounded open set of $[a, b] \times \mathbb{R}^{m+1}$ that meets all sets $\{t\} \times \mathbb{R}^{m+1}$. Let $H : \overline{D} \to \mathbb{R}^{m+1}$ be a continuous function and let $c \in \mathbb{R}^{m+1} \setminus H(\partial D)$. For $a \leq t \leq b$ and $X \subset [a, b] \times \mathbb{R}^{m+1}$ we denote

$$X_t = \{x \in \mathbb{R}^{m+1} : (t, x) \in X\}.$$

Show that $D_t \subset \mathbb{R}^{m+1}$ is a bounded open set, that $\partial D_t \subset (\partial D)_t$ (equality not true in general), and that the degree $d(H_t, D_t, c)$ is well defined and does not depend on t.

Number 5. Let D be a bounded open subset of $[0, 1] \times \mathbb{R}^{m+1}$ meeting all sets $\{t\} \times \mathbb{R}^{m+1}$, and let $H : \overline{D} \to \mathbb{R}^{m+1}$ be a continuous function. Let $a \in \mathbb{R}^{m+1}$ off $H_t(\partial D_t)$ for all $t \in [0, 1]$. Is it true that $d(H_t, D_t, a)$ depends on t?

Number 6. Write $\mathbb{R}^m = \mathbb{R}^p \times \mathbb{R}^q$ and consider the closed balls

$$\overline{D}^p_\varepsilon = \{x \in \mathbb{R}^p : \|x\| \leq \varepsilon\} \quad \text{and} \quad \overline{D}^q_\delta = \{y \in \mathbb{R}^q : \|y\| \leq \delta\}.$$

Let there be a continuous mapping

$$g : \overline{D}^p_\varepsilon \times \overline{D}^q_\delta \to \mathbb{R}^m : (x, y) \mapsto (x, g_y(x))$$

with $g(0, y) = y$. Show that

$$\bigcup_y \operatorname{graph}(g_y) = \{(x, g_y(x)) : x \in \overline{D}^p_\varepsilon, y \in \overline{D}^q_\delta\}$$

is a neighborhood of $(0, 0)$ in $\mathbb{R}^m$.

Number 7. Let D denote the unit open ball in $\mathbb{R}^{m+1}$, and let $f : \overline{D} \to \mathbb{R}^{m+1}$ be a continuous function without fixed points. Prove that the angle $\alpha(z)$ determined by the two vectors $z, z - f(z)$ reaches all values from 0 to π for $z \in \mathbb{S}^m$.

3. The degree of a differentiable mapping

Only after all properties of the Euclidean degree are available, can we deduce that for an arbitrary differentiable mapping degree can be computed by the same regular value formula that gives the degree of a smooth mapping. In any case, it is interesting to remark how simple this is if we compare it to the Brouwer-Kronecker theory (III.4.1, p. 114). We need the following lemma, interesting in its own right:

Lemma 3.1. *Let $D \subset \mathbb{R}^{m+1}$ be the open ball $\|x - c\| < \varepsilon$, and let $f : \Omega \to \mathbb{R}^{m+1}$ be a $\mathcal{C}^1$ mapping defined on a convex open neighborhood Ω of $\overline{D}$. Then the mapping*

$$H_t(x) = \begin{cases} \dfrac{f(tx + (1-t)c) - f(c)}{t} & \text{for } t \neq 0, \\ d_c f(x - c) & \text{for } t = 0 \end{cases}$$

is a continuous homotopy on Ω of $f - f(c)$ and $d_c f - d_c f(c)$.

Proof. Since Ω is convex, we can write

$$\begin{aligned} f(x) - f(c) &= \int_0^1 \frac{d}{dt} f(tx + (1-t)c)dt \\ &= \int_0^1 \sum_{i=1}^{m+1} \frac{\partial f}{dx_i}(tx + (1-t)c)(x_i - c_i)dt = \sum_{i=1}^{m+1} (x_i - c_i) f_i(x), \end{aligned}$$

where the f_i's are the continuous functions

$$f_i(x) = \int_0^1 \frac{\partial f}{dx_i}(tx + (1-t)c)dt.$$

We see that $f_i(c) = \frac{\partial f}{dx_i}(c)$, and after a little computation we get

$$H_t(x) = \sum_{i=1}^{m+1} (x_i - c_i) f_i(tx + (1-t)c),$$

which shows that H is continuous. □

Now we can prove the following:

Proposition 3.2. *Let $f : \overline{D} \to \mathbb{R}^{m+1}$ be a continuous mapping whose restriction $f|_D$ is $\mathcal{C}^1$, and let $a \in \mathbb{R}^{m+1} \setminus f(X)$ be a regular value of $f|_D$. Then $f^{-1}(a)$ is finite, and*

$$d(f, D, a) = \sum_{x \in f^{-1}(a)} \operatorname{sign} \det \Big(\frac{\partial f_i}{\partial x_j}(x) \Big).$$

Proof. That $f^{-1}(a)$ is finite follows as usual from the Inverse Function Theorem, say $f^{-1}(a) = \{c_1, \dots, c_r\}$. We pick disjoint open balls D_i centered at the points c_i with $\overline{D}_i \subset D$, such that each restriction $f|_{D_i}$ is bijective. Since $a \notin f(\overline{D} \setminus \bigcup_i D_i)$, from additivity (IV.2.5(3), p. 149), we deduce that

$$d(f, D, a) = \sum_i d(f, D_i, a).$$

Then, by the previous lemma and the invariance of homotopy, we have

$$\begin{aligned} d(f, D_i, a) &= d(f - f(c_i), D_i, 0) \\ &= d(d_{c_i} f - d_{c_i} f(c_i), D_i, 0) = d(d_{c_i} f, D_i, d_{c_i} f(c_i)). \end{aligned}$$

As $\big(\frac{\partial f_i}{\partial x_j}(c_i)\big)$ is the matrix of the linear isomorphism $d_{c_i} f$, we get what we wanted. □

Exercises and problems

Number 1. Let $h_1, \dots, h_r : \mathbb{R} \to \mathbb{R}$ be differentiable functions, and define

$$f : \mathbb{R}^2 \to \mathbb{R}^2 : (x, y) \mapsto (xy, xy^2 + x^2 \sum_i h_i(xy)^2).$$

Compute all posible values of $d(f, D, a)$.

Number 2. Let $f : \overline{D} \to \mathbb{R}^{m+1}$ be a continuous mapping and consider $a \in \mathbb{R}^{m+1} \setminus f(X)$. Suppose that f is differentiable on an open set containing $f^{-1}(a)$. Prove that

$$d(f, D, a) = \sum_{x \in f^{-1}(b)} \operatorname{sign}\det\Big(\frac{\partial f_i}{\partial x_j}(x)\Big)$$

for all regular values b close enough to a.

Number 3. Let $f : \overline{D} \to \mathbb{R}^{m+1}$ be a continuous mapping whose restriction $f|_D$ is differentiable, with Jacobian determinant ≥ 0 all through D. Let $a \in \mathbb{R}^{m+1}$ have finite preimage $f^{-1}(a) \subset D$. Is it true that

$$d(f, D, a) = \sum_{x \in f^{-1}(a)} \operatorname{sign}\det\Big(\frac{\partial f_i}{\partial x_j}(x)\Big)$$

(where $\operatorname{sign} 0 = 0$)?

Number 4. Let $f : \overline{D} \to \mathbb{R}^{m+1}$ be a continuous mapping whose restriction $f|_A$ to an open set $A \subset D$ is differentiable, and let $a \in \mathbb{R}^{m+1} \setminus f(X)$ be a regular value of $f|_A$. Find conditions on the fiber $f^{-1}(a)$ for the following formula to hold true:

$$d(f, D, a) = \sum_{x \in f^{-1}(a) \cap A} \operatorname{sign}\det\Big(\frac{\partial f_i}{\partial x_j}(x)\Big).$$

Number 5. Let $f : \overline{D} \to \mathbb{R}^{m+1}$ be a continuous mapping whose restriction $f|_D$ is differentiable. Suppose that the open set A consisting of all regular values of f in $\mathbb{R}^{m+1} \setminus f(X)$ is connected. Prove that

$$\sum_{x \in f^{-1}(a)} \operatorname{sign}\det\Big(\frac{\partial f_i}{\partial x_j}(x)\Big) = \sum_{y \in f^{-1}(b)} \operatorname{sign}\det\Big(\frac{\partial f_i}{\partial x_j}(y)\Big),$$

for any two values $a, b \in A$.

Number 6. Let $D^k \subset \mathbb{R}^k$ denote the unit open ball as usual. Let $f : \overline{D}^p \to \mathbb{R}^p$ be a continuous mapping whose restriction $f|_{D^p}$ is differentiable with $0 \notin f(\overline{D}^p \setminus D^p)$ a regular value. Consider $g : \overline{D}^{p+q} \to \mathbb{R}^{p+q}$ defined by

$$g(x_1, \dots, x_p, x_{p+1}, \dots, x_{p+q}) = (f_1(x_1, \dots, x_p), \dots, f_p(x_1, \dots, x_p), x_{p+1}, \dots, x_{p+q}).$$

Prove that $d(g, D^{p+q}, 0)$ is well defined and that it coincides with $d(f, D^p, 0)$.

Number 7. ([Führer 1971]; I.5, p. 39) We already know that Euclidean degree theory verifies all properties of Führer's characterization. Prove now that those properties give uniqueness of the theory, by the following steps:

(1) By approximation and homotopy invariance, reduce uniqueness to the case of differentiable functions.

(2) By additivity, reduce to mappings defined on open balls and regular values with a single preimage.

(3) By IV.3.1, p. 153, reduce to linear forms, and then use paths in the space of regular matrices.

4. Winding number

Here we look at the relationship between the Brouwer-Kronecker degree and the Euclidean degree, where our X in this chapter is a hypersurface. The basic notion on which the whole comparison stands is that of *winding number,* and thus we come back to the origins of the theory (I.1, p. 5, and I.2, p. 19).

First of all, for $D \subset \mathbb{R}^{m+1}$ bounded open and $X = \overline{D} \setminus D$ as usual, we have the following:

Proposition and Definition 4.1. *Let $f : X \to \mathbb{R}^{m+1}$ be a continuous mapping and consider a point $a \in \mathbb{R}^{m+1} \setminus f(X)$. Then, for all continuous extensions $\bar{f} : \overline{D} \to \mathbb{R}^{m+1}$ the degree $d(\bar{f}, D, a)$ is the same, and we define the* winding number of f around a *by*

$$w(f, a) = d(\bar{f}, D, a)$$

for any such $\bar{f}$ (which exists by the Tietze Extension Theorem).

This follows immediately from the Boundary Theorem, IV.2.6, p. 150. As is the degree (IV.2.3, p. 146), the winding number is also locally constant on the target:

Proposition 4.2. *Let $f : X \to \mathbb{R}^{m+1}$ be a continuous mapping and consider two points a, b in the same connected component of $\mathbb{R}^{m+1} \setminus f(X)$. Then $w(f, a) = w(f, b)$.*

Of course, the winding number is invariant by homotopy:

Proposition 4.3. *Let $H : [0,1] \times X \to \mathbb{R}^{m+1}$ be a continuous mapping, and let $\gamma : [0,1] \to \mathbb{R}^{m+1}$ be a continuous path such that $\gamma(t) \notin H_t(X)$ for $0 \le t \le 1$. Then $w(H_t, \gamma(t))$ does not depend on t.*

Proof. By the Tietze Extension Theorem, H extends to a homotopy $\overline{H} : [0,1] \times \overline{D} \to \mathbb{R}^{m+1}$, and IV.1.3, p. 140, applies. □

Another easy fact is the following consequence of IV.2.5(2), p. 149:

Proposition 4.4 (Boundary Theorem). *Let $f : X \to \mathbb{R}^{m+1}$ be a continuous mapping and consider a point $a \in \mathbb{R}^{m+1} \setminus f(X)$. If f has a continuous extension $\bar{f} : \overline{D} \to \mathbb{R}^{m+1}$ such that $a \notin \bar{f}(\overline{D})$, then $w(f, a) = 0$.*

Next we want to represent the Brouwer-Kronecker degree as a winding number in this Euclidean setting. The key result needed to do this follows.

Proposition 4.5. *Let X be the boundary of a compact, oriented, differentiable manifold W of dimension $m+1$. Let $\bar{f} : W \to \mathbb{R}^{m+1}$ be a differentiable mapping and let $a \in \mathbb{R}^{m+1} \setminus \bar{f}(X)$ be a regular value of $\bar{f}$, and consider the differentiable mapping*

$$g : X \to \mathbb{S}^m : x \mapsto \frac{\bar{f}(x) - a}{\|\bar{f}(x) - a\|}\,.$$

Then $\bar{f}^{-1}(a)$ is finite, and the Brouwer-Kronecker degree of g is

$$\deg(g) = \sum_{x \in \bar{f}^{-1}(a)} \operatorname{sign}_x(\bar{f})\,.$$

Proof. That $\bar{f}^{-1}(a)$ is finite follows as usual, say $\bar{f}^{-1}(a) = \{x_1, \ldots .x_r\}$. Then there are disjoint neighborhoods U_k of the points x_k such that the restrictions $\bar{f}|_{U_k} : U_k \to B$ are diffeomorphisms onto an open ball B centered at a. Then, we choose a smaller closed ball $D \subset B$ with boundary a sphere S and again restrict $\bar{f}$ to have diffeomorphisms

$$\bar{f}|_{V_k} : V_k = \bar{f}^{-1}(D) \cap U_k \to D, \quad \text{with} \quad \bar{f}(\partial V_k) = S.$$

Let $V = W \setminus \bigcup_k V_k$. This is a compact manifold with boundary $Y = X \cup \bigcup_k \partial V_k$; here X and the ∂V_k's carry the orientation as boundaries of V. This means that each ∂V_k carries the wrong orientation as boundary of V_k.

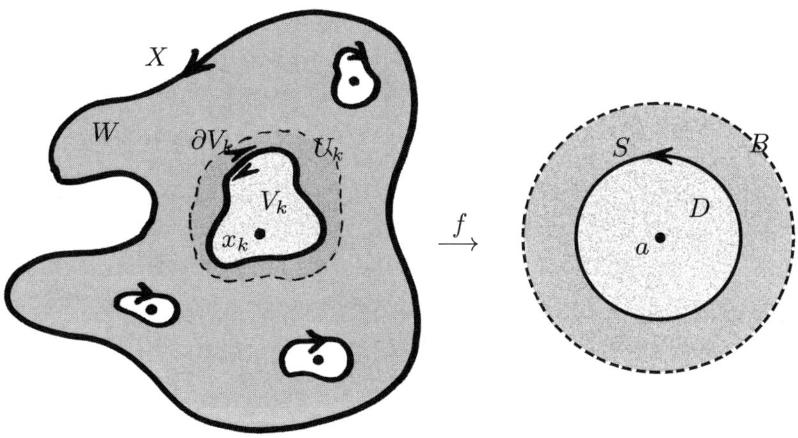

Since $a \notin \bar{f}(V)$, g is in fact defined on V, and by III.1.8, p. 101, $\deg(g|_{\partial V}) = 0$; hence

$$\deg(g|_X) - \sum_k \deg(g|_{\partial V_k}) = 0,$$

where the negative sign corresponds to the above remark on the orientations of the ∂V_k's. Moreover, if ε is the radius of D, we have

$$g|_{\partial V_k} = \tfrac{1}{\varepsilon}(\bar{f}|_{\partial V_k} - a),$$

and it follows readily that $\deg(g|_{\partial V_k}) = \deg(\bar{f}|_{\partial V_k})$. Now:

(1) If $\operatorname{sign}_{x_k}(\bar{f}) = +1$, the diffeomorphism $\bar{f}|_{V_k} : V_k \to D$ preserves orientation. Hence the restriction $\bar{f}|_{\partial V_k} : \partial V_k \to S$ preserves orientation too and has Brouwer-Kronecker degree $+1$.

(2) If $\operatorname{sign}_{x_k}(\bar{f}) = -1$, the diffeomorphism $\bar{f}|_{V_k} : V_k \to D$ reverses orientation. Hence the restriction $\bar{f}|_{\partial V_k} : \partial V_k \to S$ reverses orientation too and has Brouwer-Kronecker degree -1.

Consequently,

$$\deg(g|_X) = \sum_k \deg(g|_{\partial V_k}) = \sum_k \deg(f|_{\partial V_k}) = \sum_k \operatorname{sign}_{x_k}(\bar{f}),$$

as desired. □

Now we conclude the section with the announced representation of the Brouwer-Kronecker degree:

Proposition 4.6. *Let $X \subset \mathbb{R}^{m+1}$ be a connected, compact, boundaryless, differentiable manifold of dimension m, and let $f : X \to \mathbb{S}^m$ be a continuous mapping. Then X is the boundary of a bounded open set $D \subset \mathbb{R}^{m+1}$, and*

$$\deg(f) = w(f, 0).$$

Proof. The Jordan Separation Theorem for compact hypersurfaces (III.6.2, p. 124, and III.6.4, p. 129) says that X bounds a bounded open set D as claimed. On the other hand, since both deg and w are invariant by homotopy, we can suppose f is smooth, and then, by the smooth Tietze Extension Theorem, f has a smooth extension $\bar{f} : \overline{D} \to \mathbb{R}^{m+1}$. Finally, we pick a regular point a of $\bar{f}$ such that $\|a\| < 1$. By IV.4.2, p. 156, $w(f, 0) = w(f, a)$; hence we are reduced to showing that

$$\deg(f) = w(f, a).$$

But in this situation $w(f,a) = \sum_{x\in\bar{f}^{-1}(a)} \mathrm{sign}_x(\bar{f})$, and from IV.4.5, p. 157, we get $w(f,a) = \deg(g)$, where

$$g : X \to \mathbb{S}^m : x \mapsto \frac{f(x) - a}{\|f(x) - a\|}.$$

Finally, the homotopy

$$H_t(x) = \frac{f(x) - ta}{\|f(x) - ta\|}$$

is well defined, and by III.3.4, p. 112, we conclude that

$$\deg(f) = \deg(H_0) = \deg(H_1) = \deg(g) = w(f,a). \qquad \square$$

Exercises and problems

Number 1. ([Poincaré 1886] and [Bohl 1904]; I.2, p. 19) Let X be a compact smooth hypersurface of $\mathbb{R}^{m+1}$ and let

$$f = (f_1, \dots, f_{m+1}),\ g = (g_1, \dots, g_{m+1}) : X \to \mathbb{R}^{m+1}$$

be two continuous mappings with $0 \notin f(X), g(X)$. Prove:

(1) If $w(f,0) \neq w(g,0)$, then there is at least one point $x \in X$ such that

$$\frac{f_1(x)}{g_1(x)} = \cdots = \frac{f_{m+1}(x)}{g_{m+1}(x)} < 0.$$

(2) If $w(f,0) \neq (-1)^{m+1} w(g,0)$, then there is at least one point $y \in X$ such that

$$\frac{f_1(y)}{g_1(y)} = \cdots = \frac{f_{m+1}(y)}{g_{m+1}(y)} > 0.$$

(3) If $w(f,0) \neq \pm w(g,0)$, then the function $f_1 g_1 + \cdots + f_{m+1} g_{m+1}$ cannot have constant sign on X.

Number 2. ([Kronecker 1869a] and [Hadamard 1910]; I.1, p. 10, and I.2, p. 19) Let $f : X \to \mathbb{R}^{m+1}$ be a smooth function on a compact smooth hypersurface X of $\mathbb{R}^{m+1}$, such that $0 \notin f(X)$. Prove the following assertions:

(1) $w(f,0) = \deg\left(\frac{f}{\|f\|}\right)$.

(2) Denote by Ω the volume element of $\mathbb{S}^m$, and set $r(x) = x/\|x\|$. Then

$$w(f,0) = \frac{1}{\mathrm{vol}(\mathbb{S}^m)} \int_X (r \circ f)^* \Omega.$$

(3) Let $x = (x_1, \dots, x_m)$ be local coordinates on X. Then

$$w(f,0) = \frac{1}{\mathrm{vol}(\mathbb{S}^m)} \int_X \|f(x)\|^{-m-1} \det\left(f(x), \frac{\partial f(x)}{\partial x_1}, \dots, \frac{\partial f(x)}{\partial x_m}\right) dx.$$

Number 3. ([Cauchy 1855]; I.1, p. 6) Fix $a \in \mathbb{C} \equiv \mathbb{R}^2$. Let $f : \mathbb{S}^1 \to \mathbb{C} \setminus \{a\}$ be a smooth mapping that has a holomorphic extension to some open neighborhood of $\mathbb{S}^1$ in $\mathbb{C}$. Prove the Argument Principle:

$$w(f, a) = \frac{1}{2\pi\sqrt{-1}} \int_{\mathbb{S}^1} \frac{f'(z)}{f(z) - a} dz.$$

Check in this case the geometrical idea that the winding number tells how many times f wraps $\mathbb{S}^1$ around the point a (I.1, p. 5): *Given a half-line from the origin that meets $f(\mathbb{S}^1)$ transversally at finitely many points, $w(f, 0)$ is the difference between the number of them at which the argument of f increases and the number of them at which the argument of f decreases.*

Number 4. ([Gauss 1813]; I.1, p. 11) Let $S \subset \mathbb{R}^3$ be a compact surface. Compute the winding number at a point $a \notin S$ of the inclusion $S \subset \mathbb{R}^3$, and deduce the Gauss Theorem: *Let $v(x) = \frac{x-a}{\|x-a\|^3}$ be the electric field generated by a unit charge placed at a point a. Then the flow of v through S is*

$$\int_S \langle v, \nu_S \rangle \Omega_S = \begin{cases} 4\pi & \text{if } a \text{ is inside } S, \\ 0 & \text{if } a \text{ is outside } S \end{cases}$$

(ν_S and Ω_S stand for a normal vector field and the corresponding volume element of S).

Number 5. [Gauss 1833] Let $f, g : [0, 1] \to \mathbb{R}^3$ be two smooth disjoint knots ($f(0) = f(1)$, $g(0) = g(1)$, $f(s) \neq g(t)$ for all s, t). Consider the torus $T \subset \mathbb{R}^3$ parametrized by

$$T : x = (2 + \cos(2\pi t)) \cos(2\pi s),\ y = (2 + \cos(2\pi t)) \sin(2\pi s),\ z = \sin(2\pi t),$$

and define $h : T \to \mathbb{R}^3 : (x, y, z) \mapsto f(s) - g(t)$. Prove that

$$w(h, 0) = -\frac{1}{4\pi} \int_0^1 \int_0^1 \|f(s) - g(t)\|^{-3} \det(f(s) - g(t), f'(s), g'(t)) ds dt,$$

which is the link number of the two knots (I.2, p. 28).

Number 6. Let A be an orthogonal matrix of order $m + 1$. Use a winding number for the computation of the degree of the smooth mapping $f : \mathbb{S}^m \to \mathbb{S}^m : x \mapsto Ax$. Also do the direct computation in the Brouwer-Kronecker definition. Which is quicker?

5. The Borsuk-Ulam Theorem

Our primary goal here is to obtain the important Borsuk-Ulam Theorem. Recall that a mapping f defined in some subset of a Euclidean space is called *even* if $f(x) = f(-x)$ and *odd* if $f(x) = -f(-x)$. The famous Borsuk-Ulam Theorem concerns the degree of such mappings. Afterwards, we will deduce several purely topological results, including the Invariance of Domain Theorem.

We start with a technical statement:

Lemma 5.1. (1) *Let $K \subset M$ be compact subsets of $\mathbb{R}^n$ and let $f : K \to \mathbb{R}^m$, $m > n$, be a continuous function such that $0 \notin f(K)$. Then f has a continuous extension $\bar{f} : M \to \mathbb{R}^m$ such that $0 \notin \bar{f}(M)$.*

(2) *Let $D \subset \mathbb{R}^n$ be a bounded open set, and let $X = \overline{D} \setminus D$. We suppose that $0 \notin \overline{D}$ and D is symmetric with respect to 0 (that is, $x \mapsto -x$ induces a homeomorphism on D). Now let $f : X \to \mathbb{R}^m$, $m > n$, be an even (resp., odd) continuous mapping such that $0 \notin f(X)$. Then f has an even (resp., odd) continuous extension $\bar{f} : \overline{D} \to \mathbb{R}^m$ such that $0 \notin \bar{f}(\overline{D})$.*

(3) *Assume the same hypotheses as in (2), except that here $m = n$. Then f has an even (resp., odd) continuous extension $\bar{f} : \overline{D} \to \mathbb{R}^n$ such that $0 \notin \bar{f}(\overline{D} \cap \{x_n = 0\})$.*

Proof. (1) Henceforth, norms of mappings are always intended on the compact set M. Let $\varepsilon = \operatorname{dist}(0, f(K))$. First, by the Tietze Extension Theorem, f extends to a continuous mapping $g : \mathbb{R}^n \to \mathbb{R}^m$, and by the Weierstrass Approximation Theorem there is a polynomial mapping $h = h(x)$ such that $\|g - h\| < \frac{1}{4}\varepsilon$. Since $m > n$, the Easy Sard Theorem (II.3.4, p. 63) says that the image $h(\mathbb{R}^n)$ has empty interior in $\mathbb{R}^m$, and we can find a point $a \notin h(\mathbb{R}^n)$ such that $\|a\| < \frac{1}{4}\varepsilon$. Now consider $h' = h - a : \mathbb{R}^n \to \mathbb{R}^m$. Clearly $0 \notin h'(\mathbb{R}^n)$, and furthermore

$$\|h' - g\| \leq \|h' - h\| + \|h - g\| < \tfrac{1}{2}\varepsilon.$$

Next consider the continuous mapping

$$h''(x) = \begin{cases} \dfrac{\varepsilon h'(x)}{2\|h'(x)\|} & \text{for } \|h'(x)\| \leq \frac{1}{2}\varepsilon, \\ h'(x) & \text{otherwise.} \end{cases}$$

We have $\|h''\| \geq \frac{1}{2}\varepsilon$. On the other hand, $f(x) = g(x)$ for $x \in K$, and

$$\|h'(x)\| \geq \|g(x)\| - \|g(x) - h'(x)\| > \varepsilon - \tfrac{1}{2}\varepsilon = \tfrac{1}{2}\varepsilon,$$

so that $h''(x) = h'(x)$, and we conclude

$$\|h''(x) - f(x)\| < \tfrac{1}{2}\varepsilon \quad \text{for } x \in K.$$

Now, we can apply the Tietze Extension Theorem to the function

$$h'' - f : K \to B,$$

where B is the open ball of radius $\frac{1}{2}\varepsilon$ (homeomorphic to $\mathbb{R}^m$), and $h'' - f$ extends to a continuous mapping $h''' : M \to B$. We claim that $\bar{f} = h'' - h'''$ is the mapping we seek. Indeed, for $x \in K$ we have

$$\bar{f}(x) = h''(x) - h'''(x) = h''(x) - (h''(x) - f(x)) = f(x),$$

and taking norms on M,

$$\|\bar{f}\| \geq \|h''\| - \|h'''\| \geq \tfrac{1}{2}\varepsilon - \|h'''\| > \tfrac{1}{2}\varepsilon - \tfrac{1}{2}\varepsilon = 0.$$

This completes the proof of (1).

The proof of (2) goes by induction on n. For $n = 1$ the set D is a union of disjoint open intervals $I = (a, b)$ with $a > 0$ and their symmetric copies $-I = (-b, -a)$, and X consists of the end points of those intervals. To extend f to $\overline{D}$, we map each I onto a polygonal path in $\mathbb{R}^m \setminus \{0\}$ joining $f(a)$ to $f(b)$ (here we use $m > 1$), and for $x \in -I$ define $f(x) = f(-x)$ in the even case and $f(x) = -f(-x)$ in the odd case.

Now suppose the result holds for $n-1$. We identify $\mathbb{R}^{n-1} \equiv \{x_n = 0\} \subset \mathbb{R}^n$ and consider the set

$$D_1 = D \cap \mathbb{R}^{n-1},$$

which is symmetric. Clearly, the set $X_1 = \overline{D}_1 \setminus D_1 \subset \mathbb{R}^{n-1}$ is contained in X, and by induction we find a continuous mapping $\bar{f}_1 : \overline{D}_1 \to \mathbb{R}^m$ that extends $f_1 = f|_{X_1}$ and is never 0. Now, since $D_1 \cap X = \emptyset$, we can define a nowhere zero continuous mapping $f_2 : \overline{D}_1 \cup X \to \mathbb{R}^m$ by

$$\begin{cases} f_2 \equiv \bar{f}_1 & \text{on } \overline{D}_1, \\ f_2 \equiv f & \text{on } X. \end{cases}$$

After this preparation we can apply (1). We already have one compact set, namely

$$K = \overline{D}_1 \cup X,$$

and to define $M \supset K$, set

$$D_+ = D \cap \{x_n > 0\}, \quad D_- = D \cap \{x_n < 0\},$$

and take

$$M = \overline{D}_1 \cup X \cup D_+ = \overline{D}_1 \cup X \cup \overline{D}_+ .$$

Now (1) gives a continuous mapping $\bar{f}_2 : M \to \mathbb{R}^m$ that extends $f_2|_K$ and is never 0. To conclude the proof of (2), one extends $\bar{f}_2$ to D_- by symmetry.

Finally, the proof of (3) is analogous, but one uses (2) instead of induction. The argument ends with an application of the Tietze Extension Theorem instead of (1), which gives the condition $0 \notin \bar{f}(\overline{D} \cap \mathbb{R}^{n-1})$ instead of $0 \notin \bar{f}(\overline{D})$. □

After these extension lemmas, we are ready to prove the famous Borsuk-Ulam Theorem:

Theorem 5.2 (Borsuk-Ulam Theorem). *Let $D \subset \mathbb{R}^n$ be a symmetric bounded open set, with $0 \in D$. Set $X = \overline{D} \setminus D$, and let $f : X \to \mathbb{R}^n$ be a continuous mapping with $0 \notin f(X)$.*

(1) *Assume*
$$\frac{f(x)}{\|f(x)\|} \neq \frac{f(-x)}{\|f(-x)\|} \quad \text{for all } x \in X.$$
Then the winding number $w(f, 0)$ is odd, hence not zero.

(2) *Assume*
$$\frac{f(x)}{\|f(x)\|} \neq -\frac{f(-x)}{\|f(-x)\|} \quad \text{for all } x \in X.$$
Then the winding number $w(f, 0)$ is even, and zero for n odd.

In particular, (1) *applies if f is an odd function, and* (2) *applies if f is an even function*

Proof. We start with case (1). By the Tietze Extension Theorem, f extends to a continuous function on $\overline{D}$, again denoted $f : \overline{D} \to \mathbb{R}^n$. But we need that extension to be odd. To amend that, consider the function
$$g : \overline{D} \to \mathbb{R}^n : x \mapsto f(x) - f(-x),$$
which is clearly an odd function. Then we define the homotopy
$$H_t(x) = f(x) - tf(-x), \quad H_0 = f,\ H_1 = g.$$
We claim that $0 \notin H([0,1] \times X)$. Indeed, suppose $0 = f(x) - tf(-x)$ for some $x \in X$, $0 \leq t \leq 1$. Since $0 \notin f(X)$, it must be true that $t > 0$, and from $f(x) = tf(-x)$ it follows that $\frac{f(x)}{\|f(x)\|} = \frac{f(-x)}{\|f(-x)\|}$, contrary to the hypothesis.

We conclude that f and g have the same degree, and consequently we can substitute g for f, or merely suppose that $f : \overline{D} \to \mathbb{R}^n$ is odd, which we do henceforth.

Pick $\varepsilon > 0$ small enough so that $U = \{\|x\| \leq \varepsilon\} \subset D$ and define the mapping
$$f_1 : U \cup X \to \mathbb{R}^n : x \mapsto \begin{cases} f(x) & \text{if } x \in X, \\ x & \text{if } x \in U. \end{cases}$$
Set $D_1 = D \setminus U$, $X_1 = \overline{D}_1 \setminus D_1$, and $f_2 = f_1|_{X_1}$; note that $X_1 = X \cup \{\|x\| = \varepsilon\}$. The following figure depicts the setting:

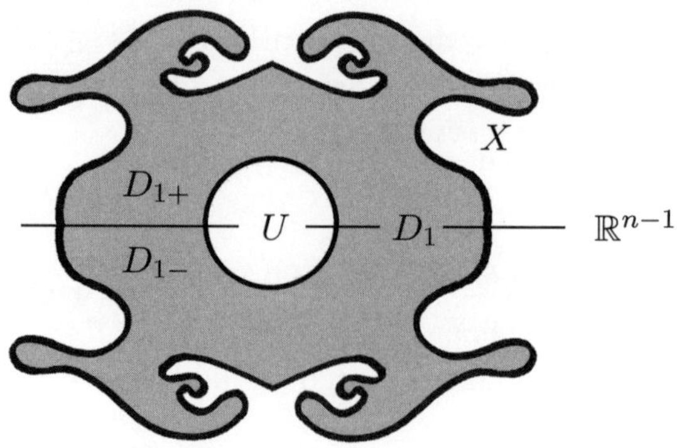

We can apply IV.5.1, p. 161, and obtain an odd continuous extension $\bar{f}_2 : \overline{D}_1 \to \mathbb{R}^n$ of f_2, such that $0 \notin \bar{f}_2(\overline{D}_1 \cap \mathbb{R}^{n-1})$. Next, we set

$$f_3 : \overline{D} \to \mathbb{R}^n : x \mapsto \begin{cases} \bar{f}_2(x) & \text{if } x \in \overline{D}_1, \\ x & \text{if } x \in U. \end{cases}$$

It is clear that $f_3|_X = f|_X$; hence by IV.2.6, p. 150,

$$d(f, D, 0) = d(f_3, D, 0).$$

Now we will use the additive property of the degree, IV.2.5(3), p. 149. We write (see the figure above)

$$D_{1+} = D_1 \cap \{x_n > 0\}, \quad D_{1-} = D_1 \cap \{x_n < 0\}.$$

As $0 \notin \bar{f}_2(\overline{D}_1 \setminus D_{1+} \cup D_{1-})$, from additivity it follows that

$$d(\bar{f}_2, D_1, 0) = d(\bar{f}_2, D_{1+}, 0) + d(\bar{f}_2, D_{1-}, 0).$$

We make the change of variables (a linear isometry in fact) $\varphi : D_{1+} \to D_{1-} : x \mapsto -x$, to get

$$\varphi^{-1} \bar{f}_2 \, \varphi(x) = \varphi^{-1} \bar{f}_2(-x) = \varphi^{-1}(-\bar{f}_2(x)) = \bar{f}_2(x).$$

Then, by direct computation,

$$d(\bar{f}_2, D_{1-}, 0) = d(\varphi^{-1} \bar{f}_2 \, \varphi, D_{1+}, 0) = d(\bar{f}_2, D_{1+}, 0)$$

and consequently,

$$d(\bar{f}_2, D_1, 0) = 2d(\bar{f}_2, D_{1+}, 0) = 2N.$$

To conclude, we look at the interior $U_0 = \{\|x\| < \varepsilon\}$ of U. We have $D \supset U_0 \cup (D \setminus U)$, and $0 \notin f_3(\overline{D} \setminus U_0 \cup (D \setminus U))$, which again by additivity, gives

$$\begin{aligned} d(f, D, 0) &= d(f_3, D, 0) = d(f_3|_{\overline{U}_0}, U_0, 0) + d(f_3, D \setminus U, 0) \\ &= d(f_3, U_0, 0) + d(f_3, D_1, 0) \\ &= d(\mathrm{Id}, U_0, 0) + d(\bar{f}_2, D_1, 0) = 1 + 2N, \end{aligned}$$

which is an odd number, as desired.

Thus we have proved part (1) of the theorem. For (2), the same argument works, with the modifications we describe next.

To reduce to the case where f is defined on the whole $\overline{D}$, the function g needed is $g(x) = f(x) + f(-x)$. The function f_1 must be defined as

$$f_1 : U \cup X \to \mathbb{R}^n : x \mapsto \begin{cases} f(x) & \text{if } x \in X, \\ |x| = (|x_1|, \dots, |x_n|) & \text{if } x \in U. \end{cases}$$

Accordingly,

$$f_3 : \overline{D} \to \mathbb{R}^n : x \mapsto \begin{cases} \bar{f}_2(x) & \text{if } x \in \overline{D}_1, \\ |x| & \text{if } x \in U. \end{cases}$$

Then,

$$\varphi^{-1} \bar{f}_2\, \varphi(x) = \varphi^{-1} \bar{f}_2(-x) = \varphi^{-1}(\bar{f}_2(x)) = -\bar{f}_2(x).$$

Thus,

$$d(\bar{f}_2, D_{1-}, 0) = d(\varphi^{-1} \bar{f}_2\, \varphi, D_{1+}, 0) = d(-\bar{f}_2, D_{1+}, 0) = (-1)^n d(\bar{f}_2, D_{1+}, 0),$$

and consequently,

$$d(\bar{f}_2, D_1, 0) = d(\bar{f}_2, D_{1+}, 0) + (-1)^n d(\bar{f}_2, D_{1+}, 0) = (1 + (-1)^n)N.$$

Finally

$$\begin{aligned} d(f, D, 0) &= d(f_3, D, 0) = d(f_3, U_0, 0) + d(f_3, D \setminus U, 0) \\ &= d(f_3, U_0, 0) + d(f_3, D_1, 0) \\ &= d(|\mathrm{Id}|, U_0, 0) + d(\bar{f}_2, D_1, 0) = 0 + (1 + (-1)^n)N, \end{aligned}$$

which is the required number in part (2). Note here that $d(|\mathrm{Id}|, U_0, 0) = 0$ by IV.2.5(2), p. 149. Indeed, $d(|\mathrm{Id}|, U_0, 0) = d(|\mathrm{Id}|, U_0, a) = 0$ for $a = (-\varepsilon, 0, \dots, 0) \notin |\mathrm{Id}|(X)$. □

The above theorem has a very nice converse:

Theorem 5.3 (Hirsch Theorem). *Let $D \subset \mathbb{R}^n$ be a symmetric bounded open set, with $0 \in D$. Set $X = \overline{D} \setminus D$, and let $f : X \to \mathbb{R}^n$ be a continuous mapping with $0 \notin f(X)$.*

(1) *If the winding number $w(f, 0)$ is odd, then there is some $x \in X$ such that*

$$\frac{f(x)}{\|f(x)\|} = \frac{-f(-x)}{\|f(-x)\|}.$$

(2) *If the winding number $w(f, 0)$ is even, then there is some $x \in X$ such that*

$$\frac{f(x)}{\|f(x)\|} = \frac{f(-x)}{\|f(-x)\|}.$$

Proof. (1) Assume by way of contradiction that

$$\frac{f(x)}{\|f(x)\|} \neq \frac{-f(-x)}{\|f(-x)\|}$$

for all $x \in X$. We extend f to a continuous mapping $\bar{f} : \overline{D} \to \mathbb{R}^n$ and define a homotopy by

$$H_t(x) = (1-t)\bar{f}(-x) + t\bar{f}(x).$$

By assumption, $H_t(x) \neq 0$ for $x \in X$, and by IV.2.4, p. 147,

$$w(f, 0) = d(\bar{f}, D, 0) = d(H_{\frac{1}{2}}, D, 0).$$

Now note that

$$H_{\frac{1}{2}}(x) = \tfrac{1}{2}\bar{f}(-x) + \tfrac{1}{2}\bar{f}(x)$$

is an even mapping, and by the Borsuk-Ulam Theorem, IV.5.2(2), p. 163, the degree $d(H_{\frac{1}{2}}, D, 0)$ must be even. This is a contradiction, because $w(f, 0)$ is odd.

Part (2) is proven similarly: one assumes

$$\frac{f(x)}{\|f(x)\|} \neq \frac{f(-x)}{\|f(-x)\|}$$

for all $x \in X$, extends f to $\bar{f}$, and defines

$$H_t(x) = (1-t)\bar{f}(-x) - t\bar{f}(x).$$

Then the odd mapping $H_{\frac{1}{2}}$ has even degree, contrary to the Borsuk-Ulam Theorem, IV.5.2(1), p. 163. □

Note that from the Borsuk-Ulam Theorem we see readily that a function like f in case (1) on p. 163 (for instance an odd f) must have some zero in $\overline{D}$ after any extension. It is also easy to obtain a fixed point result:

Corollary 5.4. *Let $D \subset \mathbb{R}^n$ be a symmetric bounded open set, with $0 \in D$. Set $X = \overline{D} \setminus D$, and let $f : \overline{D} \to \mathbb{R}^n$ be a continuous mapping whose restriction $f|_X$ is odd. Then there is some $x \in \overline{D}$ such that $f(x) = x$.*

Proof. The function $g : \overline{D} \to \mathbb{R}^n : x \mapsto x - f(x)$ also has an odd restriction $g|_X$:

$$g(-x) = -x - f(-x) = -x + f(x) = -g(x) \quad \text{for } x \in X.$$

If $0 \in g(X)$, we are done, so we assume $0 \notin g(X)$. Then, by the Borsuk-Ulam Theorem, IV.5.2(1), p. 163, $d(g, D, 0) \neq 0$; hence there is some $x \in D$ such that $g(x) = 0$ and $f(x) = x$. □

Here is a second interesting consequence:

Corollary 5.5. *Let $D \subset \mathbb{R}^n$ be a symmetric bounded open set, with $0 \in D$. Set $X = \overline{D} \setminus D$, and let $f : X \to \mathbb{R}^m$ be a continuous mapping with $m < n$. Then there is an $x \in X$ such that $f(x) = f(-x)$.*

Proof. Set $\mathbb{R}^m \equiv \{x_{m+1} = \cdots = x_n = 0\} \subset \mathbb{R}^n$, and consider the odd mapping $g : X \to \mathbb{R}^m : x \mapsto f(x) - f(-x)$. We assume $0 \notin g(X)$ (otherwise there would be nothing to prove). By the Tietze Extension Theorem g has a continuous extension $\bar{g} : \overline{D} \to \mathbb{R}^m$, and we then define $\tilde{g} : \overline{D} \to \mathbb{R}^n : x \mapsto (\bar{g}(x), 0)$. Like g, the preceding function is odd on X, and by assumption $0 \notin \tilde{g}(X)$. Thus, by the Borsuk-Ulam Theorem, IV.5.2(1), p. 163, the degree of $\tilde{g}$ is not zero: $d(\tilde{g}, D, 0) \neq 0$. However, pick $\varepsilon > 0$ small enough so that $a_\varepsilon = (0, \dots, 0, \varepsilon) \in \mathbb{R}^n \setminus \tilde{g}(X)$. Then, by IV.2.3, p. 146,

$$d(\tilde{g}, D, 0) = d(\tilde{g}, D, a_\varepsilon),$$

and the degree in the right-hand side is zero, because $a_\varepsilon \notin \tilde{g}(\overline{D})$ (IV.2.5(2), p. 149).

This contradiction comes from our assumption that $0 \notin g(X)$; thus there must be some $x \in X$ with $f(x) = f(-x)$. □

Now we give a very important theorem (which for dimension 1 is an immediate consequence of the *Bolzano Theorem*):

Theorem 5.6 (Invariance of Domain Theorem). *Let $W \subset \mathbb{R}^n$ be an open set, and let $g : W \to \mathbb{R}^n$ be a locally injective continuous mapping. Then g is an open mapping.*

Proof. Let $a \in W$ and consider the translated open set

$$D = W - a = \{x - a : x \in W\}$$

(which contains 0) and the continuous mapping

$$f : D \to \mathbb{R}^n : x \mapsto g(a + x) - g(a).$$

By hypothesis, for $\varepsilon > 0$ small enough, f is injective on the closure $\overline{B}$ of the open ball $B = \{\|x\| < \varepsilon\} \subset D$; we set $X = \overline{B} \setminus B$. Then we define the following homotopy:

$$H_t(x) = f\Big(\frac{x}{1+t}\Big) - f\Big(\frac{-tx}{1+t}\Big), \quad x \in \overline{B}.$$

We see that $0 \notin H([0,1] \times X)$: if $H(t,x) = 0$, since $f|\overline{B}$ is injective, $\frac{x}{1+t} = \frac{-tx}{1+t}$ and $x = 0 \notin X$. Thus, by homotopy invariance, IV.2.4, p. 147,

$$d(f, B, 0) = d(H_0, B, 0) = d(H_1, B, 0).$$

But the mapping

$$H_1(x) = f(\tfrac{1}{2}x) - f(-\tfrac{1}{2}x)$$

is odd, and by the Borsuk-Ulam Theorem, IV.5.2(1), p. 163, its degree is not zero. Let V denote the connected component of $\mathbb{R}^n \setminus f(X)$ that contains 0. We deduce (IV.2.3, p. 146) that

$$d(f, B, y) \neq 0 \quad \text{for every } y \in V,$$

which by IV.2.5(2), p. 149, means $y \in f(B)$. Consequently,

$$0 \in V \subset f(B).$$

Thus, there is an $\eta > 0$ such that

if $\|y'\| < \eta$, *then* $y' = f(x')$ *for some* $x' \in \mathbb{R}^n$ *with* $\|x'\| < \varepsilon$.

Setting $x' = x - a$ and $y' = y - g(a)$, we deduce that

if $\|y - g(a)\| < \eta$, *then* $y = g(x)$ *for some* $x \in \mathbb{R}^n$ *with* $\|x - a\| < \varepsilon$.

This means that $g(a + B)$ is a neighborhood of $g(a)$, and since the $a + B$'s for ε small enough form a neighborhood basis of a, we conclude that g maps every neighborhood of a onto a neighborhood of $g(a)$. Since $a \in W$ is arbitrary, g is open. □

Remark 5.7. Suppose $W = \mathbb{R}^n$ in the above theorem. Then, *if*

$$\lim_{\|x\| \to +\infty} \|f(x)\| = +\infty,$$

the mapping f is surjective.

Indeed, that condition implies that f is proper, and then $f(\mathbb{R}^n)$ is not only open as the theorem says, but also closed in $\mathbb{R}^n$. □

Variations of the previous theorem follow:

Corollary 5.8. (1) (Invariance of Domain Theorem) *Let $W \subset \mathbb{R}^n$ be an open set, and let $g : W \to \mathbb{R}^n$ be an injective continuous mapping. Then $V = g(W) \subset \mathbb{R}^n$ is open, and g is a homeomorphism onto V.*

(2) (Invariance of interiors) *Let $S, T \subset \mathbb{R}^n$ be arbitrary sets, and let $g : S \to T$ be a homeomorphim. Then*

$$f(\mathrm{Int}_n(S)) = \mathrm{Int}_n(T) \quad \textit{and} \quad f\big(S \cap \overline{\mathbb{R}^n \setminus S}\big) = T \cap \overline{\mathbb{R}^n \setminus T}$$

(of course, Int_n *means interior in $\mathbb{R}^n$).*

(3) (Invariance of dimension) *Let $W \subset \mathbb{R}^n$ and $V \subset \mathbb{R}^m$ be non-empty open sets. If W and V are homeomorphic, then $n = m$.*

Proof. Assertion (1) is an immediate consequence of the previous theorem.

For (2), note that since f is injective, $f(\mathrm{Int}_n(S))$ is an open subset of $\mathbb{R}^n$; hence $f(\mathrm{Int}_n(S)) \subset \mathrm{Int}_n(T)$. Arguing for f^{-1}, we get the other inclusion. Equality concerning closures follows by complementation, because

$$A \cap \overline{\mathbb{R}^n \setminus A} = A \setminus \mathrm{Int}_n(A) \quad \text{for any } A.$$

For (3) suppose $n < m$, and put $W' = W \times \{0\} \subset \mathbb{R}^n \times \mathbb{R}^{m-n} = \mathbb{R}^m$. Of course, W' is homeomorphic to W, hence to V by hypothesis. Then, by (2), $\mathrm{Int}_m(W')$ is homeomorphic to $\mathrm{Int}_m(V) = V$. But this is impossible, because $W' \subset \{x_m = 0\}$ has empty interior in $\mathbb{R}^m$. □

We conclude the section with a common improvement to the different fixed point theorems seen so far (III.7.2, p. 134, and IV.2.9, p. 152):

Proposition 5.9. *Let $D \subset \mathbb{R}^n$ be a bounded open set whose closure $\overline{D}$ is homeomorphic to a closed ball, and denote $X = \overline{D} \setminus D$ as usual. Then every continuous mapping $f : \overline{D} \to \mathbb{R}^n$ such that $f(X) \subset \overline{D}$ has some fixed point.*

Proof. Let $h : \overline{D} \to \overline{D}^n$ be a homeomorphism onto the unit closed ball $\overline{D}^n \subset \mathbb{R}^n$; by IV.5.8(2), p. 169, $h(X)$ is the unit sphere $\mathbb{S}^{n-1}$. Since $h(f(X)) \subset h(\overline{D}) = \overline{D}^n$, we can define a homotopy $H_t : X \to \mathbb{R}^n$ by

$$H_t(x) = x - h^{-1}(th(f(x))), \quad 0 \le t \le 1.$$

Now suppose $H_t(x) = 0$ for some $x \in X$. Then $x = h^{-1}(th(f(x))) \in X$, so that

$$th(f(x)) = h(x) \in \mathbb{S}^{n-1}$$

and as $h(f(x)) \in \overline{D}^n$, it must be true that $t = 1$. Consequently,

$$x = h^{-1}(th(f(x))) = h^{-1}(h(f(x))) = f(x),$$

and we have a fixed point in X. Otherwise, we in fact have $H_t : X \to \mathbb{R}^n \setminus \{0\}$ and by the definition and homotopy invariance of winding numbers,

$$\begin{aligned} d(\mathrm{Id} - f, D, 0) &= w(\mathrm{Id} - f|_X, 0) = w(H_1, 0) = w(H_0, 0) \\ &= w(\mathrm{Id} - h^{-1}(0)|_X, 0) = d(\mathrm{Id} - h^{-1}(0), D, 0) = 1, \end{aligned}$$

because $h^{-1}(0) \in D$. Thus, $d(\mathrm{Id} - f, D, 0) \ne 0$, and by IV.2.5(2), p. 149, there is an $x \in D$ such that $x - f(x) = 0$; that is, we have a fixed point in D. □

Exercises and problems

Number 1. Suppose we have a decomposition

$$\mathbb{S}^m = (A_1 \cup \cdots \cup A_{m+1}) \cup (-A_1 \cup \cdots \cup -A_{m+1})$$

into non-empty closed subsets with $A_i \cap -A_i = \emptyset$.

(1) Show that for every $i = 1, \dots, m+1$ there is an odd continuous function $f_i : \mathbb{S}^m \to \mathbb{R}$ such that $f_i(x) = 2$ if and only if $x \in A_i$ and $f_i(x) = -2$ if and only if $x \in -A_i$.

(2) Deduce that $\bigcap_{i=1}^{m+1} A_i \ne \emptyset$.

Number 2. Let $A_1, \dots, A_{m+1}$ be non-empty closed subsets of the sphere $\mathbb{S}^m$. Prove the following:

(1) If $\mathbb{S}^m = A_1 \cup \cdots \cup A_{m+1}$, there is a point $x \in \mathbb{S}^m$ such that $x \in A_i$ and $-x \in A_i$ for some i.

(2) If $\mathbb{S}^m = (A_1 \cup \cdots \cup A_m) \cup (-A_1 \cup \cdots \cup -A_m)$, there is a point $x \in \mathbb{S}^m$ such that $x \in A_i$ and $-x \in A_i$ for some i.

Number 3. Let $f_1, \dots, f_m : \mathbb{R}^{m+1} \to \mathbb{R}$ be homogeneous continuous mappings of odd degree (that is, $f_i(\lambda x) = \lambda^{p_i} f_i(x)$ with p_i odd). Prove that the system $f_1 = 0, \dots, f_m = 0$ has some non-trivial solution.

Number 4. Let $f : \mathbb{R}^n \to \mathbb{R}^p$ be a homogeneous continuous mapping of odd degree, without zeros. Prove that every linear subspace $L \subset \mathbb{R}^p$ of dimension $n - 1$ is perpendicular to $f(x)$ for some $x \in \mathbb{R}^n$.

Number 5. An odd continuous mapping $f : \mathbb{S}^m \to \mathbb{S}^n$ is also called *equivariant*, expressing that it commutes with the antipodal involution. Prove the following:

(1) There are equivariant mappings $f : \mathbb{S}^m \to \mathbb{S}^n$ if and only if $m \leq n$.

(2) Every equivariant mapping $f : \mathbb{S}^m \to \mathbb{S}^m$ has odd degree; hence it is not null-homotopic.

(3) Every equivariant mapping $f : \mathbb{S}^m \to \mathbb{S}^n$, $m < n$, is not surjective, hence null-homotopic.

Number 6. Here we consider topological spaces X equipped with a continuous involution σ_X (that is, $\sigma_X \circ \sigma_X = \mathrm{Id}_X$); in particular, $\sigma_{\mathbb{S}^m}$ is the antipodal involution. A continuous mapping $f : X \to Y$ of spaces with involution is called *equivariant* when $\sigma_Y \circ f = f \circ \sigma_X$. For a space with involution X, define the following:

(i) the *level* $s(X)$, as the infimum (possibly ∞) of all integers $m \geq 1$ such that there is an equivariant continuous mapping $X \to \mathbb{S}^m$,

(ii) the *colevel* $s'(X)$, as the supremum (possibly ∞) of all integers $m \geq 1$ such that there is an equivariant continuous mapping $\mathbb{S}^m \to X$.

Prove that $s'(X) \leq s(X)$. Compute both invariants when X is the boundary of a symmetric bounded open neighborhood D of the origin in $\mathbb{R}^n$, equipped with the antipodal involution.

Number 7. Consider the polynomial algebra $A = \mathbb{R}[x_1, \dots, x_n]/(1 + x_1^2 + \cdots + x_n^2)$. Complete the details of the following proof that -1 *is not a sum of less than n squares in* A.

(1) Otherwise, let $-1 = f_1(x)^2 + \cdots + f_{n-1}^2(x) + f_0(x)(1 + x_1^2 + \cdots + x_n^2)$ for some polynomials $f_k(x) \in \mathbb{R}[x]$, and write

$$f_k(\sqrt{-1}\,x) = p_k(x) + \sqrt{-1}q_k(x), \quad p_k(x), q_k(x) \in \mathbb{R}[x], \ p_k(x) \text{ even and } q_k(x) \text{ odd.}$$

(2) Replacing x with $\sqrt{-1}\,x$ in the equation for -1 above and comparing real parts, obtain

$$-1 = \sum_{k=1}^{n-1}(p_k(x)^2 - q_k(x)^2) + p_0(x)(1 - x_1^2 - \cdots - x_n^2).$$

(3) Show that the mapping $q = (q_1, \dots, q_{n-1}) : \overline{D}^n \to \mathbb{R}^{n-1}$ collapses two antipodal points in $\partial\overline{D}^n = \mathbb{S}^{n-1}$, and being odd, it has a zero $a \in \mathbb{S}^{n-1}$.

(4) Substituting $x = a$ in the equation for -1 in (2), get a contradiction.

Number 8. Invariance of Domain holds for manifolds, but no further:

(1) Prove that a locally injective continuous mapping $f : M \to N$ of boundaryless manifolds of the same dimension is open.

(2) Consider the *lemniscate* $X \subset \mathbb{R}^2$ with polynomial equation $(x^2 + y^2)^2 = xy$. Define a continuous bijection $f : \mathbb{R} \to X$ which is not a homeomorphism.

6. The Multiplication Formula

In this section we prove the general formula for the computation of the degree of a composite mapping. This is much more delicate than the corresponding result for the Brouwer-Kronecker degree, but always elementary. Here is the statement of the formula:

Proposition 6.1 (Multiplication Formula). *Let $D, E \subset \mathbb{R}^n$ be bounded open sets with boundaries $X = \overline{D} \setminus D$ and $Y = \overline{E} \setminus E$. Let $f : \overline{D} \to \mathbb{R}^n$ and $g : \overline{E} \to \mathbb{R}^n$ be continuous mappings such that $f(\overline{D}) \subset E$. The open set $E \setminus f(X)$ decomposes into countably many connected components E_ℓ, and we arbitrarily choose one point $c_\ell \in E_\ell$ in each. Now, let $a \in \mathbb{R}^n \setminus g(Y \cup f(X))$. Then*

$$d(g \circ f, D, a) = \sum_\ell d(g, E_\ell, a) \cdot d(f, D, c_\ell).$$

Proof. We split the argument into several steps.

Step I: The formula is well defined. First of all, the degree $d(g \circ f, D, a)$ exists, because $\mathbb{R}^n \setminus g(f(X)) \supset \mathbb{R}^n \setminus g(Y \cup f(X))$. Next, since E_ℓ is a connected subset of $E \setminus f(X) \subset \mathbb{R}^n \setminus f(X)$, the degree $d(f, D, \cdot)$ is constant on it (IV.2.3, p. 146). For the other factor $d(g, E_\ell, a)$ we must see that $a \notin g(\overline{E}_\ell \setminus E_\ell)$, but we have

$$\overline{E}_\ell \setminus E_\ell \subset (\overline{E} \setminus E) \cup f(X) = Y \cup f(X)$$

and $a \notin g(Y \cup f(X))$. Finally, the sum $\sum_\ell$ is finite. Indeed, since $g^{-1}(a) \subset E \setminus f(X)$ is a compact set, it is contained in the union of finitely many E_ℓ's, which are disjoint; hence $g^{-1}(a) \cap E_\ell = \emptyset$ for, say, $\ell \geq \ell_0$. Thus, by IV.2.5(2), p. 149, $d(g, E_\ell, a) = 0$ for $\ell \geq \ell_0$.

Step II: Approximation data. By the Weierstrass Approximation Theorem, there is a polynomial mapping $\hat{f}$ such that on $\overline{D}$

$$\|f - \hat{f}\| < \begin{cases} \operatorname{dist}(f(\overline{D}), \mathbb{R}^n \setminus E) & \text{and} \\ \operatorname{dist}(g^{-1}(a), f(X)). \end{cases}$$

From the first bound we deduce that $\hat{f}(\overline{D}) \subset E$, and the composite mapping $g \circ \hat{f}$ does exist. From the second bound, we see that $a \notin g(\hat{f}(X))$, hence $a \notin \mathbb{R}^n \setminus g(Y \cup \hat{f}(X))$, and $\varepsilon = \operatorname{dist}(a, g(Y \cup \hat{f}(X))) > 0$. Again by the Weierstrass Approximation Theorem, there is a polynomial mapping g_1 such that $\|g - g_1\| < \frac{1}{2}\varepsilon$ on $\overline{E}$, and by the Sard-Brown Theorem, there is a regular value b of the mapping $g_1 \circ \hat{f}|_D$ with $\|a - b\| < \frac{1}{2}\varepsilon$. We consider now the polynomial mapping $\hat{g} = g_1 + a - b$. We have:

(i) On $\overline{E}$,

$$\|g - \hat{g}\| \leq \|g - g_1\| + \|g_1 - \hat{g}\| < \tfrac{1}{2}\varepsilon + \tfrac{1}{2}\varepsilon = \varepsilon;$$

hence $a \in \mathbb{R}^n \setminus \hat{g}(Y \cup \hat{f}(X))$.

(ii) By translation, a is a regular value of $\hat{g} \circ \hat{f}|_D$.

Step III: Reduction to the approximation data. We claim that

$$d(g \circ f, D, a) = d(g \circ \hat{f}, D, a) = d(\hat{g} \circ \hat{f}, D, a).$$

For the first equality, we consider the homotopy

$$\Phi_t(x) = (1-t)f(x) + t\hat{f}(x), \quad x \in \overline{D}.$$

We have

$$\|f(x) - \Phi_t(x)\| = t\|f(x) - \hat{f}(x)\| < \operatorname{dist}(f(\overline{D}), \mathbb{R}^n \setminus E);$$

hence $\Phi_t(x) \in E$, and the homotopy $g \circ \Phi_t$ is well defined. Furthermore, $a \notin g(\Phi([0,1] \times X))$. Indeed, suppose there is an $x \in X$ such that $\Phi_t(x) \in g^{-1}(a)$. Then

$$\|\Phi_t(x) - f(x)\| = t\|\hat{f}(x) - f(x)\| < \operatorname{dist}(g^{-1}(a), f(X)),$$

which is impossible. Thus, we can use homotopy invariance (IV.2.4, p. 147) and

$$d(g \circ f, D, a) = d(g \circ \Phi_0, D, a) = d(g \circ \Phi_1, D, a) = d(g \circ \hat{f}, D, a).$$

For the second equality of our claim we use the homotopy

$$\Psi_t(x) = (1-t)g(\hat{f}(x)) + t\hat{g}(\hat{f}(x)), \quad x \in \overline{D}.$$

Again, the key fact is that $a \notin \Psi([0,1] \times X)$. But if $x \in X$, we have

$$\begin{aligned} \|a - \Psi_t(x)\| &= \|(a - g\hat{f}(x)) + t(g\hat{f}(x) - \hat{g}\hat{f}(x))\| \\ &\geq \|a - g\hat{f}(x)\| - t\|g\hat{f}(x) - \hat{g}\hat{f}(x)\| \\ &\geq \operatorname{dist}(a, g(\hat{f}(X))) - \|g - \hat{g}\| > \operatorname{dist}(a, g(\hat{f}(X))) - \varepsilon \geq 0, \end{aligned}$$

by (i) in the preceding step. Thus, we can indeed apply homotopy invariance.

Step IV: Multiplication. By IV.2.1, p. 145, we know that

$$\begin{aligned} d(g \circ f, D, a) &= \sum_{x \in (\hat{g}\circ\hat{f})^{-1}(a)} \operatorname{sign}_x(\hat{g} \circ \hat{f}) = \sum_{x \in (\hat{g}\circ\hat{f})^{-1}(a)} \operatorname{sign}_{\hat{f}(x)}(\hat{g}) \operatorname{sign}_x(\hat{f}) \\ &= \sum_{y \in \hat{g}^{-1}(a)} \sum_{x \in \hat{f}^{-1}(y)} \operatorname{sign}_y(\hat{g}) \cdot \operatorname{sign}_x(\hat{f}) \\ &= \sum_{y \in \hat{g}^{-1}(a)} \operatorname{sign}_y(\hat{g}) \Big(\sum_{x \in \hat{f}^{-1}(y)} \operatorname{sign}_x(\hat{f}) \Big) = \sum_{y \in \hat{g}^{-1}(a)} \operatorname{sign}_y(\hat{g}) \cdot d(\hat{f}, D, y). \end{aligned}$$

Now, for every k the set

$$\widehat{W_k} = \{y \in E \setminus \hat{f}(X) : d(\hat{f}, D, y) = k\}$$

is a union of connected components of $E \setminus \hat{f}(X)$, whose boundaries are contained in $Y \cup \hat{f}(X)$ (as for the E_ℓ's). Hence, by (ii) in Step II, $a \notin \hat{g}\big(\overline{\widehat{W_k}} \setminus \widehat{W_k}\big)$, and we can continue the above sequence of equalities as follows:

$$\begin{aligned} d(g \circ f, D, a) &= \cdots \\ &= \sum_{y \in \hat{g}^{-1}(a)} \operatorname{sign}_y(\hat{g}) \cdot d(\hat{f}, D, y) \\ &= \sum_k k\Big(\sum_{y \in \hat{g}^{-1}(a) \cap \widehat{W_k}} \operatorname{sign}_y(\hat{g})\Big) = \sum_k k \cdot d(\hat{g}, \widehat{W_k}, a). \end{aligned}$$

Step V: Second multiplication. We repeat the same computation for f. We set

$$W_k = \{y \in E \setminus f(X) : d(f, D, y) = k\} = \bigcup_i E_{\ell_i},$$

where the E_{ℓ_i}'s are the connected components of $E \setminus f(X)$ on which $d(f, D, \cdot) \equiv k$. Since $\overline{W}_k \setminus W_k \subset Y \cup f(X)$, we have $a \notin g(\overline{W}_k \setminus W_k)$, and by additivity (IV.2.5(3), p. 149) we get

$$d(g, W_k, a) = \sum_i d(g, E_{\ell_i}, a).$$

Indeed, as remarked in Step I, the sum is in fact finite:

$$\sum_i d(g, E_{\ell_i}, a) = \sum_{i:\, \ell_i < \ell_0} d(g, E_{\ell_i}, a)$$

because $g^{-1}(a) \cap E_\ell = \emptyset$ for $\ell \geq \ell_0$; moreover, in this situation

$$a \notin g(\overline{W}_k \setminus W_k) = g\Big(\overline{W}_k \setminus \bigcup_{i:\, \ell_i < \ell_0} E_{\ell_i}\Big)$$

and the additivity property does apply.

Notice now that the E_{ℓ_i}'s in this sum are the E_ℓ's contained in W_k. We conclude that

$$\begin{aligned} \sum_\ell d(g, E_\ell, a) \cdot d(f, D, c_\ell) &= \sum_k k \cdot \sum_{\ell:\, E_\ell \subset W_k} d(g, E_\ell, a) \\ &= \sum_k k \cdot d(g, W_k, a). \end{aligned}$$

Step VI: Simplification of mappings. Comparing the formulas in the preceding two steps, we are reduced to showing that

$$d(g, W_k, a) = d(\hat{g}, \widehat{W_k}, a)$$

for every k. But by (i) in Step II,

$$\|g - \hat{g}\| < \varepsilon = \text{dist}\,\big(a, g(Y \cup \hat{f}(X))\big) \leq \text{dist}\,\big(a, g\big(\overline{\widehat{W_k}} \setminus \widehat{W_k}\big)\big).$$

Hence

$$d(\hat{g}, \widehat{W_k}, a) = d(g, \widehat{W_k}, a)$$

and we must show the equality of the following two degrees:

$$d(g, W_k, a) = d(g, \widehat{W_k}, a).$$

Step VII: Computation of the left-hand side degree. We claim that

$$a \notin g(\overline{W}_k \setminus \widehat{W_k}).$$

Suppose $a = g(z)$ with $z \in \overline{W}_k \setminus \widehat{W_k}$. Since $a \notin g(Y \cup f(X))$ and $\overline{W}_k \setminus W_k \subset Y \cup f(X)$, it must be true that $z \in W_k$, so that $d(f, D, z) = k$. Moreover,

$$\|f - \hat{f}\| < \text{dist}(g^{-1}(a), f(X)) \leq \text{dist}(z, f(X))$$

and we get $d(\hat{f}, D, z) = d(f, D, z) = k$, that is, $z \in \widehat{W_k}$, a contradiction.

Thus the claim is proved, and, by additivity (IV.2.5(3), p. 149),

$$d(g, W_k, a) = d(g, \widehat{W_k} \cap W_k, a).$$

Step VIII: Computation of the right-hand side degree and conclusion. The argument is analogous. First, $a \notin g\big(\overline{\widehat{W_k}} \setminus W_k\big)$, because $a \notin g(Y \cup \hat{f}(X))$ and $\overline{\widehat{W_k}} \setminus \widehat{W_k} \subset Y \cup \hat{f}(X)$. Then, additivity gives

$$d(g, \widehat{W_k}, a) = d(g, \widehat{W_k} \cap W_k, a).$$

Thus, we have the equality of degrees stated at the end of Step VI, and the proof of the Multiplication Formula is finished. □

Exercises and problems

Number 1. Find two continuous functions $f : \overline{D} \to \mathbb{R}^n$ and $g : \overline{E} \to \mathbb{R}^n$ verifying the hypothesis of the Multiplication Formula (IV.6.1, p. 172), with D and E connected, such that the sum in that formula has at least two different summands.

Number 2. Let D (resp., E) be a bounded open subset of $\mathbb{R}^n$, and let $f : \overline{D} \to \mathbb{R}^n$ (resp., $g : \overline{E} \to \mathbb{R}^n$) be a continuous function. Suppose E is connected and that $f(\overline{D}) \subset \overline{E}$, $f(\partial D) \subset \partial E$. Prove that

$$d(g \circ f, D, a) = d(g, E, a) \cdot d(f, D, b)$$

for $a \in \mathbb{R}^n \setminus g(\partial E)$ and $b \in f(D)$.

Number 3. Let D (resp., E) be a bounded open subset of $\mathbb{R}^n$, and let $f : \overline{D} \to \mathbb{R}^n$ (resp., $g : \overline{E} \to \mathbb{R}^n$) be a continuous function. Suppose D is connected and that $f(\partial D) \cap f(D) = \emptyset$ and $f(\overline{D}) \subset \overline{E}$. Prove that

$$d(g \circ f, D, a) = d(g, E, a) \cdot d(f, D, b)$$

for $a \in \mathbb{R}^n \setminus g(\partial E \cup f(\partial D))$ such that $g^{-1}(a) \subset f(D)$ and $b \in E$. Show that the condition $g^{-1}(a) \subset f(D)$ is necessary for this conclusion.

Number 4. For $\rho > 0$ we denote by $D^k_\rho \subset \mathbb{R}^k$ the open ball of radius ρ centered at the origin and by $\overline{D}^k_\rho$ and S^{k-1}_ρ the corresponding closed ball and sphere; if $\rho = 1$, we just get D^k, $\overline{D}^k$, and $\mathbb{S}^{k-1}$. Fix $n \geq 1$, and set $E_\rho = (-\rho, \rho) \times D^n_\rho$.

(1) Compute the degree of the mapping

$$\varphi_\rho : \overline{E}_\rho \to \overline{D}^{n+1}_\rho : (t, x) \mapsto \big(t, \tfrac{1}{\rho}\sqrt{\rho^2 - t^2}\, x\big).$$

(2) Show that for every continuous mapping $f : \overline{D}^n \to \mathbb{R}^n$ such that $f(\mathbb{S}^{n-1}) \subset \mathbb{S}^{n-1}$ and $(\mathrm{Id}_{[-1,1]} \times f)(\overline{E}_1) \subset E_r$ with $r > 1$,

$$d(f, D^n, 0) = d(\varphi_r \circ (\mathrm{Id}_{[-1,1]} \times f), E_1, 0).$$

(3) Prove that for every continuous mapping $g : \mathbb{S}^{n-1} \to \mathbb{S}^{n-1}$ there is a unique continuous mapping $\Sigma(g) : \mathbb{S}^n \to \mathbb{S}^n$ (the *suspension* of g) such that

$$\varphi_1 \circ (\mathrm{Id}_{[-1,1]} \times g) = \Sigma(g) \circ \varphi_1,$$

and prove that $\deg(\Sigma(g)) = \deg(g)$.

Use the above construction to produce by induction on k a mapping $\mathbb{S}^k \to \mathbb{S}^k$ of any given degree d.

7. The Jordan Separation Theorem

The core of this section is to prove the following purely topological statement:

Theorem 7.1 (Jordan Separation Theorem). *Let $K, L \subset \mathbb{R}^n$, $n \geq 2$, be two compact sets. If K and L are homeomorphic, then their complements $\mathbb{R}^n \setminus K$ and $\mathbb{R}^n \setminus L$ have the same number $N \leq \infty$ of connected components.*

Proof. Note first that both complements $\mathbb{R}^n \setminus K$ and $\mathbb{R}^n \setminus L$ are open sets with exactly one unbounded connected component each, which we denote

Δ_0 and Γ_0, respectively. In fact, we consider the decompositions into connected components

$$\begin{cases} \mathbb{R}^n \setminus K = \bigcup_{i\geq 0} \Delta_i, & \alpha = \text{number of bounded } \Delta_i\text{'s} \leq \infty, \\ \mathbb{R}^n \setminus L = \bigcup_{j\geq 0} \Gamma_j, & \beta = \text{number of bounded } \Gamma_j\text{'s} \leq \infty. \end{cases}$$

Pick points $a_i \in \Delta_i$ $(i \geq 1)$ and $b_j \in \Gamma_j$ $(j \geq 1)$. Also, we consider as usual the boundaries

$$X_i = \overline{\Delta}_i \setminus \Delta_i \subset K \quad \text{and} \quad Y_j = \overline{\Gamma}_j \setminus \Gamma_j \subset L.$$

By hypothesis, we have a homeomorphism $h : K \to L$, and by the Tietze Extension Theorem we find continuous mappings $\varphi, \psi : \mathbb{R}^n \to \mathbb{R}^n$ such that $\varphi|_K = h$ and $\psi|_L = h^{-1}$ and both φ and ψ are the identity off an open ball containing K and L. In this situation, we easily compute the degrees

$$d(\varphi \circ \psi, \Gamma_i, b_j) \quad \text{and} \quad d(\psi \circ \varphi, \Delta_i, a_j).$$

Indeed, by construction

$$\begin{cases} b_j \notin L = \varphi(\psi(L)), \\ L \supset Y_i = \overline{\Gamma}_i \setminus \Gamma_i \end{cases} \quad \text{and} \quad \begin{cases} a_j \notin K = \psi(\varphi(K)), \\ K \supset X_i = \overline{\Delta}_i \setminus \Delta_i\,. \end{cases}$$

Finally, since $\varphi \circ \psi|_{Y_i} = \mathrm{Id}\,|_{Y_i}$ and $\psi \circ \varphi|_{X_i} = \mathrm{Id}\,|_{X_i}$, by the Boundary Theorem, IV.2.6, p. 150,

$$d(\varphi \circ \psi, \Gamma_i, b_j) = d(\mathrm{Id}, \Gamma_i, b_j) = \left\{ \begin{matrix} 1 & \text{if } b_j \in \Gamma_i, \text{ that is, if } i = j \\ 0 & \text{otherwise} \end{matrix} \right\} = \delta_{ij}$$

and

$$d(\psi \circ \varphi, \Delta_i, a_j) = d(\mathrm{Id}, \Delta_i, a_j) = \left\{ \begin{matrix} 1 & \text{if } a_j \in \Delta_i, \text{ that is, if } i = j \\ 0 & \text{otherwise} \end{matrix} \right\} = \delta_{ij}.$$

Next, we compute $d(\varphi \circ \psi, \Gamma_i, b_j)$ by using the Multiplication Formula.

Step I: Multiplication setting. We decompose the complement of the compact set $\psi(Y_i)$ into connected components, the first one unbounded:

$$\mathbb{R}^n \setminus \psi(Y_i) = D_0 \cup D_1 \cup D_2 \ \cdots .$$

Now, this complement contains

$$\mathbb{R}^n \setminus K = \Delta_0 \cup \Delta_1 \cup \Delta_2 \ \cdots ;$$

hence each connected component of the last union is contained in a unique D_ℓ, and we can write

$$\begin{cases} D_0 \supset U_0 = \Delta_0 \cup \Delta_{k_1(0)} \cup \Delta_{k_2(0)} \cdots & \text{and } \overline{D}_0 \setminus U_0 \subset K, \\ D_1 \supset U_1 = \qquad \Delta_{k_1(1)} \cup \Delta_{k_2(1)} \cdots & \text{and } \overline{D}_1 \setminus U_1 \subset K, \\ \quad \cdots & \qquad \cdots \end{cases}$$

Next we choose a big enough open ball $E \subset \mathbb{R}^n$ to have the following conditions:

(i) $b_j \notin \varphi(\overline{E} \setminus E)$ (recall that φ is the identity off an open ball).

(ii) $E \supset \psi(\overline{\Gamma}_i) \cup (\mathbb{R}^n \setminus D_0)$; hence $E \supset \bigcup_{\ell \geq 1} D_\ell$.

(iii) $E_0 = E \cap D_0$ is connected, and $E_0 \not\subset \psi(\overline{\Gamma}_i)$.

(iv) $E \setminus \psi(Y_i) = E_0 \cup E_1 \cup E_2 \ \cdots$, where $E_\ell = D_\ell$ for $\ell \geq 1$.

In this situation we can apply the Multiplication Formula, IV.6.1, p. 172, to the mappings $\psi|_{\overline{\Gamma}_i} : \overline{\Gamma}_i \to \mathbb{R}^n$ and $\varphi|_{\overline{E}} : \overline{E} \to \mathbb{R}^n$ to compute $d(\varphi \circ \psi, \Gamma_i, b_j)$.

Step II: Computation by multiplication. Using the Multiplication Formula, we obtain

$$d(\varphi \circ \psi, \Gamma_i, b_j) = \sum_{\ell \geq 0} d(\varphi, E_\ell, b_j) d(\psi, \Gamma_i, c_\ell),$$

for any chosen $c_\ell \in E_\ell$. Here we can immediately discard the $\ell = 0$ term by picking $c_0 \in E_0 \setminus \psi(\overline{\Gamma}_i)$ (condition (iii) above). Thus,

$$d(\varphi \circ \psi, \Gamma_i, b_j) = \sum_{\ell \geq 1} d(\varphi, E_\ell, b_j) d(\psi, \Gamma_i, c_\ell),$$

and $c_\ell \in D_\ell = E_\ell$.

Step III: Additional expansion. Here we further decompose some terms of the last sum. In fact, we are to show that

$$d(\varphi, E_\ell, b_j) = \sum_r d(\varphi, \Delta_{k_r(\ell)}, b_j).$$

First, this is a well-defined finite sum, for, since

$$b_j \in \Gamma_j \subset \mathbb{R}^n \setminus L = \mathbb{R}^n \setminus \varphi(K) \subset \mathbb{R}^n \setminus \varphi(X_{k_r(\ell)}),$$

every degree $d(\varphi, \Delta_{k_r(\ell)}, b_j)$ exists. Furthermore it is zero for r big enough. Indeed, since $b_j \notin L = \varphi(K)$, $\varphi^{-1}(b_j) \cap K = \emptyset$; hence

$$\varphi^{-1}(b_j) \setminus \Delta_0 \subset \bigcup_{k \geq 1} \Delta_k.$$

But $\varphi^{-1}(b_j) \setminus \Delta_0$ is compact (closed and bounded), so that for r large, say $r > r_0$, $\varphi^{-1}(b_j) \cap \Delta_{k_r(\ell)} = \emptyset$; hence $d(\varphi, \Delta_{k_r(\ell)}, b_j) = 0$, and

$$\sum_r d(\varphi, \Delta_{k_r(\ell)}, b_j) = \sum_{r=1}^{r_0} d(\varphi, \Delta_{k_r(\ell)}, b_j).$$

Now, by construction we know that

$$b_j \notin \varphi\Big(\overline{D}_\ell \setminus \bigcup_r \Delta_{k_r(\ell)}\Big),$$

and since $\varphi^{-1}(b_j)$ does not meet $\Delta_{k_r(\ell)}$ for any $r > r_0$, we get

$$b_j \notin \varphi\Big(\overline{D}_\ell \setminus \bigcup_{r=1}^{r_0} \Delta_{k_r(\ell)}\Big).$$

Thus we can apply additivity (IV.2.5(3), p. 149), to see that the degree $d(\varphi, E_\ell, b_j)$ is the finite sum above.

Step IV: End of the computation. Putting Steps II and III together, we have

$$\begin{aligned} d(\varphi \circ \psi, \Gamma_i, b_j) &= \sum_{\ell \geq 1} \Big(\sum_r d(\varphi, \Delta_{k_r(\ell)}, b_j)\Big) d(\psi, \Gamma_i, c_\ell) \\ &= \sum_{\ell, r \geq 1} d(\varphi, \Delta_{k_r(\ell)}, b_j) \cdot d(\psi, \Gamma_i, a_{k_r(\ell)}). \end{aligned}$$

The last equality follows by rearranging the sums and taking $c_\ell = a_{k_r(\ell)} \in \Delta_{k_r(\ell)} \subset D_\ell$ for every ℓ, i. But the last sum can also be rewritten as

$$\begin{aligned} \cdots &= \sum_{\Delta_k \not\subset D_0} d(\varphi, \Delta_k, b_j) \cdot d(\psi, \Gamma_i, a_k) \\ &= \sum_{k \geq 1} d(\varphi, \Delta_k, b_j) \cdot d(\psi, \Gamma_i, a_k). \end{aligned}$$

This is because if $\Delta_k \subset D_0$, then

$$d(\psi, \Gamma_i, a_k) = d(\psi, \Gamma_i, a'_k)$$

for any $a'_k \in D_0$ (IV.2.3, p. 146). But we can find a'_k far from $\psi(\Gamma_i)$ so that the equation $\psi(x) = a'_k$ has no solution in Γ_i and the last degree is zero (IV.2.5(2), p. 149).

With this, we have finished the computation of $d(\varphi \circ \psi, \Gamma_i, b_j)$.

All that has been done so far for $\varphi\circ\psi$ can be done analogously for $\psi\circ\varphi$, and one ends up with the following formulas:

$$(*)\qquad \sum_{k\geq 1} d(\varphi,\Delta_k,b_j)\cdot d(\psi,\Gamma_i,a_k) = d(\varphi\circ\psi,\Gamma_i,b_j) = \delta_{ij},$$

$$(**)\qquad \sum_{\ell\geq 1} d(\psi,\Gamma_\ell,a_j)\cdot d(\varphi,\Delta_i,b_\ell) = d(\psi\circ\varphi,\Delta_i,a_j) = \delta_{ij}.$$

With these formulas, we can finally prove $\alpha=\beta$.

To that end, we define two linear mappings. Let $\mathbb{R}^{(\gamma)}$ stand for the direct sum of $\gamma\leq\infty$ copies of $\mathbb{R}$, and let $\{e_i\}$ stand for its canonical basis. Then

$$\begin{cases} A:\mathbb{R}^{(\alpha)}\to\mathbb{R}^{(\beta)}: e_k\mapsto (d(\psi,\Gamma_i,a_k): i\geq 1) = (A_{ik}: i\geq 1),\\ B:\mathbb{R}^{(\beta)}\to\mathbb{R}^{(\alpha)}: e_\ell\mapsto (d(\varphi,\Delta_i,b_\ell): i\geq 1) = (B_{i\ell}: i\geq 1).\end{cases}$$

Now, we can rewrite the formulas above as

$$(*)\quad \sum_{k\geq 1} B_{kj}\cdot A_{ik} = \delta_{ij}\,,\qquad\qquad (**)\quad \sum_{\ell\geq 1} A_{\ell j}\cdot B_{i\ell} = \delta_{ij}$$

and we see that A and B are inverse to each other; hence both linear spaces have the same dimension, and $\alpha=\beta$, as desired. □

Examples 7.2. (1) A compact set $K\subset\mathbb{R}^n$ that is homeomorphic to a differentiable hypersurface disconnects $\mathbb{R}^n$.

(2) A compact set $K\subset\mathbb{R}^n$ that is homeomorphic to a *connected* differentiable hypersurface disconnects $\mathbb{R}^n$ into two open connected components, one bounded, whose boundaries are both K. No proper subset $T\subset K$ disconnects $\mathbb{R}^n$.

(3) An arc (that is, a compact set homeomorphic to an interval) never disconnects $\mathbb{R}^n$ for $n\geq 2$.

(4) A topological closed ball never disconnects $\mathbb{R}^n$ for $n\geq 2$.

(5) A compact set $K\subset\mathbb{R}^n$ homeomorphic to a proper subset of $\mathbb{S}^{n-1}$ does not disconnect $\mathbb{R}^n$.

(6) A sphere is not homeomorphic to any of its proper subsets (otherwise we would have two homeomorphic compact subsets of a Euclidean space, one disconnecting it and the other not). □

The Jordan Separation Theorem, IV.7.1, p. 176, also holds when we replace Euclidean spaces by spheres:

Proposition 7.3. *Let $K, L \subset \mathbb{S}^n$, $n \geq 2$, be two compact sets. If K and L are homeomorphic, then their complements $\mathbb{S}^n \setminus K$ and $\mathbb{S}^n \setminus L$ have the same number $N \leq \infty$ of connected components.*

Proof. If one of the two compact sets is the sphere, then the other is the sphere also, by IV.7.2(6) above. Hence we can pick two points $a \in \mathbb{S}^n \setminus K$ and $b \in \mathbb{S}^n \setminus L$ and consider the stereographic projections from them, denoted by $\mathtt{x} : \mathbb{S}^n \setminus \{a\} \to \mathbb{R}^n$ and $\mathtt{y} : \mathbb{S}^n \setminus \{b\} \to \mathbb{R}^n$, which are diffeomorphisms. Then $\mathtt{x}(K)$ and $\mathtt{y}(L)$ are compact homeomorphic subsets of $\mathbb{R}^n$; hence they disconnect it in the same number of connected components. Pulling them back to $\mathbb{S}^n$ via $\mathtt{x}$ and $\mathtt{y}$, we see that $\mathbb{S}^n \setminus K$ and $\mathbb{S}^n \setminus L$ have the same number of connected components too. Indeed, it suffices to remark that the unbounded component U of $\mathbb{R}^n \setminus \mathtt{x}(K)$ (resp., V of $\mathbb{R}^n \setminus \mathtt{y}(L)$) corresponds to the connected component $U' = \{a\} \cup \mathtt{x}^{-1}(U)$ of $\mathbb{S}^n \setminus K$ that contains a (resp., $V' = \{b\} \cup \mathtt{y}^{-1}(V)$ of $\mathbb{S}^n \setminus L$ that contains b). □

We can finally generalize IV.7.1, p. 176, to closed sets:

Proposition 7.4. *Let $C, C' \subset \mathbb{R}^n$, $n \geq 2$, be two closed sets. If C and C' are homeomorphic, then their complements $\mathbb{R}^n \setminus C$ and $\mathbb{R}^n \setminus C'$ have the same number $N \leq \infty$ of connected components.*

Proof. Let $j : C \subset \mathbb{R}^n$ be the canonical inclusion. Since j is a proper mapping, it extends to the one-point compactifications C^* of C and $\mathbb{S}^n$ of $\mathbb{R}^n$, by mapping $\infty \mapsto \infty$. Similarly, $j' : C' \subset \mathbb{R}^n$ extends to $C'^* \to \mathbb{S}^n$. Of course, C^* and C'^* are homeomorphic as C and C' are. Thus, we have two homeomorphic compact subsets C^* and C'^* of the sphere $\mathbb{S}^n$, and by the preceding proposition, they disconnect $\mathbb{S}^n$ in the same number of connected components. Note here that none of those components contains ∞; hence they pull back to $\mathbb{R}^n$ without surprise, and the result is proved. □

For instance, we can improve the statement of IV.7.2(1), p. 180, to the following: *a closed set $C \subset \mathbb{R}^n$ that is homeomorphic to a closed differentiable hypersurface disconnects $\mathbb{R}^n$.*

We close this section with a new proof of the Invariance of Domain Theorem:

Proposition 7.5. *Let $W \subset \mathbb{R}^n$ be an open set. Every locally injective continuous mapping $f : W \to \mathbb{R}^n$ is open.*

Proof. Pick any $a \in W$. Consider an open ball B centered at a of radius $\varepsilon > 0$ and the compact sets

$$K = \{\|x - a\| \le \varepsilon\} \quad \text{and} \quad L = \{\|x - a\| = \varepsilon\};$$

by hypothesis, for ε small enough, f is defined and injective on K. Then, the restrictions $f|_K$ and $f|_L$ are injective closed mappings (by compactness of their domains of definition), hence homeomorphisms onto their images $f(K)$ and $f(L)$. Consequently, $\mathbb{R}^n \setminus f(K)$ is connected (and unbounded) and $\mathbb{R}^n \setminus f(L)$ splits into two connected components D_0, D_1, with D_1 bounded. It follows that $\mathbb{R}^n \setminus f(K) \subset D_0$. Hence

$$D_1 \subset f(K) = f(B) \cup f(L),$$

and we conclude that $D_1 \subset f(B)$. But $f(B)$ is a connected subset of $\mathbb{R}^n \setminus f(L)$, so it cannot be bigger than D_1. Thus $f(B) = D_1$ is an open set. This shows that f maps all small enough open balls centered at a onto open sets. Since a is arbitrary, f is open. □

Exercises and problems

Number 1. Prove the following general facts for a compact set K in $\mathbb{R}^n$.

(1) The complement $\mathbb{R}^n \setminus K$ has exactly one unbounded connected component.

(2) The set K contains the boundaries of all connected components of $\mathbb{R}^n \setminus K$.

(3) If $\mathbb{R}^n \setminus K$ is not connected, but $\mathbb{R}^n \setminus A$ is connected for every closed subset $A \subsetneq K$, then K is the boundary of all connected components of $\mathbb{R}^n \setminus K$.

Number 2. Exhibit two homeomorphic compact subsets of $\mathbb{R}^3$ whose complements are not homeomorphic.

Number 3. Let C and C' be closed sets of $\mathbb{R}^p$ and $\mathbb{R}^q$, respectively. Show that if C and C' are homeomorphic, then $\mathbb{R}^p \times \mathbb{R}^q \setminus C \times \{0\}$ and $\mathbb{R}^p \times \mathbb{R}^q \setminus \{0\} \times C'$ are homeomorphic.

Number 4. Let $\gamma, \gamma' : [0,1] \to \mathbb{D}^2$ be two injective continuous mappings with

$$\gamma(0) = (1,0), \ \gamma(1) = (-1,0); \quad \gamma'(0) = (0,1), \ \gamma'(1) = (0,-1).$$

Prove that they meet at some point.

Number 5. Let C be a closed subset of the closed disk $\overline{D}^2$. If C and $\overline{D}^2 \setminus C$ are both connected, then $C \cap \partial\overline{D}^2$ and $\partial\overline{D}^2 \setminus C$ are both connected.

Number 6. Let M be a boundaryless connected smooth manifold, $\dim(M) \ge 2$.

(1) Find some smooth hypersurface $H \subset M$, diffeomorphic to a sphere, such that $M \setminus H$ has two connected components.

(2) Show that there may be two hypersurfaces $H_1, H_2 \subset M$, both diffeomorphic to a sphere, such that $M \setminus H_1$ has two connected components and $M \setminus H_2$ is connected.

Chapter V

The Hopf Theorems

In this chapter we formulate and prove the theorems that fully complete degree theory: the degree is the only homotopy invariant for spheres. This must be formulated for both the Brouwer-Kronecker and the Euclidean degrees, although in essence the two cases are the same. These are the *Hopf Theorems.* As these theorems refer to mappings into spheres, we start the chapter by constructing in §1 various basic examples of them. Then in §2 we prove the first Hopf Theorem, that *two mappings into a sphere of the same (Brouwer-Kronecker) degree are homotopic;* hence, in particular, the cohomotopy group $\pi^m(M)$ of any m-manifold and the homotopy group $\pi_m(\mathbb{S}^m)$ of the m-sphere are $\mathbb{Z}$. In §3 we deduce the same result for the Euclidean degree, which has the virtue of determining the cohomotopy groups $\pi^m(X)$ for some more exotic spaces X of dimension m, like the topological circle. These Hopf Theorems require the use of diffeotopies. Next, in §4 we come back to the *Hopf invariant* and define the *Hopf fibrations* to study some homotopy groups $\pi_k(\mathbb{S}^m)$ for $k > m$. Then we turn to the theory of tangent vector fields, and in §5 we discuss the essential notion of *index of a tangent vector field at an isolated zero,* using both the Brouwer-Kronecker and the Euclidean degrees. This notion is best illustrated by the *gradients of Morse functions,* which we describe in §6. Finally, we conclude the chapter and the book by presenting in §7 another deep theorem also named after Hopf, the *Poincaré-Hopf Index Theorem,* that computes the index of a tangent vector field. We also see how the *Gauss-Bonnet Formula* follows from this.

1. Mappings into spheres

In this section we discuss several methods for constructing smooth mappings into spheres, or to modify one already given.

To start with, we exhibit simple examples of mappings of arbitrary degree.

Example 1.1. *Let $\mathbb{S}^m \subset \mathbb{R}^{m+1}$ be the standard unit sphere. Then, for every integer d there are mappings $f_d : \mathbb{S}^m \to \mathbb{S}^m$ of degree d.*

Since all constant mappings have degree 0, we assume $d \neq 0$.

Case $d \geq 1$. We will generalize to dimension m the intuitive idea of winding a circle k times around another (III.1.6(3), p. 100). Any point $x \in \mathbb{S}^m$ can be parametrized as

$$x = (x_1, x_2, x') = (\rho\cos\theta, \rho\sin\theta, x') \quad \text{where} \quad \rho = \sqrt{1 - \|x'\|^2}.$$

Then, we define the winding by

$$g_d(x) = (\rho\cos(d\cdot\theta), \rho\sin(d\cdot\theta), x').$$

To eliminate the parameter θ, recall that

$$\cos(d\cdot\theta) = P_d(\cos\theta, \sin\theta), \quad \sin(d\cdot\theta) = Q_d(\cos\theta, \sin\theta),$$

for suitable homogeneous polynomials P_d and Q_d of degree d. We have

$$\begin{aligned} g_d(x) &= \big(\rho P_d\big(\tfrac{1}{\rho}x_1, \tfrac{1}{\rho}x_2\big), \rho Q_d\big(\tfrac{1}{\rho}x_1, \tfrac{1}{\rho}x_2\big), x'\big) \\ &= \big(\tfrac{1}{\rho^{d-1}}P_d(x_1, x_2), \tfrac{1}{\rho^{d-1}}Q_d(x_1, x_2), x'\big). \end{aligned}$$

This shows that the map g_d is differentiable except when $\|x'\| = 1$. To amend this, we use the deformation $\widetilde{\rho}$ of ρ defined by

$$\widetilde{\rho} = \sqrt{1 - \lambda(\|x'\|^2)},$$

for a smooth function $\lambda(t)$ which is $= t$ for $t \leq \frac{1}{2}$ and $< t$ for $t > \frac{1}{2}$, as in the figure below:

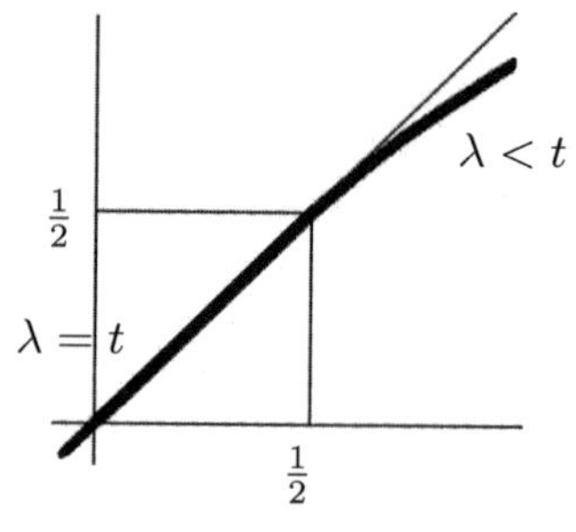

Thus, we have the smooth mapping

$$\widetilde{g}_d(x) = \Big(\frac{1}{\widetilde{\rho}^{d-1}}P_d(x_1, x_2), \frac{1}{\widetilde{\rho}^{d-1}}Q_d(x_1, x_2), x'\Big).$$

For $\|x'\| \leq \frac{1}{2}$ this coincides with $g_d(x) \in \mathbb{S}^m$, but otherwise, we must normalize to get $f_d = \widetilde{g}_d/\|\widetilde{g}_d\| : \mathbb{S}^n \to \mathbb{S}^n$. We claim that $\deg(f_d) = d$.

Indeed, pick $a = (1, 0, \ldots, 0) \in \mathbb{S}^m$. Suppose $f_d(x) = a$. Then $x' = 0$ and $\rho = 1$, so that $f_d(x) = g_d(x) = (\cos(d\cdot\theta), \sin(d\cdot\theta), 0)$, and we obtain

$$x = (\cos\theta_k, \sin\theta_k, 0), \quad \text{where} \quad \theta_k = 2\pi\frac{k}{d}, 0 \leq k < d.$$

Near such a point we can use the parametrization $x = (\rho\cos\theta, \rho\sin\theta, x')$, which is compatible with the standard orientation of the sphere (straightforward computation). The local expression of our mapping is then

$$f_d \equiv g_d : (\theta, x') \mapsto (d{\cdot}\theta, x').$$

This is clearly a diffeomorphism that preserves orientation, which shows that a is a regular value of f_d, and

$$\deg(f_d) = \sum_{x\in f_d^{-1}(a)} \operatorname{sign}_x(f_d) = \sum_{k=1}^{d}(+1) = d.$$

One can use the same method for d negative, just noting that one winds around clockwise, instead of counterclockwise. However, here we have the following alternative argument.

Case $d = -1$. As we saw in III.1.6(2), p. 100, the symmetry

$$f_{-1} : x \mapsto (x_1, \dots, x_m, -x_{m+1})$$

has degree $\deg(f_{-1}) = -1$.

Case $d < 0$. Define $f_d = f_{-1} \circ f_{-d}$, so that by III.1.7, p. 100,

$$\deg(f_d) = \deg(f_{-1}) \cdot \deg(f_{-d}) = (-1)(-d) = d. \qquad \square$$

We can extend the preceding result to arbitrary compact manifolds (compactness is required because the target $\mathbb{S}^m$ is compact):

Proposition 1.2. *Let M be a compact, oriented, boundaryless, differentiable manifold of dimension m. Then for every integer d there is a differentiable mapping $h_d : M \to \mathbb{S}^m$ of degree d.*

Proof. First we remark that it is enough to find $h : M \to \mathbb{S}^m$ of degree ± 1. Indeed, given such an h, we can use the mappings $f_{\pm d}$ of the preceding example to define $h_d = f_{\pm d} \circ h$, which has degree d (III.1.7, p. 100). In fact, let us define a differentiable mapping $h : M \to \mathbb{S}^m$ such that the north pole $(0, \dots, 0, 1) \in \mathbb{S}^m$ is a regular value with a single preimage.

Let U be an open subset of M diffeomorphic to $\mathbb{R}^m$, and consider a smooth function

$$\mu(t) = \begin{cases} 1 & \text{for } t \le \frac{1}{4}, \\ 0 & \text{for } t \ge 1. \end{cases}$$

We identify $U \equiv \mathbb{R}^m$ to define $h: M \to \mathbb{S}^m$ by

$$h(x) = \begin{cases} (0,\dots,0,-1) & \text{the south pole, for } x \notin U, \\ \Big(\dfrac{2\tau(x)x}{\tau(x)^2+\|x\|^2}, \dfrac{\tau(x)^2-\|x\|^2}{\tau(x)^2+\|x\|^2}\Big) & \text{where } \tau(x)=\mu(\|x\|^2), \text{ for } x \in U. \end{cases}$$

This mapping is well defined because for $x \in U$ with $\|x\| \geq 1$ we have $\tau(x) = 0$ and $h(x) = (0,\dots,0,-1)$. On the other hand, using the equations of the stereographic projection π from the south pole (II.2.4, p. 55), for $x \in U$ with $\tau(x) \neq 0$, we get

$$h(x) = \pi^{-1}\Big(\frac{x}{\tau(x)}\Big).$$

Thus, h is the diffeomorphism π^{-1} for $\tau(x) = 1$, that is, for $\|x\| \leq \frac{1}{2}$. On the other hand, $x = 0$ is the only preimage of $(0,\dots,0,1)$. Indeed, suppose $(0,\dots,0,1) = h(x)$. Then $\tau(x) \neq 0$, and

$$(0,\dots,0,1) = h(x) = \pi^{-1}\Big(\frac{x}{\tau(x)}\Big);$$

hence $x = 0$. We are done. □

Remark 1.3. The existence of non-homotopic maps $M \to N$ between two manifolds depends on the nature of the given manifolds. In general we cannot expect the richness shown in V.1.2 above. To illustrate the opposite situation we recall that *all continuous mappings* $\mathbb{S}^2 \to \mathbb{S}^1 \times \mathbb{S}^1$ *are null-homotopic,* hence have degree zero.

Indeed, the reason is that any continuous mapping $f: \mathbb{S}^2 \to \mathbb{S}^1 \times \mathbb{S}^1$ can be lifted to $\widetilde{f}: \mathbb{S}^2 \to \mathbb{R} \times \mathbb{R}$, through the universal covering

$$\pi: \mathbb{R} \times \mathbb{R} \to \mathbb{S}^1 \times \mathbb{S}^1 : (\alpha, \beta) \mapsto (\cos\alpha, \sin\alpha; \cos\beta, \sin\beta),$$

so that $f = \pi \circ \widetilde{f}$. Then the homotopy $\widetilde{H}_t(x) = (1-t)\widetilde{f}(x)$ goes down to $H_t = \pi \circ \widetilde{H}_t$ with $H_0 = f$ and H_1 constant. □

Now we want to modify mappings already given. The modification must keep the homotopy type. For mappings into spheres this comes easily, as follows:

Lemma 1.4. *Let M be a possibly non-compact manifold. Let $f, g: M \to \mathbb{S}^m$ be two mappings that do not have antipodal images: $f(x) \neq -g(x)$ for all $x \in M$. Then*

$$H_t(x) = \frac{tg(x) + (1-t)f(x)}{\|tg(x) + (1-t)f(x)\|}$$

is a well-defined homotopy with $H_0 = f$, $H_1 = g$.

In particular, if f *is not surjective, it is null-homotopic.*

Proof. Let $x \in M$. Then $p = tg(x) + (1-t)f(x)$ is a point in the segment joining $f(x)$ and $g(x)$. As both points belong to the sphere, $p = 0$ only if $f(x)$ and $g(x)$ are antipodal, which is excluded by hypothesis. Thus H_t is indeed well defined.

Finally, if there is a $p \notin f(M)$, f is homotopic to the constant mapping $g \equiv -p$. □

An immediate application of the lemma is that every differentiable mapping $M \to \mathbb{S}^m$ with $\dim(M) < m$ is null-homotopic. Indeed, with that hypothesis, no smooth mapping $M \to \mathbb{S}^m$ is surjective, by the Easy Sard Theorem.

Now that we have this homotopy lemma, we will further explore the idea in the proof of V.1.2, p. 185:

(1.5) Antipodal extension, linearization, and normalization. Let M be an oriented, boundaryless, differentiable manifold of dimension m, possibly non-compact.

(1) Suppose we fix some parametrization $\mathbb{R}^m \equiv U \subset M$ and a smooth mapping $h : \mathbb{R}^m \to \mathbb{R}^m$ such that $h^{-1}(0) = \{c\}$. Then we use the stereographic projection $\pi : \mathbb{S}^m \to \mathbb{R}^m$ from the south pole $a_S = (0, \dots, 0, -1) \in \mathbb{S}^m$ to identify $\mathbb{R}^m$ with the open set $V = \mathbb{S}^m \setminus \{a_S\} \subset \mathbb{S}^m$ and $0 \in \mathbb{R}^m$ with the north pole $a_N = (0, \dots, 0, 1) \in \mathbb{S}^m$. Thus, $h \equiv \pi^{-1} \circ h$ is seen as a mapping into $\mathbb{R}^m \equiv V \subset \mathbb{S}^m$:

$$\begin{array}{ccccc} M & & & & \mathbb{S}^m \\ \cup & c & \longmapsto & 0 \equiv a_N & \cup \\ U \equiv \mathbb{R}^m & & \xrightarrow{h} & \mathbb{R}^m & \equiv V \end{array}$$

What we want is to extend $h : U \to V$ from a neighborhood of c to $\bar{h} : M \to \mathbb{S}^m$ by sending everything off another neighborhood of c to the south pole.

For such an extension we modify h by a trick analogous to the one used in the proof mentioned above:

$$\bar{h}(x) = \begin{cases} (0, \dots, 0, -1) & \text{for } x \notin U, \\ \left(\dfrac{2\tau(x)h(x)}{\tau(x)^2 + \|h(x)\|^2}, \dfrac{\tau(x)^2 - \|h(x)\|^2}{\tau(x)^2 + \|h(x)\|^2} \right) & \text{for } x \in U, \end{cases}$$

with a smooth bump function

$$\tau(x) = \mu(\|x - c\|^2) = \begin{cases} 1 & \text{for } \|x - c\| \leq \frac{1}{2}, \\ 0 & \text{for } \|x - c\| \geq 1. \end{cases}$$

This defines well a smooth mapping, because $\|h(x)\| \neq 0$ on a neighborhood of $\|x - c\| \geq 1$, and $\bar{h}(x) = (0, \dots, 0, -1)$ for $\|x - c\| \geq 1$. After some straightforward computation, for $x \in U$ with $\tau(x) \neq 0$ we find

$$\bar{h}(x) = \pi^{-1}\Big(\frac{h(x)}{\tau(x)}\Big);$$

hence $h \equiv \pi^{-1} \circ h$ on the ball of radius $\frac{1}{2}$. Note also that $\bar{h}(x) = (0, \dots, 0, 1)$ if and only if $x \in U$ and $h(x) = 0$, if and only if $x = c$. We will call this construction the *antipodal extension.*

In view of the homotopy lemma, V.1.4, p. 186, we are also interested in possible antipodal images of $\bar{h}$ and h in $\mathbb{S}^m$. Suppose this is the case for some point $x \in U$. Then

$$\Big(\frac{2\tau(x)h(x)}{\tau(x)^2 + \|h(x)\|^2}, \frac{\tau(x)^2 - \|h(x)\|^2}{\tau(x)^2 + \|h(x)\|^2}\Big) = -\Big(\frac{2h(x)}{1 + \|h(x)\|^2}, \frac{1 - \|h(x)\|^2}{1 + \|h(x)\|^2}\Big).$$

Looking at the last component of these two points, we find that $\tau(x) = \|h(x)\|^2$ and $h(x) \neq 0$. Then, looking at the other components, we get

$$\frac{\tau(x)h_i(x)}{\tau(x)^2 + \|h(x)\|^2} = -\frac{h_i(x)}{1 + \|h(x)\|^2},$$

which is impossible. Hence we conclude that *the mapping h and its extension $\bar{h}$ have no antipodal images.*

(2) Next, suppose h is a diffeomorphism, and denote $L = d_c h$, which is a linear isomorphism. Consider the homotopy of IV.3.1, p. 153, which here simplifies to

$$H_t(x) = \begin{cases} \dfrac{h(tx + (1 - t)c)}{t} & \text{for } t \neq 0, \\ L(x - c) & \text{for } t = 0. \end{cases}$$

Clearly $H_t(x) = 0$ if and only if $x = c$, and we can apply the same antipodal extension to H_t replacing h by H_t all through the construction. We get a homotopy $\bar{H}_t$ such that $\bar{H}_1 = \bar{h}$ and $\pi \circ \bar{H}_0 \equiv L - L(c)$ near c. We have thus *linearized* the mapping $\bar{h}$.

(3) Now, the determinant of the linear isomorphism L can be either positive or negative. In the first case, let A denote the identity, and in the

second, the symmetry with respect to the first variable in $\mathbb{R}^m$. In any case, the determinants of L and A have the same sign. Now, $m \times m$ matrices are points in $\mathbb{R}^{m\times m}$, and the closed hypersurface $\det = 0$ decomposes $\mathbb{R}^{m\times m}$ into two connected components: $\det > 0$ and $\det < 0$. Since L and A are in the same one of those two components, there is a continuous mapping $t \mapsto A_t$ with $\det(A_t) \neq 0$ for all t, $A_0 = L$ and $A_1 = A$. Again we apply the antipodal extension to $B_t = A_t - A_t(c)$ and find a homotopy $\bar{B}_t$. We remark here that the construction is possible because A_t is bijective ($\det(A_t) \neq 0$): this guarantees that $B_t(x) = A_t(x - c) = 0$ if and only if $x = c$. This *normalization* shows that the mapping $\bar{h}$ is homotopic to $g = \bar{B}_1$, which is the antipodal extension of $A - A(c)$.

The construction we have just described is the method for classifying smooth mappings $f : M \to \mathbb{S}^m$ by homotopy. Indeed, suppose the north pole $a \in \mathbb{S}^m$ is a regular value of f with finitely many preimages $c_1, \dots, c_r$. As usual, we find local coordinates around each c_i, on whose domains U_i the restrictions $f|_{U_i}$ are diffeomorphisms. Choosing the U_i's disjoint and small enough, we can glue the antipodal extensions of the restrictions $f|_{U_i}$ into a smooth mapping $\bar{f} : M \to \mathbb{S}^m$. By (1) above, f and $\bar{f}$ have no antipodal images, and by V.1.4, p. 186, f and $\bar{f}$ are homotopic. This mapping $\bar{f}$ depends only on the preimages of the regular value and the signs of f at them: exactly what we use to compute the degree.

Actually, this is almost a proof of the Hopf Theorem, and what remains is quite evident: (i) to move the preimages freely and (ii) to dispose of those pairs whose signs cancel each other. We devote the following section to settling these two matters, by means of diffeotopies (II.6, p. 80).

Exercises and problems

Number 1. Points (x, y, t) in the sphere $\mathbb{S}^2 \subset \mathbb{R}^3 \equiv \mathbb{C} \times \mathbb{R}$ can be written as pairs (z, t), $z = x + iy$. For every integer $k > 0$ define a continuous mapping $f : \mathbb{S}^2 \to \mathbb{S}^2$ by a formula $f(z, t) = (z^k, th(t))$ involving a suitable function $h > 0$. Compute the degree of f.

Number 2. It is a fact that all smooth mappings $\mathbb{CP}^1 \times \mathbb{CP}^1 \to \mathbb{CP}^2$ have even degree. Mappings of every possible degree can be obtained as follows.

(1) Use the *Segre embedding*

$$\mathbb{CP}^1 \times \mathbb{CP}^1 \to \mathbb{CP}^3 : (x_0 : x_1; y_0 : y_1) \mapsto (x_0y_0 : x_0y_1 : x_1y_0 : x_1y_1)$$

and a conic projection in $\mathbb{CP}^3$ to produce a smooth mapping $\mathbb{CP}^1 \times \mathbb{CP}^1 \to \mathbb{CP}^2$ of degree 2.

(2) Compose the mapping in (1) with mappings $\mathbb{CP}^1 \times \mathbb{CP}^1 \to \mathbb{CP}^1 \times \mathbb{CP}^1$ of arbitrary degree.

Number 3. Let $f : \mathbb{S}^m \to \mathbb{S}^m$ be a continuous mapping. Show that:

(1) If $\deg(f) \neq +1$, there is an $x \in \mathbb{S}^m$ with $f(x) = -x$.

(2) If $\deg(f) \neq (-1)^{m+1}$, there is an $x \in \mathbb{S}^m$ with $f(x) = x$.

(3) If f is null-homotopic, then it fixes a point and sends another to its antipode.

Number 4. Let $f, g : \mathbb{S}^{2n} \to \mathbb{S}^{2n}$ be two continuous mappings. Prove the following assertions:

(1) At least one of the mappings f, g, or $g \circ f$ has a fixed point.

(2) If f has no fixed point, there are $x, y \in \mathbb{S}^{2n}$ such that $f(x) = y$ and $f(y) = x$.

(3) If f has no fixed point, then it sends some point to its antipode.

Number 5. Let $\tau : \mathbb{S}^m \to \mathbb{S}^m$ be a continuous involution ($\neq \mathrm{Id}_{\mathbb{S}^m}$). Show that τ sends some point to its antipode by way of contradiction, as follows:

(1) Prove that, otherwise,

$$f : \mathbb{S}^m \to \mathbb{S}^m : x \mapsto \frac{x + \tau(x)}{\|x + \tau(x)\|}$$

is a well-defined continuous map of degree 1, hence surjective.

(2) Let $F \subset \mathbb{S}^m$ denote the set of fixed points of τ, and write $V = \mathbb{S}^m \setminus F$, $U = f^{-1}(V) \subset \mathbb{S}^m$. Show that $f|_U : U \to V$ is a well-defined proper map of the same degree as f.

(3) Use the identification $x \equiv \tau(x)$ to define a surjective local difeomorphism $\pi : U \to M$ onto an abstract smooth manifold M (just as $x \equiv -x$ gives $M = \mathbb{RP}^m$).

(4) Show that $f|_U$ factorizes through M, and use that factorization to conclude that f has even degree.

Number 6. Let G be a non-trivial group of homeomorphisms acting on a sphere of even dimension. Suppose that no $h \in G$ distinct from the identity has fixed points. Then $G \equiv \mathbb{Z}_2$.

Number 7. Let $f, g : \mathbb{S}^m \to \mathbb{S}^m$ be two continuous mappings which are never perpendicular (that is, $\langle f(x), g(x) \rangle \neq 0$ for all $x \in \mathbb{S}^m$). Then $\deg(f) = \pm \deg(g)$.

Number 8. Let P, Q denote two complex polynomials of degrees p and q, respectively, and consider the rational function $F = P/Q : \mathbb{C} \equiv \mathbb{R}^2 \to \mathbb{C} \equiv \mathbb{R}^2$. This function is defined except at finitely many poles $c_1, \dots, c_r$: the zeros of Q which do not cancel with those of P. Show that via the sterographic projection from the north pole $a_N = (0,0,1)$, F extends to a smooth mapping $f : \mathbb{S}^2 \to \mathbb{S}^2$ by setting

$$f(c_1) = \cdots = f(c_r) = f(a_N) = a_N.$$

Compute the degree of f.

Number 9. Let $M \subset \mathbb{R}^p$ be a connected, boundaryless, smooth manifold of codimension m. Show that there is a differentiable mapping $f : \mathbb{R}^p \to \mathbb{S}^m$ such that the north pole $a_N = (0, \dots, 0, 1) \in \mathbb{S}^m$ is a regular value of f and $M = f^{-1}(a_N)$.

2. The Hopf Theorem: Brouwer-Kronecker degree

Here we will show that an example like III.3.5, p. 112, cannot occur for mappings into spheres.

Theorem 2.1 (Hopf Theorem). *Let M be a connected, compact, oriented, boundaryless, differentiable manifold of dimension m. Two continuous mappings $M \to \mathbb{S}^m$ with the same Brouwer-Kronecker degree are homotopic.*

Proof. We will prove the theorem by showing that all mappings $f : M \to \mathbb{S}^m$ of a given degree have a common normal form, in the sense described in the previous section. Suppose $\deg(f) = d \geq 0$ (resp., $d \leq 0$) and fix a coordinate domain $U \equiv \mathbb{R}^m$, whose local coordinates preserve orientation, and any d points $c_1, \dots, c_d \in U$ (none if $d = 0$). We are to built up a mapping homotopic to f that solely depends on these points and d.

We can assume f is differentiable and choose a regular value $a \in \mathbb{S}^m$, which by a diffeotopy (II.6.5, p. 82) of the sphere we move to the north pole. Then a has $2r + d$ inverse images, so that

(i) at $d + r$ of the images the sign of f is positive (resp., negative),

(ii) at r of the images the sign of f is negative (resp., positive).

Then, choose $2r + d$ points $p_1, q_1, \dots, p_r, q_r, c_1, \dots, c_d \in U$, and use a diffeotopy to move the points in $f^{-1}(a)$ as follows (we do nothing if there are no points to move):

(1) r of them at which f is negative (resp., positive) to $q_1, \dots, q_r$,

(2) some r at which f is positive (resp., negative) to $p_1, \dots, p_r$,

(3) the remaining d of them at which f is positive (resp., negative) to $c_1, \dots, c_d$.

Using the chosen coordinates in U, we identify $U \equiv \mathbb{R}^m$, and our points q_k, p_k, c_i, are points in $\mathbb{R}^m$. Then we use a diffeotopy of $\mathbb{R}^m$ which is the identity off a big enough ball (II.6.6, p. 83) to move those points so that

(1) $q_k = (-3k, 0, \dots, 0)$,

(2) $p_k = (3k, 0, \dots, 0)$,

(3) $c_i = (3(s + i), 0, \dots, 0)$, for some fixed $s \geq r$.

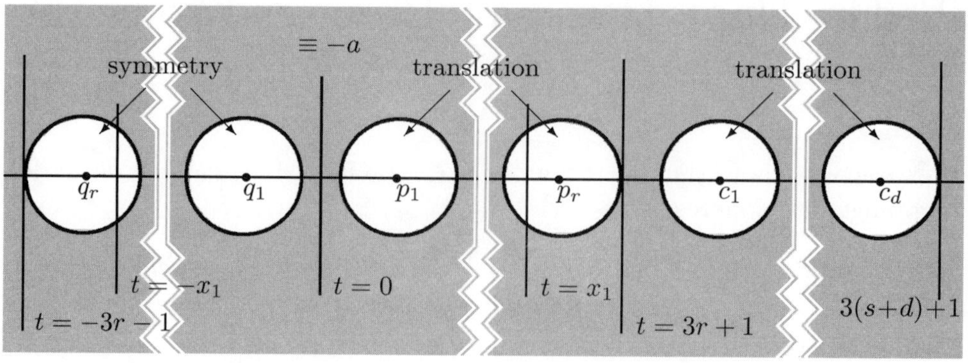

Now, using the antipodal extension, linearization, and normalization (V.1.5, p. 187), we find a mapping $\bar{f}$ homotopic to f, which is $\equiv -a$ off the open balls of radius 1 around the points q_k, p_k, c_i, and inside those balls it is, respectively, the antipodal extension of

(1) the symmetry: $(x_1, y) \mapsto (-x_1, y) + q_k$,
(2) the translation: $(x_1, y) \mapsto (x_1, y) - p_k$, and
(3) the translation: $(x_1, y) \mapsto (x_1, y) - c_i$.

What is important here is that $\bar{f}(-x_1, y) = \bar{f}(x_1, y)$ for $|x_1| \leq 3r + 1$. This implies that the following homotopy is well defined for $0 \leq t \leq 3r + 1$:

$$H_t(x_1, y) = \begin{cases} \bar{f}(x_1, y) & \text{for } |x_1| \geq t, \\ \bar{f}(t, y) & \text{for } |x_1| \leq t. \end{cases}$$

Note that this homotopy on $\mathbb{R}^m \equiv U \subset M$ extends to M, because $H_t \equiv -a$ off $-3r - 1 \leq x_1 \leq 3(s + d) + 1$, $\|y\| \leq 1$. We have $H_0 = \bar{f}$, and for $t = 3r + 1$:

$$H_t(x) = \begin{cases} \bar{f}(x) & \text{on the balls centered at the points } c_i, \\ -a & \text{off those balls.} \end{cases}$$

This $g = H_t$ is the mapping we seek. Of course, it is homotopic to $\bar{f}$, hence to f, and, in addition, the north pole a is a regular value of g, $g^{-1}(a) = \{c_1, \ldots, c_d\}$, and g is the antipodal extension of the translation $(x_1, y) \mapsto (x_1, y) - c_i$ on the unit ball around each c_i. Note that if $d = 0$, this is the constant mapping $\equiv -a$.

Clearly, this construction can be carried along simultaneously for two mappings of degree d taking enough points p_k, q_k, and a common s, and the final form is the same for both. □

Corollary 2.2. *Let M be a connected, compact, oriented, boundaryless, differentiable manifold of dimension m. Then*

$$\pi^m(M) = \mathbb{Z}.$$

In particular, $\pi_m(\mathbb{S}^m) = \pi^m(\mathbb{S}^m) = \mathbb{Z}$.

Proof. This follows immediately from V.2.1, p. 191, and V.1.2, p. 185, because $\pi^m(M) = [M, \mathbb{S}^m]$. □

Remarks 2.3. (1) The condition that M is boundaryless is essential to have different homotopy classes. In fact, if M has a boundary $\partial M \neq \emptyset$, then *all mappings* $M \to \mathbb{S}^m$ *are null-homotopic.*

We only present a quick sketch of the argument. First one glues along their boundaries M and a copy $-M$ with the opposite orientation, to get a new oriented, compact, boundaryless manifold $M^* = M \cup -M$ of the same dimension (some smoothing is required around $\partial M \equiv \partial(-M)$ by means of a bump function). Then every continuous mapping $f : M \to \mathbb{S}^m$ extends in an obvious unique way to $f^* : M^* \to \mathbb{S}^m$, and this f^* has degree zero (pick a close smooth approximation g of f^* and look at a regular value of g off $g(\partial M)$). By the Hopf Theorem, the mapping f^* is null-homotopic, and so is its restriction $f = f^*|M$.

(2) After the above example, it is natural to ask for a degree for manifolds with boundary. This can indeed be done, using the ideas behind Euclidean degree. If $f : M \to N$ is a proper mapping from an $(m+1)$-manifold with boundary into a connected, boundaryless, $(m+1)$-manifold, there is a degree $d(f, M, a)$ for $a \in N \setminus f(\partial M)$. In this setting homotopies do not touch the boundary and the situation in (1) does not occur.

(3) In case M is non-compact, there is no proper mapping $M \to \mathbb{S}^m$, so that there is no notion of degree for a mapping $f : M \to \mathbb{S}^m$. But still, one can ask about the homotopy type of f. The dramatic answer to this question is that f *is always null-homotopic.* Actually, the argument is a kind of infinite degree computation. We only sketch the idea.

First, replace f by some close smooth mapping, which will be homotopic. Then, pick a regular value $a \in \mathbb{S}^m$, and move it to the north pole. Replace f with a homotopic mapping which is $\equiv -a$ off a compact neighborhood of $f^{-1}(a)$ (the antipodal extension is still posible because $f^{-1}(a)$ is discrete). Pick a countable discrete family of coordinate domains in M far from that neighborhood, and reversing the method used in the proof of the Hopf Theorem to cancel points with opposite sign, create a pair of them on each such domain. From this we are sure that $f^{-1}(a)$ consists of an infinite sequence (p_k) of points at which the sign of f is positive and another sequence (q_k) at which the sign is negative. Then we can move each pair p_k, q_k into a coordinate domain far from the other points, using a diffeotopy that does not move anything else. Thus they cancel each other separately as in the proof of the Hopf Theorem, and in the end, $f^{-1}(a)$ is empty. Thus, f is not surjective, hence null-homotopic.

(4) For non-orientable manifolds, the mod 2 Brouwer-Kronecker degree (III.6.1, p. 124) provides the following parallel version of the above Hopf Theorem, V.2.1, p. 191:

Let M be a connected, compact, non-orientable, boundaryless, differentiable manifold of dimension m. Two continuous mappings $M \to \mathbb{S}^m$ with the same mod 2 Brouwer-Kronecker degree are homotopic.

In particular, we see that for such an M, $\pi^m(M) = \mathbb{Z}_2$.

The proof is no surprise: one can cancel pairs of points p, q in the inverse image of any regular value. Indeed, after some diffeotopy, non-orientability gives two parametrizations $\varphi, \psi : \mathbb{R}^m \to U, V$ with $p, q \in U \cap V$ that induce different orientations at p, but the same at q. Hence the localization of the given mapping via one of those parametrizations must preserve orientation at p and reverse orientation at q (or the opposite), and then we eliminate the pair as in the orientable case. □

From the Hopf Theorem we easily deduce for spheres the converse to the Boundary Theorem, III.3.3, p. 111:

Proposition 2.4. *Let X be an oriented differentiable C^r manifold of dimension $m+1$ with connected boundary $\partial X = M$, and let $f : M \to \mathbb{S}^m$ be a proper (resp., proper C^r) mapping of degree 0. Then f has a continuous (resp., C^r) extension $\bar{f} : X \to \mathbb{S}^m$.*

Proof. As f is proper, M is compact. By V.2.1, p. 191, there is a homotopy $H_t : M \to \mathbb{S}^m$ with $H_0 \equiv a \in \mathbb{S}^m$ and $H_1 = f$. Then, the set $Z = (\{0\} \times X) \cup ([0,1] \times M)$ is closed in $[0,1] \times X$, and we have the continuous mapping

$$F : Z \to \mathbb{S}^m : (t, x) \mapsto \begin{cases} a & \text{for } t = 0, \\ H_t(x) & \text{for } t \neq 0. \end{cases}$$

Now, by the Tietze Extension Theorem, this extends to a continuous mapping $\widetilde{F} : [0,1] \times X \to \mathbb{R}^{m+1}$. Consider the set $A = \widetilde{F}^{-1}(\mathbb{R}^{m+1} \setminus \{0\})$, which is an open neighborhood of Z in $[0,1] \times X$. In particular, there is an open set $U \subset X$ such that $U \supset M$ and $[0,1] \times U \subset A$, and we choose a bump function θ which is $\equiv 0$ off U and $\equiv 1$ on M. We define a continuous mapping $\bar{f} : X \to \mathbb{S}^m$ by

$$\bar{f}(x) = \frac{\widetilde{F}(\theta(x), x)}{\|\widetilde{F}(\theta(x), x)\|}.$$

This is well defined, because $\widetilde{F}(\theta(x), x) = 0$ implies $(\theta(x), x) \notin A$; hence $x \notin U$ and $\theta(x) = 0$, so that

$$0 = \widetilde{F}(\theta(x), x) = \widetilde{F}(0, x) = a \in \mathbb{S}^m,$$

a contradiction. Furthermore $\bar{f}$ is an extension of f, since $\theta(x) = 1$ on M, so that $\bar{f}|_M = H_1 = f$.

We have thus proved the assertion for proper mappings. For proper $\mathcal{C}^r$ mappings the argument is the same using the $\mathcal{C}^r$ Tietze Extension Theorem, II.1.5, p. 52. The only requirement is to change the first definition:

$$F : Z \to \mathbb{S}^m : (t, x) \mapsto \begin{cases} a & \text{for } t = 0, \\ H_{\eta(t)}(x) & \text{for } t \neq 0, \end{cases}$$

where η is a smooth bump function $\equiv 0$ for $t \leq \frac{1}{3}$ and $\equiv t$ for $t \geq \frac{1}{2}$. This guarantees F is $\mathcal{C}^r$, and the proof follows readily. □

In fact the above proof is an adapted version of the general argument used to extend homotopies from closed subsets of a metric space to the whole space (the *Borsuk Theorem*). This then implies that *the property that a continuous mapping on such a closed subset extends to the whole space only depends on the homotopy type of the mapping.*

Exercises and problems

Number 1. Let $f, g : \mathbb{S}^m \to \mathbb{S}^m$ be two continuous mappings. Prove that $f \circ g$ and $g \circ f$ are homotopic.

Number 2. Let $f, g : \mathbb{S}^1 \to \mathbb{S}^1$ be smooth mappings. Then the (complex) multiplication $f \cdot g : \mathbb{S}^1 \to \mathbb{S}^1$ is well defined, and

$$\deg(f \cdot g) = \deg(f) + \deg(g).$$

Prove this directly from the Hopf Theorem and compare with Problem Number 8 of III.3. Deduce that *if a complex polynomial $P(z)$ has m roots (counted with multiplicities) inside the circle $\mathbb{S}^1 \subset \mathbb{R}^2 \equiv \mathbb{C}$ and none on it, then the mapping $f : \mathbb{S}^1 \to \mathbb{S}^1 : z \mapsto \frac{P(z)}{|P(z)|}$ has degree m.*

Number 3. Use the preceding formula $\deg(f \cdot g) = \deg(f) + \deg(g)$ to review Problem Number 4 of III.4, noting that for the g and h there, $h = \mathrm{Id} \cdot g$.

Number 4. Let X be a space with involution for which $s(X) = s'(X) = m < \infty$ (Problem Number 6 of IV.5). Then $\pi_m(X)$ contains $\mathbb{Z}$.

Number 5. Let M and N be smooth manifolds, N boundaryless and connected, and fix a point $c \in N$. For every proper smooth mapping $f : M \to N$ such that $f(\partial M) = \{c\}$ define

$$\deg(f) = \sum_{x \in f^{-1}(a)} \mathrm{sign}_x(f),$$

where $a \neq c$ is a regular value of f. Show that this degree is well defined and that it is invariant by homotopies H_t such that $H_t(\partial M) = \{c\}$.

Number 6. Two spaces X and Y have the same homotopy type if there are continuous mappings $f : X \to Y$ and $g : Y \to X$ such that $g \circ f$ and $f \circ g$ are homotopic to Id_X and Id_Y. Prove the following:

(1) Two spheres of different dimensions do not have the same homotopy type.

(2) A sphere does not have the same homotopy type as the product of two others.

Number 7. Let $N \subset M \subset \mathbb{R}^p$ be differentiable manifolds, N closed in M. Let X be a topological space and let $H : [0,1] \times N \to X$ be a continuous homotopy. Suppose that $f = H_0 : N \to X$ extends to a continuous mapping $\bar{f} : M \to X$. Then H extends to a continuous homotopy $\overline{H} : [0,1] \times M \to X$ such that $\bar{f} = \overline{H}_0$.

Number 8. Let X be a metric space, $C \subset X$ a closed subset, and $H : [0,1] \times C \to M$ a continuous mapping into a differentiable manifold M. Suppose that $f = H_0 : C \to M$ extends to a continuous mapping $\bar{f} : X \to M$. Then H extends to a continuous homotopy $\overline{H} : [0,1] \times X \to M$ such that $\bar{f} = \overline{H}_0$.

3. The Hopf Theorem: Euclidean degree

In this section we prove the suitable version of the Hopf Theorem, V.2.1, p. 191, in the Euclidean degree setting:

Theorem 3.1 (Hopf Theorem). *Let $X \subset \mathbb{R}^{m+1}$ be a compact set that disconnects $\mathbb{R}^{m+1}$ into two connected components, D_1 bounded and D_0 unbounded, whose topological boundary is X. Let $f, g : X \to \mathbb{R}^{m+1} \setminus \{0\}$ be two continuous mappings with the same winding number: $w(f,0) = w(g,0)$. Then f and g are homotopic.*

Proof. The proof is a recollection of various results already at hand. After a translation, $0 \in D_1$, and we choose an open ball B of radius ρ containing $\overline{D}_1$, whose boundary (the sphere of radius ρ) we denote by $M = \overline{B} \setminus B$. Now recall that

$$w(f,0) = d(\bar{f}, D_1, 0) \quad \text{and} \quad w(g,0) = d(\bar{g}, D_1, 0),$$

where $\bar{f}, \bar{g}$ are arbitrary continuous extensions of f, g to $\overline{D}_1 = D_1 \cup X$, which we can further extend to $\overline{B}$, *without zeros in* M (Tietze Extension Theorem). By the same definition of degree (IV.2.1, p. 145, and IV.2.2, p. 146), we can assume that $\bar{f}$ and $\bar{g}$ are smooth on $\overline{B}$ and that 0 is a regular value for both on B. Actually, $\bar{f}$ and $\bar{g}$ are polynomials; hence they are defined on the whole Euclidean space $\mathbb{R}^{m+1}$.

Let us now concentrate on $\bar{f}$. The discrete set $\bar{f}^{-1}(0) \cap B$ is finite, and its points split into those in D_1 and those in $D_0 \cap B$ (note that $0 \notin f(X) \cup f(M)$): say $c_1, \dots, c_r \in D_1$, $b_1, \dots, b_s \in D_0 \cap B$. Now pick an open ball $B' \subset D_1$ centered at $0 \in D_1$. Since D_1 is connected, we can move the c_j's into B' by a diffeotopy of $\mathbb{R}^{m+1}$ that fixes X and all other zeros of $\bar{f}$; then, we pick a ball $B'' \supset \overline{B}$, so that $\bar{f}$ has no zeros in $B'' \setminus B$, and since $D_1 \cap B''$ is connected, we move the b_i's into $B'' \setminus B$ by a diffeotopy of $\mathbb{R}^{m+1}$ that fixes X and all other zeros of $\bar{f}$ (II.6.6, p. 83).

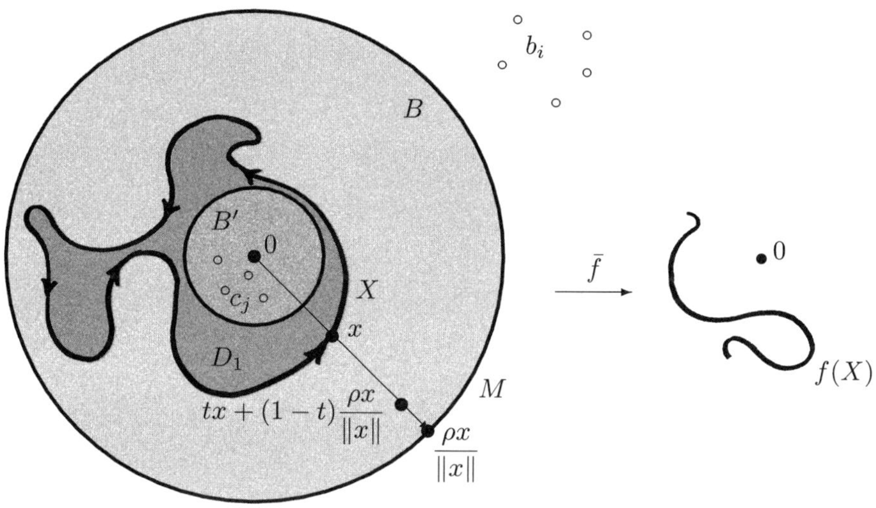

We have moved the points so that 0 has the same preimages in B and D_1; hence $d(\bar{f}, D_1, 0) = d(\bar{f}, B, 0)$. Furthermore the mapping

$$\widetilde{f} : U = \overline{B} \setminus B' \to \mathbb{S}^m : x \mapsto \frac{\bar{f}(x)}{\|\bar{f}(x)\|}$$

is well defined. Now, by IV.4.5, p. 157, the Euclidean degree $d(\bar{f}, B, 0)$ is the Brouwer-Kronecker degree of the smooth mapping $\widetilde{f}|_M$; that is,

$$d(\bar{f}, D_1, 0) = \deg(\widetilde{f}|_M).$$

The same construction for g gives a smooth mapping

$$\widetilde{g} : U = \overline{B} \setminus B' \to \mathbb{S}^m : x \mapsto \frac{\bar{g}(x)}{\|\bar{g}(x)\|}$$

with

$$d(\bar{g}, D_1, 0) = \deg(\widetilde{g}|_M).$$

Consequently $\deg(\widetilde{f}|_M) = \deg(\widetilde{g}|_M)$, and by the Hopf Theorem, V.2.1, p. 191, $\widetilde{f}|_M$ and $\widetilde{g}|_M$ are homotopic. It follows immediately that the two mappings

$$f^* : X \to \mathbb{S}^m : x \mapsto \widetilde{f}(\rho x/\|x\|) \quad \text{and} \quad g^* : X \to \mathbb{S}^m : x \mapsto \widetilde{g}(\rho x/\|x\|)$$

are also homotopic or we can simply say that $f^* : X \to \mathbb{R}^{m+1} \setminus \{0\}$ and $g^* : X \to \mathbb{R}^{m+1} \setminus \{0\}$ are homotopic. Thus, it remains to show that f^* and g^* are, respectively, homotopic to f and g.

We write the proof for f in two steps. First, since $\widetilde{f}$ is defined on $\overline{B} \setminus B'$, we can consider the homotopy

$$H_t(x) = \widetilde{f}(tx + (1-t)\rho x/\|x\|) \neq 0, \quad x \in X$$

(every segment joining $x \in X$ to $\rho x/\|x\| \in M$ is fully contained in $\overline{B} \setminus B'$). Clearly $H_0 = f^*$ and $H_1 = \widetilde{f}|_X$.

Second, we show that $\widetilde{f}|X$ is homotopic to f. To that end, consider the mapping

$$F_t(x) = \frac{\bar{f}(x)}{(1-t) + t\|\bar{f}(x)\|}.$$

If the denominator vanishes at some $x \in X$, then $t = 1$ and $\bar{f}(x) = 0$, which is impossible. Also, if $F_t(x) = 0$ for some $x \in X$, then $\bar{f}(x) = 0$, again impossible. Hence, F_t is a well-defined homotopy with $F_0 = \bar{f}|_X = f$ and $F_1 = \widetilde{f}|_X$.

For g the argument is the same, and the proof of the theorem is thus finished. □

As in the preceding section, this completely describes the cohomotopy group:

Corollary 3.2. *Let $X \subset \mathbb{R}^{m+1}$ be a compact set that disconnects $\mathbb{R}^{m+1}$ into two connected components whose topological boundary is X. Then*

$$\pi^m(X) = \mathbb{Z}.$$

Proof. The group $\pi^m(X)$ classifies by homotopy the mappings $X \to \mathbb{S}^m$. Using the inclusion $\mathbb{S}^m \subset \mathbb{R}^{m+1}$ and its canonical retraction $\mathbb{R}^{m+1} \setminus \{0\} \to \mathbb{S}^m : x \mapsto x/\|x\|$, one sees immediately that $\pi^m(X) = [X, \mathbb{R}^{m+1} \setminus \{0\}]$. Then, by the latter Hopf Theorem, V.3.1, p. 196, the winding number gives an inclusion $[X, \mathbb{R}^{m+1} \setminus \{0\}] \subset \mathbb{Z}$, and to conclude, it is enough to exhibit mappings $f : X \to \mathbb{R}^{m+1} \setminus \{0\}$ of arbitrary degree. This follows easily by the technique used in the preceding proof.

Let D_1 be the bounded component of $\mathbb{R}^{m+1} \setminus X$, and pick any big ball $B \supset X \cup D_1$. Let M denote the sphere that bounds B and choose a smooth mapping $h : M \to \mathbb{S}^m \subset \mathbb{R}^{m+1}$ of Brouwer-Kronecker degree $\deg(h) = d$ (V.1.1, p. 183); by IV.4.6, p. 158,

$$d = \deg(h) = w(h, 0).$$

Let $\bar{h} : \overline{B} \to \mathbb{R}^{m+1}$ be a smooth extension of h. Then $d = w(h, 0) = d(\bar{h}, B, 0)$ is computed through the preimages of any regular value a of $\bar{h}$. Now, those preimages off D_1 can be moved inside by a diffeotopy that leaves M invariant. Consequently, we can assume all preimages are in D_1, so that

$$d = d(\bar{h}, B, 0) = d(\bar{h}, D_1, 0) = w(\bar{h}|_X, 0).$$

We are done. □

Also, we can deduce the converse of the Boundary Theorem for winding numbers (IV.4.4, p. 157):

Proposition 3.3. *Let $f : X \to \mathbb{R}^{m+1} \setminus \{0\}$ be a continuous mapping with winding number $w(f, 0) = 0$. Then f has a continuous extension $\bar{f} : \overline{D} \to \mathbb{R}^{m+1} \setminus \{0\}$.*

The proof is a copy of that of V.2.4, p. 194, using V.3.1, p. 196, instead of V.2.1, p. 191. We leave it to the reader.

Exercises and problems

Number 1. We have shown, by a combination of various results, that the Hopf Theorem for the Euclidean degree (V.3.1, p. 196) follows from the Hopf Theorem for the Brouwer-Kronecker degree (V.2.1, p. 191). Show that to the contrary, the converse implication is quite immediate for hypersurfaces.

Number 2. Let $h : \mathbb{R} \to \mathbb{R}$ be a periodic continuous mapping of the form

$$h(x + 2\pi) = h(x) + 2k\pi, \quad \text{for a fixed integer } k.$$

Compute the winding number $w(f,0)$ of the continuous mapping $f : \mathbb{S}^1 \to \mathbb{R}^2 \setminus \{0\}$ induced by h.

Number 3. Let $f : \mathbb{S}^m \to \mathbb{R}^{m+1} \setminus \{0\}$ be a continuous mapping such that

$$f(\mathbb{S}^m \cap \{x_{m+1} \geq 0\}) \subset \{x_{m+1} \geq 0\} \quad \text{and} \quad f(\mathbb{S}^m \cap \{x_{m+1} \leq 0\}) \subset \{x_{m+1} \leq 0\}.$$

Then identify $\mathbb{S}^{m-1} \equiv \mathbb{S}^{m-1} \times \{0\}$ and $\mathbb{R}^m \equiv \mathbb{R}^m \times \{0\}$, so that f restricts to a mapping $g = f|_{\mathbb{S}^{m-1}} : \mathbb{S}^{m-1} \to \mathbb{R}^m \setminus \{0\}$. Prove the following:

(1) $w(g,0) = w(f,0)$.

(2) The suspension of $g/\|g\|$ (Problem Number 4 of IV.6) is homotopic to $f/\|f\|$.

Number 4. Let $X \subset R^2$ be the curve of Problem Number 5 of IV.1. Show that although X is not homeomorphic to a circle, it verifies the hypotheses of V.3.1, p. 196, and find simple representatives for all homotopy classes in $[X, \mathbb{R}^2 \setminus \{0\}]$.

4. The Hopf fibration

The Hopf Theorems in the preceding sections show how the degree fully determines the cohomotopy groups $[M, \mathbb{S}^m] = \mathbb{Z}$ when $\dim(M) = m$. We also know that $[M, \mathbb{S}^m] = \{0\}$ when $\dim(M) < m$ (V.1.4, p. 186). Thus the problem is the case $\dim(M) > m$, in particular for spheres $M = \mathbb{S}^k$. However, the degree can still be used to obtain important information concerning mappings $\mathbb{S}^{2m-1} \to \mathbb{S}^m$. The tool is the Hopf invariant introduced in III.5.1, p. 119. The additional key property of this invariant that we give here follows.

Proposition 4.1. *Let $g : \mathbb{S}^{2m-1} \to \mathbb{S}^{2m-1}$ and $f : \mathbb{S}^{2m-1} \to \mathbb{S}^m$ be smooth mappings. Then*

$$H(f \circ g) = H(f) \deg(g).$$

Proof. We can assume m even, because otherwise $H \equiv 0$. Since the Hopf invariant and the degree depend only on the homotopy class, we can at any moment change every mapping by any other one homotopic to it. This said, we distinguish several cases.

Case I: $\deg(g) = 0$. Then g is null-homotopic, and we can replace it with a constant function, so that $f \circ g$ is also constant, and the assertion is trivial.

Case II: $\deg(g) = 1$. Then g is homotopic to the identity, and for the identity the result is evident.

Case III: $\deg(g) = -1$. We can assume that g is the linear symmetry with respect to the hyperplane $x_1 = 0$, which is compatible with the stereographic projection from the south pole, and then $g = g^{-1}$. Pick regular values a, b to compute $H(f)$ and $H(f \circ g)$. We have the commutative diagram

$$\begin{array}{ccc} f^{-1}(a) \times f^{-1}(b) & \xrightarrow{\phi_{a,b}} & \mathbb{S}^{2m-2} \\ {\scriptstyle g\times g}\downarrow \approx & & \approx \downarrow {\scriptstyle g} \\ g(f^{-1}(a)) \times g(f^{-1}(b)) & \xrightarrow{\phi^g_{a,b}} & \mathbb{S}^{2m-2} \end{array}$$

because

$$\phi^g_{a,b}(g(x), g(y)) = \frac{\phi(g(y)) - \phi(g(x))}{\|\phi(g(y)) - \phi(g(x))\|} = g\Big(\frac{\phi(y) - \phi(x)}{\|\phi(y) - \phi(x)\|}\Big).$$

Consequently,

$$H(f \circ g) \deg(g \times g) = \deg(g) H(f).$$

Here, we know that $\deg(g) = -1$ (III.1.6(2), p. 100) and $\deg(g \times g) = +1$. Thus, we conclude $\deg(g)H(f) = H(f \circ g)$.

Case IV: $\deg(g) = d > 0$. Replacing g with a homotopic mapping of the type obtained in the proof of V.2.1, p. 191, we can assume that :

(1) The north pole p has a neighborhood V such that $g^{-1}(V)$ is a disjoint union of d open sets U_i and on each of them g is a diffeomorphism that preserves orientations.

(2) There are open sets $V_i \supset \overline{U}_i$, which do not meet the south pole, and after projection from the south pole, every two of them can be separated by a hyperplane.

(3) Off the V_i's everything goes to the south pole.

We denote by g_i the mapping that coincides with g on V_i and maps everything else to the south pole. By construction, $\deg(g_i) = +1$. Now, let $a, b \neq f(-p)$ be regular values of f. By a diffeotopy of $\mathbb{S}^{2m-1}$ that collapses everything to the north pole (II.6.3, p. 81), we can assume $f^{-1}(a)$ and $f^{-1}(b)$ are contained in V, and so a, b are also regular values of $f \circ g$, both distinct from $f(g(-p)) = f(-p)$. In this situation we have

$$(f \circ g)^{-1}(a) = \bigcup_i g_i^{-1}(f^{-1}(a)), \quad (f \circ g)^{-1}(b) = \bigcup_i g_i^{-1}(f^{-1}(b)),$$

so that

$$\begin{aligned} H(f \circ g) &= \ell\big((f \circ g)^{-1}(a), (f \circ g)^{-1}(b)\big) \\ &= \sum_{i,j} \ell\big((f \circ g_i)^{-1}(a), (f \circ g_j)^{-1}(b)\big). \end{aligned}$$

We look first at a pair $i \neq j$. Since the two inverse images involved are projected into U_i and U_j, respectively, there is a hyperplane L that separates those projections. Hence no line joining a point in one to a point in the other can be parallel to L, and the corresponding map into $\mathbb{S}^{2m-2}$ is not surjective and hence has degree 0. Thus we are left with

$$H(f \circ g) = \sum_i \ell((f \circ g_i)^{-1}(a), (f \circ g_i)^{-1}(b)) = \sum_i H(f \circ g_i).$$

As $\deg(g_i) = +1$, by Case II we know that $H(f \circ g_i) = H(f)$, and we conclude that

$$H(f \circ g) = H(f) \cdot d = H(f) \deg(g).$$

Case V: $\deg(g) = d < 0$. This runs like Case IV, using Case III instead of Case II, with the only modification that g reserves orientations on U_i and consequently $\deg(g_i) = -1$. □

By the preceding proposition, given $f : \mathbb{S}^{2m-1} \to \mathbb{S}^m$ with $H(f) \neq 0$, its compositions with mappings $g : \mathbb{S}^{2m-1} \to \mathbb{S}^{2m-1}$ of different degrees give non-homotopic mappings $f \circ g : \mathbb{S}^{2m-1} \to \mathbb{S}^m$. Thus we get an injection $\mathbb{Z} = \pi_{2m-1}(\mathbb{S}^{2m-1}) \to \pi_{2m-1}(\mathbb{S}^m)$.

Clearly, the best f's for this construction seem to be those with $H(f) = \pm 1$. The classical method to define them follows.

(4.2) Hopf fibrations. Let $F : \mathbb{R}^m \times \mathbb{R}^m \to \mathbb{R}^m$ be the multiplication of (i) complex numbers for $m = 2$, (ii) quaternions for $m = 4$, and (iii) octonions for $m = 8$, and define the *Hopf fibration* associated to F by

$$f : \mathbb{S}^{2m-1} \to \mathbb{S}^m : (x, y) \mapsto (\|x\|^2 - \|y\|^2, 2F(x, y)).$$

Thus we obtain smooth mappings

$$\mathbb{S}^3 \to \mathbb{S}^2, \quad \mathbb{S}^7 \to \mathbb{S}^4, \quad \mathbb{S}^{15} \to \mathbb{S}^8,$$

whose Hopf invariant is 1. The fibers of these mappings are, respectively, $\mathbb{S}^1$, $\mathbb{S}^3$, and $\mathbb{S}^7$; that is, the three mappings are *fibrations of spheres by spheres.* We will content ourselves with the proof of this for $m = 2$, the first case settled by Hopf in 1931 (see I.3, p. 30).

Proposition 4.3. *The Hopf invariant of the Hopf fibration* $\mathbb{S}^3 \to \mathbb{S}^2$ *is* 1.

Proof. The explicit equations of this mapping are

$$f(x_1, x_2, y_1, y_2) = (x_1^2 + x_2^2 - y_1^2 - y_2^2, 2x_1y_1 - 2x_2y_2, 2x_2y_1 + 2x_1y_2).$$

An easy computation shows that the poles of $\mathbb{S}^2$, $a = (0,0,-1)$ and $b = (0,0,1)$ (both $\neq (-1,0,0) = f(0,0,0,-1)$), are two regular values with inverse images:

$$\begin{cases} C_a = f^{-1}(a) = \mathbb{S}^3 \cap \{y_1 + x_2 = y_2 + x_1 = 0\}, \\ C_b = f^{-1}(b) = \mathbb{S}^3 \cap \{y_1 - x_2 = y_2 - x_1 = 0\}. \end{cases}$$

By projection from the south pole we get in $\mathbb{R}^3$ two ellipses E_a and E_b:

$$E_a : \begin{cases} v + w = 0, \\ (u-1)^2 + 2v^2 = 2, \end{cases} \qquad E_b : \begin{cases} v - w = 0, \\ (u+1)^2 + 2v^2 = 2. \end{cases}$$

They are depicted below, with the orientations ζ_a and ζ_b that they carry as inverse images.

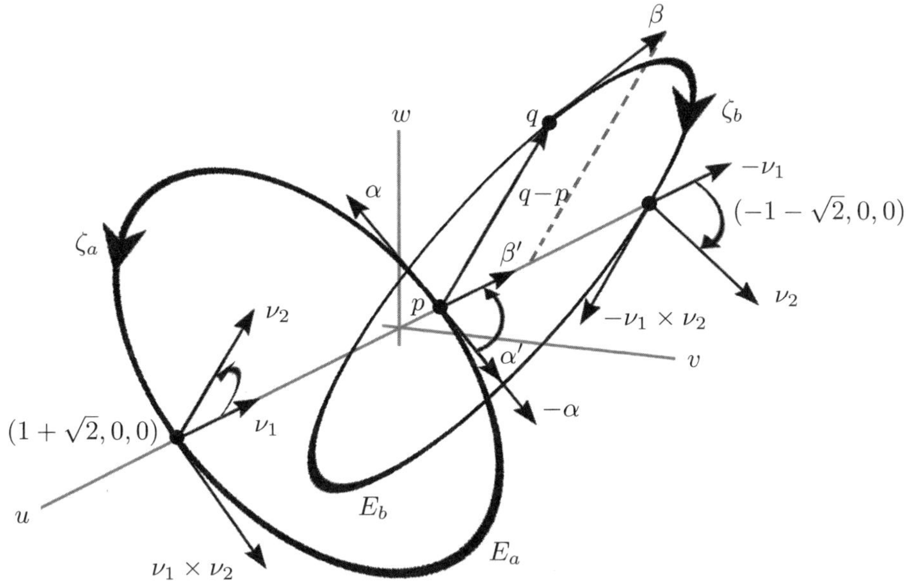

Let us describe ζ_a by means of II.7.6, p. 89. We localize f using (i) the projection from the south pole in $\mathbb{S}^3$ and (ii) the projection from the north pole in $\mathbb{S}^2$. Since both projections reverse orientation (II.7.5(3), p. 88), we get the right orientation in the ellipse E_a, which for this localization has the equations

$$\begin{cases} g_1 = \frac{1}{\delta}(4u^2 + 4v^2 - 4w^2 - t^2) = 0, \\ g_2 = \frac{1}{\delta}(8uw - 4vt) = 0, \end{cases} \quad \text{where} \quad \begin{cases} t = 1 - u^2 - v^2 - w^2, \\ \delta = (2-t)^2 - 8vw - 4ut. \end{cases}$$

The denominator δ is important (at least its sign), because it affects the derivative of the localization $g = (g_1, g_2)$ of f. Let ν_1 and ν_2 be the gradients of g_1 and g_2, respectively. Then the inverse image is oriented by the vector product $\nu_1 \times \nu_2$. Indeed, that vector product is perpendicular to the gradients, hence tangent to the inverse image, and of course $\det(\nu_1, \nu_2, \nu_1 \times \nu_2) > 0$. In addition, $d_p g$ maps ν_1, ν_2 onto a positive basis in $\mathbb{R}^2$:

$$\det(d_p g(\nu_1), d_p g(\nu_2)) = \det(\langle \nu_i, \nu_j \rangle)_{ij} = \|\nu_1\|^2 \|\nu_2\|^2 - \langle \nu_1, \nu_2 \rangle^2 > 0$$

(Cauchy-Schwartz). With this settled, as orientation is determined by one single point, we compute ν_1 and ν_2 for an easy one, namely $(1 + \sqrt{2}, 0, 0)$. For this point, $u > 0$, $v = w = 0$, $t = -2u < 0$, $\delta > 0$, and we have

$$\begin{cases} \nu_1 = \frac{1}{\delta}(8u(1-u), 0, 0) = (-, 0, 0), \\ \nu_2 = \frac{1}{\delta}(0, 8u, 8u) = (0, +, +), \end{cases}$$

so that the orientation ζ_a is that of

$$\nu_1 \times \nu_2 = \tfrac{1}{\delta^2}(0, -64u^2(1-u), 64u^2(1-u)) = (0, +, -),$$

as in the figure.

For ζ_b the argument and computations are most similar, but in this case we localize using in $\mathbb{S}^2$ the projection from the south pole, which preserves orientation. Thus $\nu_1 \times \nu_2$ does not give the right orientation, and we must take the opposite $-\nu_1 \times \nu_2$. The details are left to the reader.

Finally, we have to compute the degree of the mapping

$$\Phi : E_a \times E_b \to \mathbb{S}^2 : (p, q) \mapsto \frac{q - p}{\|q - p\|}.$$

Now, the derivative of this mapping at (p, q) is quite simple: it is the linear mapping $(\alpha, \beta) \mapsto \beta - \alpha$, followed by the orthogonal projection onto the plane perpendicular to $q - p$, plus a scaling by the factor $1/\|q - p\|$. Then we pick p, q and α, β as in the figure. It is clear that $\Phi^{-1}(\Phi(p, q)) = \{(p, q)\}$; hence we only must check that the two vectors

$$d_{(p,q)}\Phi(\alpha, 0) = \alpha', \qquad d_{(p,q)}\Phi(0, \beta) = \beta'$$

form a positive basis of $T_{\Phi(p,q)}\mathbb{S}^2$. This last linear space is the plane perpendicular to $q - p$, and α', β' form a positive basis if and only if

$$\det(q - p, \alpha', \beta') > 0.$$

But this should be clear from the picture! □

As mentioned before, once we have this, composition with the Hopf fibration gives an inclusion $\mathbb{Z} \subset \pi_3(\mathbb{S}^2)$, which shows that $\pi_3(\mathbb{S}^2)$ is infinite. As a matter of fact, it can be proved that the above inclusion is an equality; hence

$$\pi_3(\mathbb{S}^2) = \mathbb{Z}.$$

This is the first non-zero homotopy group of a sphere (excluding of course $\pi_m(\mathbb{S}^m) = \mathbb{Z}$), because $\pi_k(\mathbb{S}^1) = 0$ for $k > 1$.

Concerning the other two Hopf fibrations, it is also true that their Hopf invariant is 1; hence we have injections $\pi_{2m-1}(\mathbb{S}^{2m-1}) \to \pi_{2m-1}(\mathbb{S}^m)$, and so $\mathbb{Z} \subset \pi_{2m-1}(\mathbb{S}^m)$, but no more: it is known that

$$\pi_7(\mathbb{S}^4) = \mathbb{Z} \oplus \mathbb{Z}_4 \oplus \mathbb{Z}_3 \quad \text{and} \quad \pi_{15}(\mathbb{S}^8) = \mathbb{Z} \oplus \mathbb{Z}_8 \oplus \mathbb{Z}_3 \oplus \mathbb{Z}_5.$$

Furthermore, except for $m = 2$, the Hopf invariant does not determine the homotopy class: there are non-homotopic mappings $\mathbb{S}^7 \to \mathbb{S}^4$ (and $\mathbb{S}^{15} \to \mathbb{S}^8$) with the same Hopf invariant. On the other hand, not every integer is a Hopf invariant for arbitrary even m (see III.5.2(4)): even Hopf invariants always occur, but odd ones occur only for the crucial dimensions $m = 2, 4, 8$. All of this enters into the wide open problem of computing homotopy groups of spheres, a beautiful topic which stands far beyond our reach here.

Exercises and problems

Number 1. View the unit sphere $\mathbb{S}^{2m+1}$ in $\mathbb{C}^{m+1}$ as defined by $z_1\overline{z}_1 + \cdots + z_{m+1}\overline{z}_{m+1} = 1$. The mapping $h : \mathbb{S}^{2m+1} \to \mathbb{CP}^m : z = (z_1, \ldots, z_1) \to (z_1 : \cdots : z_{m+1})$ is also called *Hopf fibration*. Explain this name for $m = 1$. Use the fact that $\pi_{2m+1}(\mathbb{S}^1) = 0$ to show that two continuous mappings $f, g : \mathbb{S}^{2m+1} \to \mathbb{S}^{2m+1}$ such that $h \circ f = h \circ g$ must have the same degree. Compute the degree of all $\sigma : \mathbb{S}^{2m+1} \to \mathbb{S}^{2m+1}$ such that $h(\sigma(z)) = h(\overline{z})$. Compare this with Problem Number 7 of III.1.

Number 2. Let $f : \mathbb{S}^{2m-1} \to \mathbb{S}^m$ and $g : \mathbb{S}^m \to \mathbb{S}^m$ be smooth mappings ($m \geq 2$). Show that

$$H(g \circ f) = \deg(g)^2 H(f).$$

Number 3. Let $f : \mathbb{S}^{2m-1} \to \mathbb{S}^m$ be a continuous mapping with non-zero Hopf invariant (not necessarily 1). Let $\overline{g} : \mathbb{S}^{2m-1} \to \mathbb{S}^{2m-1}$ be a continuous lifting of $g : \mathbb{S}^m \to \mathbb{S}^m$ (that is, $f \circ \overline{g} = g \circ f$). Show that $\deg(\overline{g}) = \deg(g)^2$, and deduce that few continuous mappings are liftings. Deduce also that liftings of homotopic mappings are homotopic. What about the other way around?

Number 4. Let $g : \mathbb{S}^{2m-1} \to \mathbb{S}^{2m-1}$ and $f : \mathbb{S}^{2m-1} \to \mathbb{S}^m$ be smooth mappings ($m \geq 2$). Show that

$$\varrho(f \circ g) = \varrho(f) \deg(g),$$

where ϱ is the invariant defined in Problem Number 1 of III.5.

Number 5. Let $f : \mathbb{S}^{2m-1} \to \mathbb{S}^m$ and $g : \mathbb{S}^m \to \mathbb{S}^m$ be smooth mappings ($m \geq 2$). Prove the formula

$$\varrho(g \circ f) = \deg(g)^2 \varrho(f).$$

Number 6. Compute the invariant $\varrho(f)$ of the Hopf fibration $f : \mathbb{S}^3 \to \mathbb{S}^2$.

5. Singularities of tangent vector fields

In this section we describe the information borne out by tangent vector fields on a manifold. That information is disclosed by the computation of the so-called *index of a tangent vector field.*

Let $M \subset \mathbb{R}^p$ be an oriented, boundaryless, smooth manifold of dimension m.

(5.1) Non-degenerate zeros of a tangent vector field. A (smooth) tangent vector field on M is a smooth mapping $\xi : M \to \mathbb{R}^p$ such that $\xi(z) \in T_zM$ for every $z \in M$. We can represent ξ in local coordinates as follows. Let $\varphi : U \to M$ be a parametrization of M. For $x \in U$ set $z = \varphi(x) \in M$. The tangent space T_zM is generated by the basis of partial derivatives

$$\frac{\partial}{\partial \mathbf{x}_i}\Big|_z = d_x\varphi(e_i) = \frac{\partial \varphi}{\partial x_i}(x), \quad 1 \leq i \leq m.$$

Hence we can write

$$\xi(z) = \xi(\varphi(x)) = \sum_{i=1}^m \xi_i(x) \frac{\partial}{\partial \mathbf{x}_i}\Big|_z = \sum_{i=1}^m \xi_i(x) \frac{\partial \varphi}{\partial x_i}(x),$$

where the functions $\xi_i : U \to \mathbb{R}$ are uniquely determined and smooth. We will also consider the smooth mapping

$$\bar{\xi} : U \to \mathbb{R}^m : x \mapsto (\xi_1(x), \ldots, \xi_m(x)).$$

Now suppose $z = \varphi(x)$ is a *zero (or singularity)* of ξ, that is, $\xi_i(x) = 0$ for all i. Then

$$\begin{aligned} d_z\xi\Big(\frac{\partial}{\partial \mathbf{x}_j}\Big|_z\Big) &= d_x(\xi \circ \varphi)(e_j) = \frac{\partial}{\partial x_j}(\xi \circ \varphi)(x) \\ &= \sum_{i=1}^m \frac{\partial \xi_i}{\partial x_j}(x) \frac{\partial \varphi}{\partial x_i}(x) + \sum_{i=1}^m \xi_i(x) \frac{\partial^2 \varphi}{\partial x_j \partial x_i}(x) \\ &= \sum_{i=1}^m \frac{\partial \xi_i}{\partial x_j}(x) \frac{\partial \varphi}{\partial x_i}(x) = \sum_{i=1}^m \frac{\partial \xi_i}{\partial x_j}(x) \frac{\partial}{\partial \mathbf{x}_i}\Big|_z. \end{aligned}$$

Thus, we actually have a linear mapping $d_z\xi : T_zM \to T_zM \subset \mathbb{R}^p$, whose matrix with respect to the basis of partial derivatives is the Jacobian matrix $J_x\bar{\xi}$ of the localization $\bar{\xi}$. In particular, the determinant of $d_z\xi$ is a well-defined invariant of ξ:

$$\det(d_z\xi) = \det(J_x\bar{\xi}), \quad J_x\bar{\xi} = \Big(\frac{\partial \xi_i}{\partial x_j}(x)\Big)_{i,j}.$$

We will say that the zero z of ξ is *non-degenerate* when this determinant is not zero. This is equivalent to saying that the localization $\bar{\xi} : U \to \mathbb{R}^m$ is a local diffeomorphism at x, hence locally injective, so that our zero z is *isolated.* If z is non-degenerate, the sign ± 1 of $\det(d_z\xi) \neq 0$ is called the *index of* ξ *at* z and is denoted by

$$\mathrm{ind}_z(\xi) = \mathrm{sign}\det(d_z\xi) = \mathrm{sign}_x(\bar{\xi}).$$

Examples 5.2. Below there are some basic examples of vector fields on the plane, with a single non-degenerate zero, and the corresponding index. What is depicted is the *flow* generated by the field, which consists of the *integral lines* whose tangents are prescribed by the field. They come from an associated differential equation, and the investigation of their nature is a highly interesting topic beyond our purposes here.

Source (or *impulsor*)

$$\begin{cases} \xi = x\frac{\partial}{\partial x} + y\frac{\partial}{\partial y}, \quad \bar{\xi} = (x,y), \\ d_{(0,0)}\bar{\xi} = \begin{pmatrix} 1 & 0 \\ 0 & 1 \end{pmatrix}, \\ \mathrm{ind}_0\,\xi = \mathrm{sign}_0\,\bar{\xi} = +1 \end{cases}$$

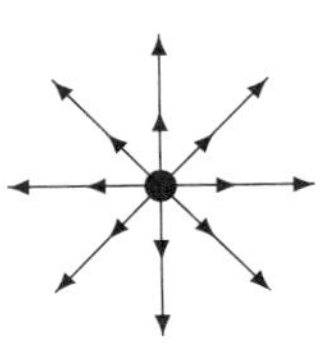

Sink (or *attractor*)

$$\begin{cases} \xi = -x\frac{\partial}{\partial x} - y\frac{\partial}{\partial y}, \quad \bar{\xi} = (-x,-y), \\ d_{(0,0)}\bar{\xi} = \begin{pmatrix} -1 & 0 \\ 0 & -1 \end{pmatrix}, \\ \mathrm{ind}_0\,\xi = \mathrm{sign}_0\,\bar{\xi} = +1 \end{cases}$$

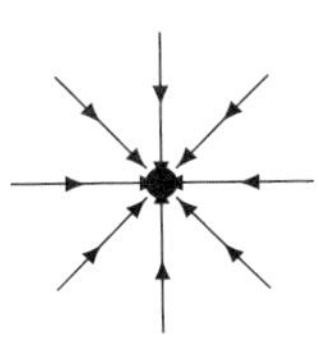

Saddle

$$\begin{cases} \xi = x\frac{\partial}{\partial x} - y\frac{\partial}{\partial y}, \quad \bar{\xi} = (x,-y), \\ d_{(0,0)}\bar{\xi} = \begin{pmatrix} 1 & 0 \\ 0 & -1 \end{pmatrix}, \\ \mathrm{ind}_0\,\xi = \mathrm{sign}_0\,\bar{\xi} = -1 \end{cases}$$

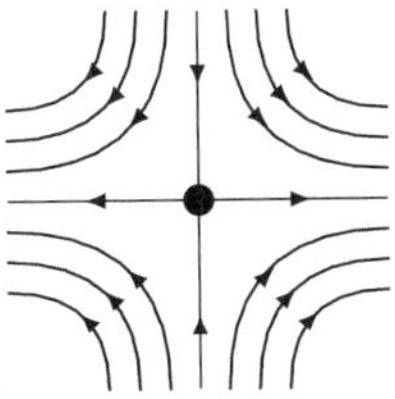

Circulation (or *center*)

$$\begin{cases} \xi = -y\frac{\partial}{\partial x} + x\frac{\partial}{\partial y}, \quad \bar{\xi} = (-y, x), \\ d_{(0,0)}\bar{\xi} = \begin{pmatrix} 0 & -1 \\ 1 & 0 \end{pmatrix}, \\ \operatorname{ind}_0 \xi = \operatorname{sign}_0 \bar{\xi} = +1 \end{cases}$$

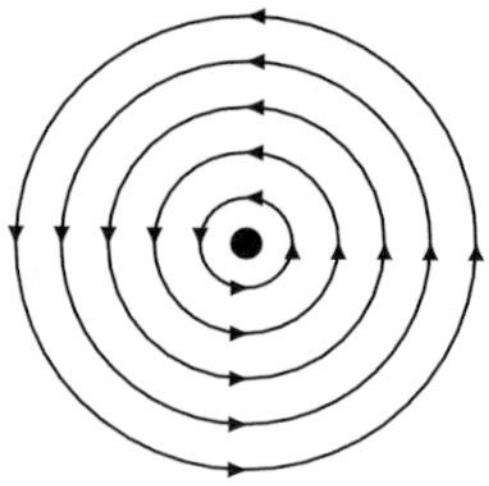

□

But we want to deal with arbitrary isolated zeros, not only the non-degenerate. This is achieved as follows.

Proposition and Definition 5.3 (Splitting of an isolated zero). *Let $\xi : M \to \mathbb{R}^p$ be a smooth tangent vector field. Let Ω be an open set where ξ has a unique zero z. Then there are an open neighborhood U of z, with $\overline{U} \subset \Omega$, and a smooth tangent vector field $\zeta : M \to \mathbb{R}^p$ such that we have the following:*

(1) *$\overline{U}$ is a smooth compact manifold with boundary $Z = \overline{U} \setminus U$.*
(2) *$\zeta \equiv \xi$ off U.*
(3) *ζ has finitely many zeros in Ω, all of them in U, and all of them non-degenerate.*

We say that ζ is a splitting of ξ at z.

Proof. The question is clearly local; hence we can assume $\Omega = \mathbb{R}^m$, $\xi = \bar{\xi}$, and $z = 0$. Then, pick a regular value a of ξ with $\|a\| < \|\xi(x)\|$ for $1 \leq \|x\| \leq 2$. Also, let θ we a smooth bump function with $0 \leq \theta \leq 1$, $\theta \equiv 1$ on $\|x\| \leq 1$, and $\theta \equiv 0$ on $\|x\| \geq 2$. Then set $\zeta = \xi - a\theta$. We have:

(i) $\zeta \equiv \xi$ off $U : \|x\| < 2$ and hence has no zeros off U.

(ii) $\|\xi\| > \|a\| \geq \|a\theta\|$ on $1 \leq \|x\| \leq 2$; hence ζ has no zeros off $V : \|x\| < 1$.

(iii) $\zeta = \xi - a$ on V; hence the zeros of ζ in V are the points in $V \cap \xi^{-1}(a)$. As a is a regular value of ξ, those zeros are finitely many and non-degenerate.

Finally, $\overline{U}$ is a closed ball in $\mathbb{R}^m$, hence a compact manifold with boundary the sphere $Z : \overline{U} \setminus U : \|x\| = 2$. □

This splitting is the key step to defining the index at isolated degenerate zeros:

Proposition and Definition 5.4. *Let $\xi : M \to \mathbb{R}^p$ be a smooth tangent vector field, and let $z \in M$ be an isolated zero of ξ. Let ζ be a splitting of ξ at z in the domain Ω corresponding to a parametrization φ. Then*

$$\sum_{z^*} \operatorname{ind}_{z^*}(\zeta) = \deg\Big(\frac{\bar{\xi}}{\|\bar{\xi}\|}\Big|_S \Big)$$

where the sum ranges over all (non-degenerate) zeros z^* of ζ and S is a small enough sphere centered at $x = \varphi^{-1}(z)$. Hence, this sum does not depend on either the splitting or the parametrization.

This sum is the index of ξ at z. *We write*

$$\operatorname{ind}_z(\xi) = \sum_{z^*} \operatorname{ind}_{z^*}(\zeta).$$

If ξ has finitely many zeros, then all of them are isolated, and the sum of the indices of ξ at the zeros is the total index of ξ. *We write*

$$\operatorname{Ind}(\xi) = \sum_{z} \operatorname{ind}_z(\xi)$$

(the sum is zero if the field has no zeros).

Proof. Consider the open neighborhood $U \subset \Omega$ associated to the splitting, and its boundary $Z = \overline{U} \setminus U$. Set $X = \varphi^{-1}(Z)$ and $x^* = \varphi^{-1}(z^*)$, so that

$$\begin{aligned}
\sum_{z^*} \operatorname{ind}_{z^*}(\zeta) &= \sum_{x^*} \operatorname{sign}_{x^*}(\bar{\zeta}) && \text{(by definition of index at a non-degenerate zero)} \\
&= \deg\Big(\frac{\bar{\zeta}}{\|\bar{\zeta}\|}\Big|_X \Big) && \text{(by IV.4.5, p. 157)} \\
&= \deg\Big(\frac{\bar{\xi}}{\|\bar{\xi}\|}\Big|_X \Big) && \text{(because } \xi \equiv \zeta \text{ on } Z) \\
&= \deg\Big(\frac{\bar{\xi}}{\|\bar{\xi}\|}\Big|_S \Big) && \text{(by the Boundary Theorem, III.3.3, p. 111)}
\end{aligned}$$

Now, the right-hand side does not depend on ζ, and the left-hand side does not depend on φ; hence none depends on either ζ or φ. □

Examples 5.5. Here we give two examples of degenerate isolated zeros with indices ± 2:

$$\textit{Dipole}\quad\begin{cases}\xi = -2xy\frac{\partial}{\partial x} + (x^2-y^2)\frac{\partial}{\partial y},\\ f = \dfrac{\bar{\xi}}{\|\bar{\xi}\|} = \dfrac{(-2xy, x^2-y^2)}{x^2+y^2},\\ \operatorname{ind}_0 \xi = \deg(f|_{\|x\|=\varepsilon}) = +2\end{cases}$$

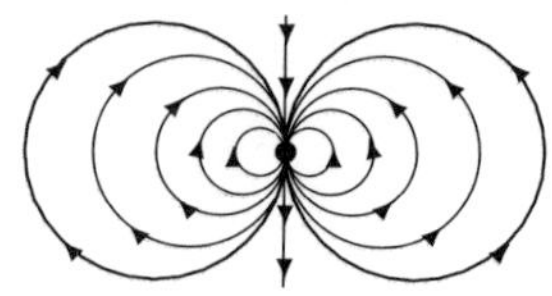

$$\begin{cases}\xi = (x^2-y^2)\frac{\partial}{\partial x} - 2xy\frac{\partial}{\partial y},\\ f = \dfrac{\bar{\xi}}{\|\bar{\xi}\|} = \dfrac{(x^2-y^2, -2xy)}{x^2+y^2},\\ \operatorname{ind}_0 \xi = \deg(f|_{\|x\|=\varepsilon}) = -2\end{cases}$$

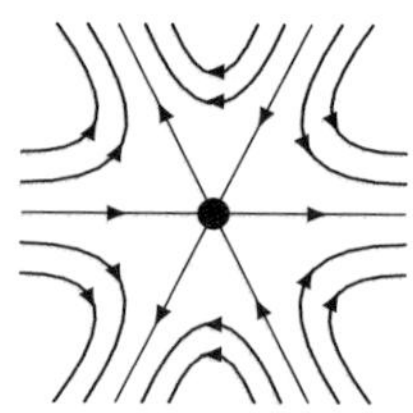

□

Exercises and problems

Number 1. Show that the only possible indices at an isolated zero of a tangent vector field on a curve are $0, +1$, and -1.

Number 2. Produce smooth tangent vector fields on the plane $\mathbb{R}^2$ with isolated zeros of arbitrary indices.

Number 3. Let M be a compact, boundaryless, smooth manifold of dimension m. Suppose M is *parallelizable*, that is, there are tangent vector fields $\xi_1, \dots, \xi_m$ such that $\{\xi_{1,x}, \dots, \xi_{m,x}\}$ is a basis of T_xM at each point $x \in M$; this defines an orientation on M. Show that:

(1) Each (smooth) tangent vector field ξ on M is defined by $\xi = \sum_{i=1}^m f_i\xi_i$, where every $f_i : M \to \mathbb{R}$ is a smooth function.

(2) A zero of ξ is a zero of the mapping $f = (f_1, \dots, f_m) : M \to \mathbb{R}^m$, and it is non-degenerate if and only if it is a regular point of f.

(3) If ξ has only non-degenerate zeros, then it has finitely many, and $\operatorname{Ind}(\xi) = \deg(f)$.

Conclude that *on a parallelizable compact manifold, all tangent fields with isolated zeros have total index zero.*

Number 4. Let ξ be a smooth tangent vector field on a smooth manifold M, and let $f : M \to \mathbb{R}$ be a smooth function. Consider the tangent vector field $\zeta_z = f(z)\xi_z$. Study the zeros of ζ in terms of those of ξ. When are they isolated? When non-degenerate? Can something be said concerning indices in the latter case?

Number 5. Use the dipole (V.5.5 above) and stereographic projection to obtain a tangent vector field on $\mathbb{S}^2$ with a unique zero of index $+2$.

Number 6. Consider the vector field on $\mathbb{R}^3$ defined by

$$\xi = xz^2\frac{\partial}{\partial x} + yz^2\frac{\partial}{\partial y} + z(z^2-1)\frac{\partial}{\partial z},$$

and let $M \subset \mathbb{R}^3$ be the surface $x^2 + y^2 = z^2 - 1$, $z > 0$. Show that ξ restricts to a tangent vector field on M with a unique zero, and compute its index.

Number 7. Let ξ (resp., ζ) be a tangent vector field on a manifold M (resp., N) with an isolated zero $x \in M$ (resp., $y \in N$). Prove that $\xi \times \zeta$ is a tangent vector field on $M \times N$, that $z = (x, y)$ is an isolated zero of $\xi \times \zeta$, and that

$$\operatorname{ind}_z(\xi \times \zeta) = \operatorname{ind}_x(\xi) \operatorname{ind}_y(\zeta).$$

6. Gradient vector fields

We will discuss here a particular case of tangent vector fields and compute their indices. We start with the following:

Proposition and Definition 6.1. *Let $M \subset \mathbb{R}^p$ be a boundaryless smooth manifold of dimension m, and let $f : M \to \mathbb{R}$ be a smooth function. Then there is a unique smooth tangent vector field ξ such that*

$$d_z f(u) = \langle u, \xi_z \rangle \quad \text{for all } u \in T_z M,\ z \in M,$$

where $\langle \cdot, \cdot \rangle$ stands for the Euclidean scalar product in $\mathbb{R}^p$.

This vector field ξ is called the gradient of f *and is denoted* $\operatorname{grad}(f)$.

Of course, if M is an open set of $\mathbb{R}^p$, then $\operatorname{grad}(f) = \left(\frac{\partial f}{\partial x_1}, \dots, \frac{\partial f}{\partial x_p}\right)$.

Proof. First recall that for every $z \in M$ the mapping $v \mapsto \langle \cdot, v \rangle$ is a linear isomorphism from $T_z M$ onto its dual space $\mathcal{L}(T_z M, \mathbb{R})$ (this is the *Riesz Representation Theorem*). Consequently, ξ is uniquely defined by the condition in the statement, and we must check that it is indeed a *smooth* tangent vector field, which is a local matter. Thus, consider any parametrization φ of an open neighborhood U of $z \in M$ and the corresponding localization

$$\xi|_U = \sum_{k=1}^{m} \xi_k \frac{\partial}{\partial \mathrm{x}_k}, \qquad \bar{\xi} = (\xi_1, \dots, \xi_m),$$

and let us show that the ξ_k's are smooth functions. We have

$$\begin{aligned}\frac{\partial (f \circ \varphi)}{\partial x_i}(x) &= d_{\varphi(x)} f\Big(\frac{\partial \varphi}{\partial x_i}(x)\Big) = \Big\langle \frac{\partial \varphi}{\partial x_i}(x), \xi_{\varphi(x)} \Big\rangle \\ &= \sum_{k=1}^{m} \Big\langle \frac{\partial \varphi}{\partial x_i}(x), \frac{\partial \varphi}{\partial x_k}(x) \Big\rangle \xi_k(x).\end{aligned}$$

Thus we obtain a linear system with unknowns the ξ_k's, and the matrix $G_x\varphi$ of this system is the Euclidean *Gramm matrix* of the basis of partial derivatives, which is the matrix of the Euclidean scalar product with respect to that basis. Clearly such a Gramm matrix is positive definite and in particular has determinant > 0. Consequently, we can solve the system and find smooth expresions for the ξ_k's. We are done. □

Note that the critical points of $f : M \to \mathbb{R}$ are exactly the zeros of its gradient (and this is why zeros of tangent vector fields are called singularities). We want to determine those which are non-degenerate zeros (V.5.1, p. 206). Such a non-degenerate zero is called a *non-degenerate critical point of f*.

(6.2) The Jacobian of a gradient vector field at a zero. Let $f : M \to \mathbb{R}$ be a smooth function as above, and set $\xi = \operatorname{grad}(f)$. Let z be a critical point of f, that is, a zero of ξ. We consider a parametrization φ of an open neighborhood U of z with, say, $z = \varphi(x)$, and as in the preceding proof we have

$$\begin{cases} \xi|_U = \sum_k \xi_k \frac{\partial}{\partial \mathrm{x}_k}, \quad \bar{\xi} = (\xi_1, \dots, \xi_m), \\ \frac{\partial(f\circ\varphi)}{\partial x_i} = \sum_k \left\langle \frac{\partial \varphi}{\partial x_i}, \frac{\partial \varphi}{\partial x_k} \right\rangle \xi_k \,. \end{cases}$$

We are interested in the following Jacobian determinant:

$$\det(d_z\xi) = \det(J_x\bar{\xi}), \quad J_x\bar{\xi} = \Big(\frac{\partial \xi_i}{\partial x_j}(x)\Big)_{i,j}.$$

Let us derive the formulas above:

$$\frac{\partial^2(f\circ\varphi)}{\partial x_j \partial x_i} = \sum_k \frac{\partial\langle \frac{\partial\varphi}{\partial x_i}, \frac{\partial\varphi}{\partial x_k}\rangle}{\partial x_j}\xi_k + \sum_k \Big\langle \frac{\partial\varphi}{\partial x_i}, \frac{\partial\varphi}{\partial x_k}\Big\rangle \frac{\partial \xi_k}{\partial x_j}.$$

Since $z = \varphi(x)$ is a zero of ξ, the first sum vanishes at x, and so

$$\frac{\partial^2(f\circ\varphi)}{\partial x_j \partial x_i}(x) = \sum_k \Big\langle \frac{\partial\varphi}{\partial x_i}(x), \frac{\partial\varphi}{\partial x_k}(x)\Big\rangle \frac{\partial \xi_k}{\partial x_j}(x).$$

Again we find the Gramm matrix $G_x\varphi$, and introducing the *symmetric* matrix

$$H_x(f\circ\varphi) = \Big(\frac{\partial^2(f\circ\varphi)}{\partial x_j \partial x_i}(x)\Big)_{i,j},$$

we rewrite the sum in matricial form as follows:

$$H_x(f\circ\varphi) = G_x\varphi \cdot J_x\bar{\xi}.$$

Since $\det(G_x\varphi) > 0$, this completely determines the matrix $J_x\bar{\xi}$, hence the linear mapping $d_z\xi$. Furthermore, we get

$$\operatorname{ind}_z(\xi) = \operatorname{sign}\det(d_z\xi) = \operatorname{sign}\det\big(H_x(f\circ\varphi)\big).$$

Here we include the case ind $=$ sign $= 0$, meaning that z is a degenerate zero of ξ, that is, z is a degenerate critical point of f.

In the previous paragraph we encountered the *Hessian* of the localization $f\circ\varphi$. In fact, this is an intrinsic notion:

Proposition and Definition 6.3. *In the setting above and with all of the notation there, consider on T_zM the quadratic form $Q_z(f)$ whose matrix with respect to the basis $\{\frac{\partial}{\partial x_i}\big|_z\}$ is $H_x(f\circ\varphi)$. Let a tangent vector $u \in T_zM$ be defined by a curve germ $\gamma(t)$ in M, that is, $\gamma(0) = z$, $\gamma'(0) = u$. Then*

$$Q_z(f)(u) = (f\circ\gamma)''(0).$$

In particular, $Q_z(f)$ does not depend on the parametrization φ. This quadratic form is the Hessian of f at z.

Proof. This is a straightforward computation. Let $(u_1,\dots,u_m)$ be the coordinates of u with respect to the basis of the partial derivatives. Then

$$Q_z(f)(u) = \sum_{ij}\frac{\partial^2(f\circ\varphi)}{\partial x_i\partial x_j}(x)\,u_i\,u_j.$$

Now, set

$$\bar{\gamma} = \varphi^{-1}\circ\gamma = (\gamma_1,\dots,\gamma_m),$$

so that the above coordinates are $u_i = \gamma_i'(0)$. We will compute $(f\circ\gamma)''(0)$. First,

$$(f\circ\gamma)'(t) = \big((f\circ\varphi)\circ\bar{\gamma}\big)'(t) = \sum_i \frac{\partial(f\circ\varphi)}{\partial x_i}(\bar{\gamma}(t))\gamma_i'(t),$$

and then

$$(f\circ\gamma)''(t) = \sum_{ij}\frac{\partial^2(f\circ\varphi)}{\partial x_j\partial x_i}(\bar{\gamma}(t))\gamma_j'(t)\gamma_i'(t) + \sum_i\frac{\partial(f\circ\varphi)}{\partial x_i}(\bar{\gamma}(t))\gamma_i''(t).$$

For $t = 0$ we obtain

$$(f\circ\gamma)''(0) = \sum_{ij}\frac{\partial^2(f\circ\varphi)}{\partial x_j\partial x_i}(x)\,u_j\,u_i$$

(the second sum vanishes because $x = \bar{\gamma}(0)$ is a zero of all the partial derivatives $\frac{\partial(f\circ\varphi)}{\partial x_i}$). □

The Hessian is used to define the index of a critical point:

Definition 6.4. *Let $f : M \to \mathbb{R}$ be a smooth function and let $z \in M$ be a non-degenerate critical point of f. The index (= the number of negative eigenvalues) of the Hessian $Q_z(f)$ is called the* index of f at z *and is denoted by* $\operatorname{ind}_z(f)$.

With this we can compute indices of gradient vector fields:

Proposition 6.5. *Let $f : M \to \mathbb{R}$ be a smooth function and let $\xi = \operatorname{grad}(f)$. Let $z \in M$ be a non-degenerate critical point of f. Then*

$$\operatorname{ind}_z(\xi) = (-1)^{\operatorname{ind}_z(f)}.$$

Proof. Keeping all preceding notation, we already know that

$$\operatorname{ind}_z(\xi) = \operatorname{sign}\det\big(H_x(f\circ\varphi)\big),$$

where $H_x(f\circ\varphi)$ is the symmetric matrix of $Q_z(f)$ with respect to a suitable base. Recall now that the determinant of a matrix is the product of its eigenvalues. As $\operatorname{ind}_x(f)$ is precisely the number of those that are negative, the assertion follows. □

(6.6) Morse functions. A *Morse function* $f : M \to \mathbb{R}$ is a smooth function whose critical points are all non-degenerate, equivalently, such that all the zeros of its gradient vector field $\xi = \operatorname{grad}(f)$ are non-degenerate. Thus, they are isolated, and in case there are finitely many (for instance, if M is compact), the total index of ξ is

$$\operatorname{Ind}(\xi) = \sum_z (-1)^{\operatorname{ind}_z(f)},$$

where the z's are the critical points of f. This can be rewritten as

$$\operatorname{Ind}(\xi) = \sum_k (-1)^k \alpha_k\,,$$

where α_k is the number of critical points with index k.

Of course, the question is whether or not Morse functions exist. The answer is that they do. In fact, we will see that *many a linear form is a Morse function.* We formulate this as follows.

(6.7) Height functions. Let $M \subset \mathbb{R}^p$ be a boundaryless smooth manifold of dimension m. For every $a \in \mathbb{R}^p$ consider the smooth function

$$f_a : M \to \mathbb{R} : x \mapsto \langle x, a \rangle.$$

This can be seen as a *height function* with respect to the linear hyperplane H perpendicular to a: the *oriented distance to* H is given by

$$\operatorname{dist}(x, H) = \tfrac{1}{\|a\|} f_a(x).$$

Also, note that after a linear change of coordinates in $\mathbb{R}^p$, we have $a = (0, \dots, 0, 1)$, and then $f_a(x) = x_p$.

Finally, denote by $\mathcal{M}$ the set of all $a \in \mathbb{R}^p$ such that f_a is a Morse function.

The set $\mathcal{M}$ is quite explicitly determined for hypersurfaces:

Proposition 6.8. *Let $M \subset \mathbb{R}^{m+1}$ be a connected, closed, boundaryless, smooth hypersurface, and let $\nu : M \to \mathbb{S}^m$ be a unitary global normal field. Let $a \in \mathbb{R}^{m+1}$. Then*

$$f_a : M \to \mathbb{R} : x \mapsto \langle x, a \rangle$$

is a Morse function if and only if $\frac{1}{\|a\|} a \in \mathbb{S}^m$ is a regular value of both ν and $-\nu$.

In particular, the set $\mathcal{M}$ is a residual, hence dense, subset of $\mathbb{R}^{m+1}$.

Proof. Recall that ν exists by III.6.4, p. 129. On the other hand, since $f_{\lambda a} = \lambda f_a$, we can assume $\|a\| = 1$. Now, we look for the critical points z of $f = f_a$. Clearly, $d_z f = \langle \cdot, a \rangle|_{T_z M}$ vanishes if and only if a is perpendicular to $T_z M$, that is, if and only if $a = \pm\nu(z)$. Suppose this is true, and let us see when the critical point z is non-degenerate. We first claim that

$$Q_z(f)(u) = \langle u, \mp d_z\nu(u) \rangle.$$

Indeed, let γ be a curve germ in M with $\gamma(0) = z$, $\gamma'(0) = u$. Then we derive

$$(f \circ \gamma)(t) = \langle \gamma(t), a \rangle$$

twice to get

$$Q_z(f)(u) = (f \circ \gamma)''(0) = \langle \gamma''(0), a \rangle.$$

But on the other hand, deriving $\langle \gamma'(t), \nu(\gamma(t)) \rangle = 0$, we see that

$$0 = \langle \gamma''(t), \nu(\gamma(t)) \rangle + \langle \gamma'(t), d_z\nu(\gamma'(t)) \rangle,$$

which for $t = 0$ gives

$$\langle \gamma''(0), \nu(z) \rangle = \langle u, -d_z\nu(u) \rangle.$$

But $\nu(z) = \pm a$; hence $\langle \gamma''(0), a \rangle = \langle u, \mp d_z\nu(u) \rangle$, and the claim follows.

Now, the claim implies that the quadratic form $Q_z(f)$ has, with respect to any orthonormal basis of T_zM, the same matrix as the *self-adjoint* endomorphism $\mp d_z\nu$ (III.2.6, p. 108), hence the same determinant. We have thus seen that z is non-degenerate if and only if it is not a critical point of $\pm\nu$. From this, the result follows readily.

The last assertion in the proposition is an easy consequence of the Sard-Brown Theorem. □

For arbitrary manifolds one shows a slightly weaker result in a less explicit way:

Proposition 6.9. *Let $M \subset \mathbb{R}^p$ be a boundaryless smooth manifold of dimension m. As above, let $\mathcal{M}$ be the set of all points $a \in \mathbb{R}^p$ such that the mapping*

$$f_a : M \to \mathbb{R} : x \mapsto \langle x, a \rangle$$

is a Morse function. Then $\mathcal{M}$ is dense in $\mathbb{R}^p$.

Proof. First we remark that every point in M has a neighborhood U on which some linear projection $\pi : x \mapsto (x_{i_1}, \ldots, x_{i_m})$ defines a local coordinate system. Indeed, $d_p\pi = \pi$ must be injective on T_pM for some choice of indices i_k, and then the Inverse Function Theorem applies. Moreover, since M has a countable basis of open sets, it has a countable covering consisting of such open coordinate domains as U.

Now pick any U as above, and, to symplify notation, suppose the projection π corresponds to the indices $i_k = k$; let $x = (x', x'') \in \mathbb{R}^p = \mathbb{R}^m \times \mathbb{R}^{p-m}$. We claim that the following statement holds true:

Let $a'' \in \mathbb{R}^{p-m}$, and denote by $\mathcal{M}_U(a'')$ the set of all $a' \in \mathbb{R}^m$ such that

$$f_a|_U : U \to \mathbb{R} : x \mapsto \langle x', a' \rangle + \langle x'', a'' \rangle$$

is a Morse function. Then the set $\mathbb{R}^m \setminus \mathcal{M}_U(a'')$ has measure zero.

Assume this for the moment, and consider the set $\mathcal{M}_U$ of all $a \in \mathbb{R}^p$ such that $f_a|_U$ is a Morse function. What the claim says is that $\mathbb{R}^p \setminus \mathcal{M}_U$ cuts every $(p-m)$-plane $x'' = a''$ in a measure zero set; hence, by the Fubini Theorem, $\mathbb{R}^p \setminus \mathcal{M}_U$ itself has measure zero. Since there are countably many U's, the union $\bigcup_U (\mathbb{R}^p \setminus \mathcal{M}_U)$ also has measure zero, hence has empty interior, and its complement $\bigcap_U \mathcal{M}_U$ is dense. As, obviously, $\mathcal{M} \supset \bigcap_U \mathcal{M}_U$, we are done.

To complete the argument, let us prove the claim. Since $\pi|_U$ is a diffeomorphism onto an open set $W \subset \mathbb{R}^m$, its inverse has the form $x' \mapsto (x', \phi(x'))$, for a suitable smooth function $\phi : W \to \mathbb{R}^{p-m}$. Then the localization of $f_a|_U$ is

$$g_a : W \to \mathbb{R} : x' \mapsto \langle x', a' \rangle + h(x'), \quad \text{where } h(x') = \langle \phi(x'), a'' \rangle,$$

and straight from the definition one sees that g_a is a Morse function if and only if $-a'$ is a regular value of the gradient $\mathrm{grad}(h) : W \to \mathbb{R}^m$. Consequently, the claim follows from the proof of the Sard-Brown Theorem (see (*) on p. 64). □

Exercises and problems

Number 1. Let $M \subset \mathbb{R}^p$ be a smooth manifold and let $F : \mathbb{R}^p \to \mathbb{R}$ be a smooth function; set $f = F|M$. Describe $\mathrm{grad}(f)$ in terms of $\mathrm{grad}(F)$ and orthogonal projections into the tangent spaces of M.

Number 2. Is there any smooth function $f : \mathbb{S}^2 \to \mathbb{R}$ with gradient $-y\frac{\partial}{\partial x} + x\frac{\partial}{\partial y}$?

Number 3. Let f (resp., g) be a Morse function on a manifold M (resp., N). Show that

$$h : M \times N \to \mathbb{R} : (x, y) \mapsto f(x) + g(y)$$

is a Morse function on $M \times N$, and describe its critical points, as well as their indices, in terms of those of f and g.

Number 4. Let f and g be Morse functions on two compact manifolds M and N, respectively. Show that for any two positive real numbers a and b large enough, the function

$$h : M \times N \to \mathbb{R} : (x, y) \mapsto (a + f(x))(b + g(y))$$

is a Morse function on $M \times N$. Describe the critical points of h and their indices.

Number 5. Let $M = SO(3) \subset \mathbb{R}^{3\times 3}$ be the *rotation group* in $\mathbb{R}^3$, consisting of all 3×3 matrices $x = (x_{ij})$ whose columns form a positive orthonormal basis of $\mathbb{R}^3$. This M is a boundaryless, compact, smooth manifold of dimension 3. Show that

$$h(x) = 2x_{11} + 3x_{22} + 4x_{33}$$

is a Morse function on M and that it has four critical points with indices $0, 1, 2, 3$.

Number 6. Study when the function $f : \mathbb{RP}^m \to \mathbb{R} : (x_0 : \ldots : x_m) \mapsto \frac{1}{\|x\|^2} \sum_k c_k x_k^2$ is a Morse function. Then, describe its critical points and their indices.

7. The Poincaré-Hopf Index Theorem

Now that the notion of total index has been described, we can turn in this last section to the main fact that the total index is an invariant of the manifold, independent of the tangent vector field used to compute it. To prove this, we must take a closer look at tubular neighborhoods.

(7.1) Normal vector fields on tubes. Let $M \subset \mathbb{R}^p$ be a compact boundaryless smooth manifold of dimension m. Then, M has a tubular retraction $\rho : U \to M$ (II.4.3, p. 71) such that $x - \rho(x)$ is perpendicular to $T_{\rho(x)}M$. Consider the smooth function

$$\tau : U \to \mathbb{R} : x \mapsto \|x - \rho(x)\|^2.$$

As $x - \rho(x)$ is perpendicular to $d_x\rho(\mathbb{R}^p) \subset T_{\rho(x)}M$, it follows easily that

$$\operatorname{grad}_x(\tau) = 2(x - \rho(x)).$$

This gradient vanishes exactly for $x = \rho(x)$, so that $x \in M$ and $\tau(x) = 0$. Thus, any $\varepsilon > 0$ is a regular value of τ, so that $N : \tau(x) \leq \varepsilon$ is a smooth manifold with boundary $\partial N : \tau(x) = \varepsilon$. We also know that $\operatorname{grad}_x(\tau)$ is perpendicular to ∂N at x, and consequently

$$\eta(x) = \frac{\operatorname{grad}_x(\tau)}{\|\operatorname{grad}_x(\tau)\|} = \tfrac{1}{\sqrt{\varepsilon}}(x - \rho(x))$$

is a unitary normal vector field on ∂N (III.2.6, p. 108).

Now recall that (i) $\|x - \rho(x)\| = \operatorname{dist}(x, M)$ for every $x \in U$ and (ii) $\operatorname{dist}(M, \mathbb{R}^p \setminus U) > 0$. Consequently, for $\varepsilon > 0$ small enough

$$N = \{x \in \mathbb{R}^p : \operatorname{dist}(x, M) \leq \sqrt{\varepsilon}\}, \quad \partial N = \{x \in \mathbb{R}^p : \operatorname{dist}(x, M) = \sqrt{\varepsilon}\}.$$

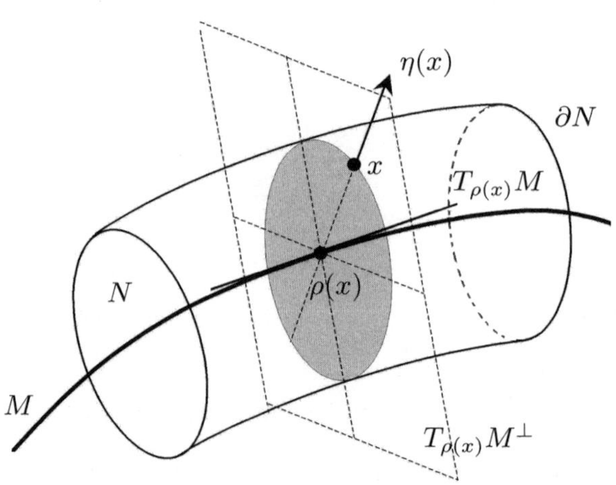

Summing up, N is a compact manifold, and η is a normal vector field on its boundary ∂N, which we call the *normal vector field on a tube around* M.

Using the above normal vector fields, we can state the main theorem:

Theorem 7.2 (Poincaré-Hopf Index Theorem). *Let $M \subset \mathbb{R}^p$ be a compact oriented boundaryless smooth manifold, and let ξ be a smooth tangent vector field with isolated zeros. Let $\eta : \partial N \to \mathbb{S}^{p-1}$ be a normal vector field on a tube around M. Then*

$$\operatorname{Ind}(\xi) = \deg(\eta).$$

Proof. By V.5.3, p. 208, and V.5.4, p. 209, we can split all zeros of ξ to assume they are non-degenerate. Once this is the case, with the notation of the preceding paragraph, we define

$$f : N \to \mathbb{R}^p : x \mapsto x - \rho(x) + \xi(\rho(x)).$$

We claim that 0 is a regular value of f. Indeed, since $x - \rho(x)$ and $\xi(\rho(x))$ are perpendicular, $f(x) = 0$ if and only if $x = \rho(x)$ and $\xi(\rho(x)) = 0$. This means that $x \in M$ is a zero a of ξ. Furthermore, given such a zero a, we have:

(1) $f \equiv \xi$ on M; hence $d_a f|_{T_a M} = d_a \xi$.

(2) $\rho \equiv a$ on the orthogonal complement $a + E$ of $T_a M$; hence $d_a f|_E \equiv \operatorname{Id}_E$.

This means that $d_a f = \operatorname{Id}_E \oplus d_a \xi$, so that $\det(d_a f) = \det(d_a \xi) \neq 0$, since a is a non-degenerate zero of ξ. This shows that 0 is a regular value of f, its preimages are the zeros of ξ, and

$$\sum_a \operatorname{sign}_a(f) = \sum_a \operatorname{sign}_a(\xi) = \operatorname{Ind}(\xi).$$

On the other hand, by IV.4.5, p. 157,

$$\deg\Big(\frac{f}{\|f\|}\Big|_{\partial N}\Big) = \sum_a \operatorname{sign}_a(f);$$

hence we only must show that $\deg(\eta) = \deg\big(\frac{f}{\|f\|}\big|_{\partial N}\big)$. But these two mappings from ∂N into $\mathbb{S}^{p-1}$ are homotopic because they do not have antipodal images (V.1.4, p. 186). Indeed, if they had some, i.e., if

$$-\lambda(x - \rho(x)) = x - \rho(x) + \xi(\rho(x)), \quad \text{with } x \in \partial N \text{ and } \lambda > 0,$$

then $\xi(\rho(x)) = -(1 + \lambda)(x - \rho(x))$. This is impossible for $x \notin M$. □

Remark 7.3. If m is odd, this invariant has little significance. Indeed, first note that $\operatorname{Ind}(-\xi) = (-1)^m \operatorname{Ind}(\xi)$ (this is immediate for the index at every non-degenerate zero and then follows for the total index). Thus, if m is odd, $\operatorname{Ind}(\xi) = 0$. □

(7.4) Euler characteristic. The Poincaré-Hopf Theorem also says that *the total index is the Euler characteristic* $\chi(M)$ *of* M (hence $= 0$ if m is odd) and consequently depends only on the topological type of M:

$$\operatorname{Ind}(\xi) = \chi(M).$$

However, to bring in χ, we would need a rigorous definition, which is not our concern here. Instead we draw some pictures to see how χ enters the scene.

First notice that by the last theorem, it is enough to find a field ξ whose total index is indeed the Euler characteristic. We will describe some informal but very natural means to construct such a ξ.

(1) Suppose we have a *triangulation*. Then we mark one point a_k at each face of dimension k. These points will be the zeros of our field ξ, and all are non-degenerate. In fact, ξ can be defined on each face by a source at a_k, that is, by the local form $\xi = \sum_{i=1}^k x_i \frac{\partial}{\partial x_i}$. Of course, only on the faces of maximal dimension $k = m$ do we get a true source. On the others the flow is incoming from higher-dimensional faces. Thus we start at the a_m's with sign $+1$, and the sign changes to -1 at the a_{m-1}'s, and then again to $+1$ at the a_{m-2}'s, and so on. Thus we have

$$\operatorname{Ind}(\xi) = A_m - A_{m-1} + A_{m-2} - \cdots + (-1)^m A_0$$

where A_k stands for the number of faces of dimension k. If m is even, this sum is χ, and if m is odd, it is $-\chi$, and we can change ξ to $-\xi$ (although it does not matter much because in that case $\chi = 0$).

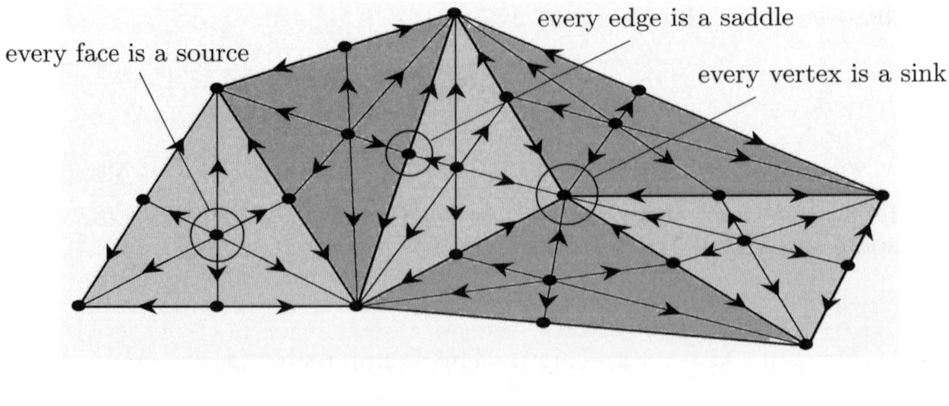

The preceding picture shows χ in dimension 2: we get a source at every face, a saddle at every edge, and a sink at every vertex, adding up to the classical Euler formula $\chi = F - E + V$.

(2) We can imagine the construction above in a physical way, as the flow of a liquid poured over the manifold from some vintage points. We can use this image to guess the Euler characteristic of a torus with g holes. The figure below shows this: sources have index $+1$, saddles -1, and sinks $+1$.

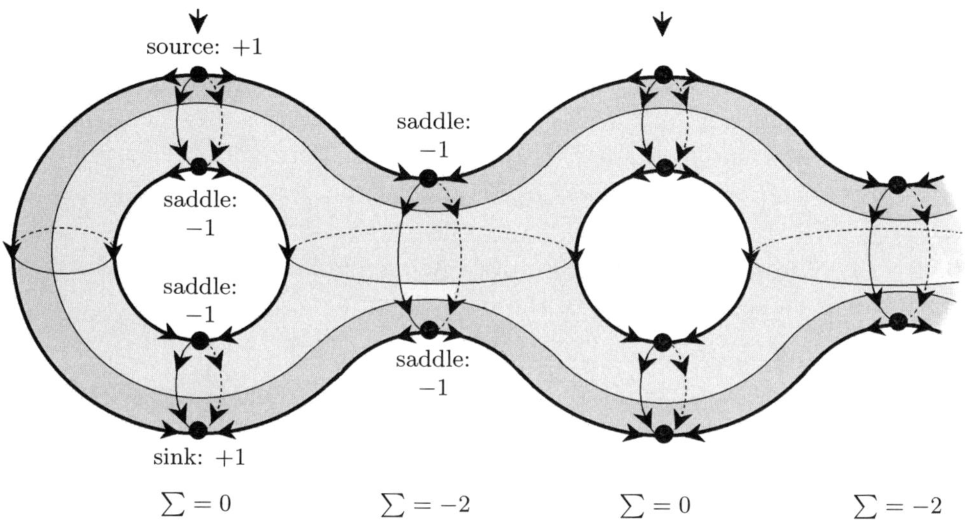

When we add up all indices, the four zeros around each hole cancel each other, and each pair of zeros in between two consecutive holes gives -2. Since there are g holes, there are $g-1$ of those pairs, and the final result is what it must be:

$$\chi = \mathrm{Ind} = \sum \mathrm{ind} = -2(g-1) = 2 - 2g.$$

This g is the *genus* of the surface. The reader can try different forms of pouring the liquid and check that, of course, the result is always the same.

(3) For the sphere $\mathbb{S}^2 \subset \mathbb{R}^3$ ($g = 0$), we have just a source and a sink, which gives index 2 and $\chi(\mathbb{S}^2) = 2 = 2 - 2g$. Or we can simply use the field $\xi(x, y, z) = (-y, x, 0)$ depicted after the Hedge-hog Theorem, III.7.4, p. 135: its two zeros are circulations and hence have index $+1$. Also, we understand now what that theorem says for arbitrary compact, boundaryless manifolds: *if there is a tangent vector field ξ without zeros, then* $\chi = \mathrm{Ind}(\xi) = 0$.

(7.5) The Morse inequalities. This is a description of the Euler characteristic by means of gradient vector fields. Again let M be an oriented, compact, boundaryless, smooth manifold of dimension m, and let $f : M \to \mathbb{R}$ be a Morse function (which always exists; V.6.8, p. 215, and V.6.9, p. 216). We have already computed the total index of the gradient $\xi = \operatorname{grad}(f)$ of f:

$$\operatorname{Ind}(\xi) = \sum_k (-1)^k \alpha_k,$$

where α_k is the number of critical points of index k of f (V.6.6, p. 214). Thus we get

$$\chi(M) = \sum_k (-1)^k \alpha_k.$$

This formula for the Euler characteristic in terms of the critical points of a Morse function is part of the so-called *Morse inequalities* (in fact this is the unique *equality* among them). Those inequalities are actually formulated in terms of the *homology groups of M* and escape our context.

(7.6) The Gauss-Bonnet Formula. As we explained earlier (III.2.6, p. 108), this is the computation of the integral curvature

$$\int_M K\Omega_M = \operatorname{vol}_m(\mathbb{S}^m)\deg(\nu) = \tfrac{1}{2}\operatorname{vol}_m(\mathbb{S}^m)\chi(M)$$

for a compact hypersurface $M \subset \mathbb{R}^{m+1}$ of even dimension m with Gauss mapping ν. There we saw how the first equality comes from degree theory by integration, and what was left was the computation

$$\deg(\nu) = \tfrac{1}{2}\chi(M).$$

But this follows from the Poincaré-Hopf Index Theorem, V.7.2, p. 219, which for hypersurfaces of even dimension has a better form. Let us sketch the argument.

First notice that as M is a hypersurface of dimension m, ∂N consists of two disjoint copies M_+ and M_- of M: for every $x \in M$ there are exactly two points x_+ and x_- at a distance $\sqrt{\varepsilon}$ from x, and the signs can be chosen by the conditions

$$\begin{cases} x_+ = x + \sqrt{\varepsilon}\nu(x), & \text{hence } \eta(x_+) = \nu(x), \\ x_- = x - \sqrt{\varepsilon}\nu(x), & \text{hence } \eta(x_-) = -\nu(x). \end{cases}$$

Thus ρ induces two diffeomorphisms $M_+, M_- \to M$ with inverses described just above. In particular, the two derivatives of those inverses are

$$\operatorname{Id} - \sqrt{\varepsilon} d_x\nu \quad \text{and} \quad \operatorname{Id} - \sqrt{\varepsilon} d_x\nu,$$

arbitrarily close to Id for ε small enough.

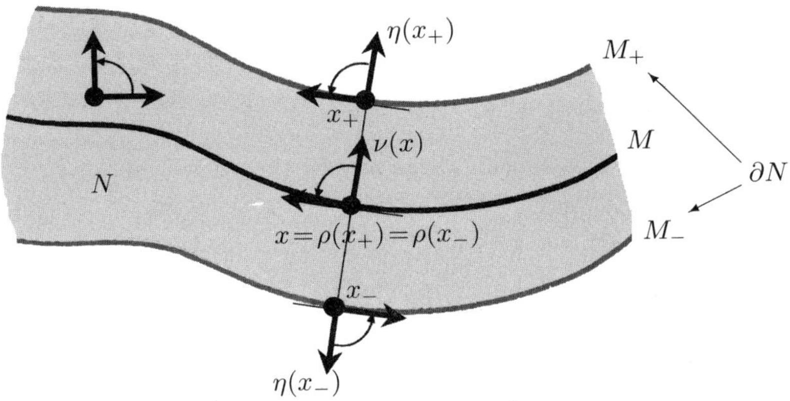

All of this readily implies that ρ preserves orientations on M_+, and it does not on M_-. Finally, write

$$\begin{cases} \eta|_{M_+} = \nu \circ \rho|_{M_+}, \\ \eta|_{M_-} = \sigma \circ \nu \circ \rho|_{M_-}, \end{cases}$$

where σ is the antipodal diffeomorphism of $\mathbb{S}^m$. Since the diffeomorphism σ has degree $+1$ if m is odd and -1 if m is even (III.1.6(1), p. 99), we get

$$\begin{aligned} \deg(\eta) &= \deg(\eta|_{M_+}) + \deg(\eta|_{M_-}) \\ &= \deg(\nu \circ \rho|_{M_+}) + \deg(\sigma \circ \nu \circ \rho|_{M_-}) \\ &= \deg(\nu)(+1) + (\pm 1)\deg(\nu)(-1) = \begin{cases} 0 & \text{if } m \text{ is odd,} \\ 2\deg(\nu) & \text{if } m \text{ is even.} \end{cases} \end{aligned}$$

As we have already explained, $\deg(\eta) = \chi(M)$, and for m odd we again get that $\chi(M) = 0$ and we get nothing about $\deg(\nu)$. But for m even we obtain $\chi(M) = 2\deg(\nu)$, and the Gauss-Bonnet Formula follows.

Exercises and problems

Number 1. Let $\mathbb{S}^m \subset \mathbb{R}^{m+1}$ be a sphere of even dimension. Use the tangent vector field

$$\xi(x_1, x_2, \dots, x_{m-1}, x_m, x_{m+1}) \mapsto (-x_2, x_1, \dots, -x_m, x_{m-1}, 0)$$

to compute the Euler characteristic $\chi(\mathbb{S}^m)$.

Number 2. Prove the following product formula for the Euler characterisitic

$$\chi(M \times N) = \chi(M) \cdot \chi(N).$$

Number 3. Show that the torus $T \subset \mathbb{R}^3$ generated by the circle $y = 0$, $(x-2)^2 + z^2 = 1$ around the axis $x = y = 0$ is diffeomorphic to the product $\mathbb{S}^1 \times \mathbb{S}^1 \subset \mathbb{R}^4$, and deduce that its Euler characteristic is 0. Confirm this in three other ways:

(1) Obtain a tangent vector field on T without zeros, and apply the Poincaré-Hopf Index Theorem.

(2) Compute the degree of the Gauss mapping $\nu : T \to \mathbb{S}^2$, and apply the Gauss-Bonnet Theorem.

(3) Exhibit a Morse function on T, and use the Morse Inequalities.

Notice that (1) and (3) can also be done for $\mathbb{S}^1 \times \mathbb{S}^1 \subset \mathbb{R}^4$, but (2) cannot, as the latter is not a hypersurface.

Number 4. Prove, using Euler characteristics, that a sphere of even dimension cannot be homeomorphic to a product of two other spheres.

Number 5. Let $M \subset \mathbb{R}^3$ be a compact smooth surface, and let $f : M \to \mathbb{R}$ be a Morse function. Show that f has at least two critical points, and if there are no more, then M has genus 0 (hence M is diffeomorphic to the sphere $\mathbb{S}^2$).

Number 6. Compute the Euler characteristic of the rotation group $M = SO(3) \subset \mathbb{R}^{3\times 3}$ using the Morse function h in Problem Number 5 of V.6.

Number 7. Let $M \subset \mathbb{R}^3$ be the *cornered sphere* $x^4 + y^4 + z^4 = 1$. Check the Gauss-Bonnet Theorem (i) through explicit computation of the integral curvature and (ii) computing the degree of the Gauss mapping.

Names of mathematicians cited

Historical references

We list below the references mentioned in Chapter I. We also include three basic papers on the history of degree theory: [Siegberg 1980a], [Siegberg 1980b], and [Mawhin 1999].

[Adams 1969] J.F. Adams: *Lectures on Lie Groups.* New York-Amsterdam: W.A. Benjamin, Inc., 1969.

[Aleksandrov-Hopf 1935] P. Aleksandrov, H. Hopf: *Topologie.* Berlin: Verlag von Julius Springer, 1935.

[Amann-Weiss 1973] H. Amann, S.A. Weiss: On the uniqueness of topological degree. *Math. Z.* **130** (1973), 39–54.

[Banach 1922] S. Banach: Sur les opérations dans les ensembles abstraites et leurs applications aux equations intégrales. *Fundamenta Math.* **3** (1922), 133–181.

[Bohl 1904] P. Bohl: Über die Bewegung eines mechanischen Systems in der Nähe einer Gleichgerwichtslage. *J. Reine Angew. Math.,* **127** (1904), 179–276.

[Brouwer 1912a] L.E.J. Brouwer: Über abbildung von Mannigfaltigkeiten. *Mat. Ann.* **71** (1912), 97–115.

[Brouwer 1912b] L.E.J. Brouwer: On looping coefficients. *Konink. Nederl. Akad. von Wet. Proc.* **15** (1912), 113–122.

[Brouwer 1912c] L.E.J. Brouwer: Continuous one-one transformations of surfaces in themselves. *CNAG Proc.* **15** (1912), 352–360.

[Brouwer 1976] L.E.J. Brouwer: *Collected works, vol. II. Geometry, analysis, topology and mechanics.* H. Freudenthal (ed.). Amsterdam-Oxford: North Holland, 1976.

[Brown 1935] A. B. Brown: Functional dependence. *Trans. Amer. Math. Soc.* **38** (1935), 379–394.

[Caccioppoli 1936] R. Caccioppoli: Sulle corrispondenze funzionalli inverse diramata: teoria generale e applicazioni ad alcune equazioni funzionali non lineari e al problema di Plateau, I & II. *Rend. Accad. Naz. Linzei* **24** (1936), 258–263 & 416–421.

[Cauchy 1837a] A.-L. Cauchy: Calcul des indices des fonctions. *J. École Polytéchnique* **XV** (1837), 176–226.

[Cauchy 1837b] A.-L. Cauchy: Extrait d'une lettre sur une mémoire publié à Turin, le 16 Juin 1833, et relatif aux racines des équations simultanées. *Comptes Rendus Acad. Sc. Paris* **IV** (1837), 672–675.

[Cauchy 1855] A.-L. Cauchy: Sur les compteurs logarithmiques. *Comptes Rendus Acad. Sc. Paris* **XL** (1855), 1009–1016.

[Čech 1932] E. Čech: Höherdimensionale Homotopiegruppen. *Verhandlungen der Internationalen Mathematiken Kongresses,* Zurich (1932), 203.

[Dancer 1983] E.N. Dancer: On the existence of zeros of perturbed operators. *Nonlinear Anal.* **7** (1983), 717–727.

[Dancer 1985] E.N. Dancer: A new degree for $\mathbb{S}^1$-invariant gradient mappings and applications. *Ann. Inst. Henri Poincaré, Anal. Non-linéaire* **2** (1985), 329–370.

[Dancer *et al.* 2005] E.N. Dancer, K. Geba, S.M. Rybicki: Classification of homotopy classes of equivariant gradient maps. *Fundamenta Math.* **185** (2005), 1–18.

[Deimling 1985] K. Deimling: *Non-linear functional analysis.* Berlin: Springer-Verlag, 1985.

[Dyck 1888] W.F.A. Dyck: Beiträge zur Analysis situs I. Aufsatz ein- und zweidimensionale Mannigfaltigkeiten. *Math. Ann.* **32** (1888), 457–512.

[Dyck 1890] W.F.A. Dyck: Beiträge zur Analysis situs II. Aufsatz Mannigfaltigkeiten von n-dimensionen. *Math. Ann.* **34** (1888), 273–316.

[Dylawerski *et al.* 1991] G. Dylawerski, K. Geba, J. Jodel, W. Marzantowicz: An $\mathbb{S}^1$-equivariant degree and the Fuller index. *Ann. Pol. Math.* **52** (3) (1991), 243–280.

[Elworthy-Tromba 1970a] K.D. Elworthy, A.J. Tromba: Differential structures and Fredholm maps on Banach manifolds. In *Global Analysis* (Berkeley, 1968) 45–94, Proc. Symp. Pure Math. **15**. Providence, R.I.: American Mathematical Society, 1970.

[Elworthy-Tromba 1970b] K.D. Elworthy, A.J. Tromba: Degree theory on Banach manifolds. In *Nonlinear Functional Analysis* (Chicago, 1968) 75–94, Proc. Symp. Pure Math. **18**, part 1. Providence, R.I.: American Mathematical Society, 1970.

[Führer 1971] L. Führer: Theorie des Abbildungsgrades in endlich-dimensionalen Räumen. *Dissertation,* Freie Universität Berlin, 1971.

[Führer 1972] L. Führer: Ein elementarer analytischer Beweis zur Eindeutigkeit des Abbildungsgrades in $\mathbb{R}^n$. *Math. Nachr.* **54** (1972), 259-267.

[Fuller 1967] B. Fuller: An index of fixed point type for periodic orbits. *American J. Math.* **89** (1967), 133–148.

[Gauss 1813] K.-F. Gauss: Theoria attractionis corporum sphaeroidicorum ellipticorum homogeneorum methodo nova tractata. *Göttingische gelehrt Anzeigen, April 5* (1813), 545–552.

[Gauss 1833] K.-F. Gauss: Zur mathematischen Theorie der electrodynamische Wirkungen. In *Werke, 5*. Göttingen: Königlichen Gesellschaft der Wissenchaften zu Göttingen (1867), 601–626.

[Geba *et al.* 1986] K. Geba, I. Massabò, A. Vignoli: Generalized topological degree and bifurcation. In *Non-linear Functional Analysis and Applications* (Proc. NATO Adv. Study Inst., Maratea, Italy, 1985) 55–73, Mathematical and Physical Sciences **173**. Dordrecht: Reidel, 1986.

[Geba *et al.* 1990] K. Geba, I. Massabò, A. Vignoli: On the Euler characteristic of equivariant gradient vector fields. *Boll. Un. Mat. Ital. A(7)* **4** (1990), 243–251.

[Hadamard 1910] J.-S. Hadamard: Sur quelques applications de l'indice de Kronecker. In [Tannery 1910], 437–477.

[Heinz 1959] E. Heinz: An elementary analytic theory of the degree of mapping in n-dimensional space. *J. Math. and Mech.* **8** (1959), 231–247.

[Hopf 1925] H. Hopf: Über die Curvatura integra geschlossener Hyperflächen. *Math. Ann.* **95** (1925), 340–367.

[Hopf 1926a] H. Hopf: Abbildungsklassen n-dimensionaler Mannigfaltigkeiten. *Math. Ann.* **96** (1926), 209–224.

[Hopf 1926b] H. Hopf: Vectorfelder in n-dimensionalen Mannigfaltigkeiten. *Math. Ann.* **96** (1926), 225–250.

[Hopf 1931] H. Hopf: Über die Abbildungen der dreidimensionalen Sphäre auf die Kugelfläche. *Math. Ann.* **104** (1931), 637–665.

[Hopf 1933] H. Hopf: Die Klassen der Abbildungen der n-dimensionalen Polyeder auf die n-dimensionale Sphäre. *Comment. Math. Helvetici* **5** (1933), 39–54.

[Hopf 1935] H. Hopf: Über die Abbildungen von Sphären auf Sphären niedrigerer Dimension. *Fundamenta Math.* **25** (1935), 427–440.

[Hopf 1966] H. Hopf: Ein Abschnitt aus der Entwicklung der Topologie. *Jber. Deutsch. Math.-Verein.* **68** (1966), 96–106.

[Hurewicz 1935a] W. Hurewicz: Beiträge zur Topologie der Deformationen I. Höherdimensionalen Homotopiegruppen. *Proc. Akad. Wetenschappen* **38** (1935), 112–119.

[Hurewicz 1935b] W. Hurewicz: Beiträge zur Topologie der Deformationen II. Homotopie und Homologiegruppen. *Proc. Akad. Wetenschappen* **38** (1935), 521–528.

[Hurewicz 1936a] W. Hurewicz: Beiträge zur Topologie der Deformationen III. Klassen und Homologietypen von Abbildungen. *Proc. Akad. Wetenschappen* **39** (1936), 117–126.

[Hurewicz 1936b] W. Hurewicz: Beiträge zur Topologie der Deformationen IV. Aspharische Räume. *Proc. Akad. Wetenschappen* **39** (1936), 215–224.

[Ize 1981] J. Ize: Introduction to bifurcation theory. In *Differential Equations* (Sao Paulo, 1981), 145–203. Lecture Notes in Math. **957**. Berlin-New York: Springer-Verlag, 1982.

[Ize *et al.* 1986] J. Ize, I. Massabò, A. Vignoli: Global results on continuation and bifurcation for equivariant maps. In *Non-linear Functional Analysis and Applications* (Maratea, Italy, 1985), 75–111. NATO Adv. Sci. Inst. Ser. C Math. Phys. Sci. **173**. Dordrecht: Reidel, 1986.

[Ize *et al.* 1989] J. Ize, I. Massabò, A. Vignoli: Degree theory for equivariant maps, I. *Trans. Amer. Math. Soc.* **315** (2) (1989), 433–510.

[Ize *et al.* 1992] J. Ize, I. Massabò, A. Vignoli: Degree theory for equivariant maps, II: The general $\mathbb{S}^1$-actions. *Memoirs Amer. Math. Soc.* **481** (1992).

[Ize-Vignoli 2003] J. Ize, A. Vignoli: *Equivariant Degree Theory.* de Gruyter Series in Nonlinear Analysis and Applications **8**. Berlin: Walter de Gruyter & Co., 2003.

[Jordan 1893] C. Jordan: *Cours d'analyse de l'École Polytechnique, I.* 1893.

[Kronecker 1869a] L. Kronecker: Über Systeme von Functionen mehrerer Variabeln, I. *Monatsberichte königlich Preuss. Akad. Wissens. Berlin, March 4* (1869), 159–193.

[Kronecker 1869b] L. Kronecker: Über Systeme von Functionen mehrerer Variabeln, II. *Monatsberichte königlich Preuss. Akad. Wissens. Berlin, August 5* (1869), 688–698.

[Leray-Schauder 1933] J. Leray, J.P. Schauder: Topologie et équations fonctionnelles. *Comptes Rendus Acad. Sc. Paris* **1197** (1933), 115–117.

[Leray-Schauder 1934] J. Leray, J.P. Schauder: Topologie et équations fonctionnelles. *Ann. Sc. École Normale Sup. Paris* **51** (1934), 45–78.

[Liouville-Sturm 1837] J. Liouville, J.-Ch.-F. Sturm: Note sur un théorème de M. Cauchy relatif aux racines des équations simultanées. *Comptes rendus Acad. Sc. Paris* **40** (1937), 720–724.

[Mawhin 1999] J. Mawhin: Leray-Schauder degree: A half century of extensions and applications. *Topol. Methods Nonlinear Anal.* **14** (1999), 195–228.

[Miranda 1940] C. Miranda: Un'osservazione su un teorema di Brouwer. *Boll. Un. Mat. Ital. (2)* **3** (1940), 5–7.

[Nagumo 1951a] M. Nagumo: A theory of degree of mapping based on infinitesimal analysis. *American J. Math.* **73** (1951), 485–496.

[Nagumo 1951b] M. Nagumo: Degree of mapping in convex linear topological spaces. *American J. Math.* **73** (1951), 497–511.

[Nirenberg 1971] L. Nirenberg: An application of generalized degree to a class of nonlinear problems. In *Troisième Colloque sur l'Analyse Fonctionnelle* (Liège, 1970), 57–74. Louvain: Vander, 1971.

[Picard 1891] Ch.-E. Picard: Sur le nombre des racines communes à plusieurs équations simultanées. *Comptes Rendus Acad. Sc. Paris* **113** (1891), 256–358.

[Picard 1891/1905] Ch.-E. Picard: *Traité d'Analyse, I* and *II*. Paris: Gauthier-Villars, 1891 and 1893/1905.

[Poincaré 1883] J.H. Poincaré: Sur certains solutions particulières du problème des trois corps. *Comptes Rendus Acad. Sc. Paris* **97** (1883), 251–252.

[Poincaré 1885a] J.H. Poincaré: Sur les courbes definies par les équations differéntielles. *J. Math. Pures Appl.* **1** (1885), 167–244.

[Poincaré 1895b] J.H. Poincaré: *Analysis Situs. J. École Polytechnique (2)* **1** (1895), 1–123.

[Poincaré 1886] J.H. Poincaré: Sur les courbes definies par une équation differéntielle. *J. Math. Pures Appl.* **2** (1886), 151–217.

[Poincaré 1899] J.H. Poincaré: Complément à l'Analysis Situs. *Rend. Circ. Mat. Palermo* **13** (1899), 285–343.

[Poincaré 1900] J.H. Poincaré: Complément à l'Analysis Situs. *Proc. London Math. Soc.* **32** (1900), 277–308.

[Pontryagin 1955] L.S. Pontryagin: Smooth manifolds and their application in homotopy theory. *Amer. Math. Soc. Transl. (2)* **11** (1959), 1–114, from *Trudy Mat. Inst. im Steklov* **45** (1955).

[de Rham 1955] G. de Rham: *Variétés différentiables.* Paris: Hermann, 1955.

[Rothe 1936] E. Rothe: Über Abbildungsklassen von Kugeln des Hilbertschen Raumes. *Compos. Math.* **4** (1936), 294–307.

[Ruiz del Portal 1991] F.R. Ruiz del Portal: Teoría del grado topológico generalizado y aplicaciones. *Dissertation.* Madrid: Universidad Complutense, 1991.

[Ruiz del Portal 1992] F.R. Ruiz del Portal: On the additivity property of the generalized degree. *Math. Japonica* **37** (1992), 657–664.

[Rybicki 1994] S. Rybicki: A degree for $\mathbb{S}^1$-equivariant orthogonal maps and its applications to bifurcation theory. *Nonlinear Anal.* **23** (1994), 83–102.

[Sard 1942] A. Sard: The measure of critical points of differentiable maps. *Bull. Amer. Math. Soc.* **48** (1942), 883–897.

[Schönflies 1902] A.M. Schönflies: Über einen grundlegenden Satz der Analysis Situs. *Gött. Nachr. Math. Phys. Kl.* (1902), 185–192.

[Siegberg 1980a] H. W. Siegberg: Brouwer degree: History and numerical computation. In *Numerical solutions of highly nonlinear problems,* W. Forster (ed.), 389–411. Amsterdam: North-Holland, 1980.

[Siegberg 1980b] H. W. Siegberg: Some historical remarks concerning degree theory. *American Math. Monthly* **88** (1981), 125–139.

[Smale 1965] S. Smale: An infinite dimensional version of Sard's Theorem. *American J. Math.* **87** (1965), 861–866.

[Sylvester 1853] J.J. Sylvester: On a theory of the syzygetic relations of two rational integral functions, comprising an application to the theory of Sturm's functions, and that of greatest algebraic common measure. *Philos. Trans. Roy. Soc. London* **143** (1853), 407-562.

[Tannery 1910] J. Tannery: *Introduction à la théorie des fonctions d'une variable, t. 2,* 2ième edition. Paris: Hermann, 1910.

[Thom 1954] R. Thom: Quelques propriétés globales des variétés différentiables. *Comment. Math. Helvetici* **28** (1954), 17–86.

[Veblen 1905] O. Veblen: Theory of plane curves in non-metrical analysis situs. *Trans. Amer. Math. Soc.* **6** (1905), 83–98.

Bibliography

For prerequisites, we recommend the quite *ad hoc* texts [4], [10], [16], and [18]. On the other hand, there is a wealth of literature on degree theory and related topics; we suggest as further reading the following books:

[1] J. Cronin: *Fixed points and topological degree in nonlinear analysis.* Mathematical Surveys **11**. Providence, R.I.: American Mathematical Society, 1964.

[2] K. Deimling: *Nonlinear Functional Analysis.* Berlin: Springer-Verlag, 1985.

[3] A. Dold: *Teoría de punto fijo* (I, II, III). México: Monografías del Instituto de Matemática, 1986.

[4] J.M. Gamboa, J.M. Ruiz: *Iniciación al estudio de las variedades diferenciales* (2ª edición revisada). Madrid: Sanz y Torres, 2006.

[5] A. Granas, J. Dugunji: *Fixed Point Theory.* Springer Monographs in Mathematics. New York: Springer-Verlag, 2003.

[6] V. Guillemin, A. Pollack: *Differential Topology.* Englewood Cliffs, N.J.: Prentice Hall, Inc.,1974.

[7] M.W. Hirsch: *Differential Topology.* Graduate Texts in Mathematics **33**. Springer-Verlag, New York-Heidelberg: Springer-Verlag, 1976.

[8] M.A. Krasnosel'skii: *Topological methods in the theory of nonlinear integral equations* (translated from Russian). New York: Pergamon Press, 1964.

[9] W. Krawcewicz, J. Wu: *Theory of degrees with applications to bifurcations and differential equations.* New York: John Wiley & Sons, 1997.

[10] S. Lang: *Differential manifolds.* Berlin: Springer-Verlag, 1988.

[11] E.L. Lima: *Introduçao à Topologia Diferencial.* Rio de Janeiro: Instituto de Matemática Pura e Aplicada, 1961.

[12] N.G. Lloyd: *Degree Theory.* Cambridge Tracts in Mathematics **73**. Cambridge: Cambridge University Press, 1978.

[13] I. Madsen, J. Tornehave: *From calculus to cohomology: de Rham cohomology and characteristic classes.* Cambridge, New York, Melbourne: Cambridge University Press, 1997.

[14] J. Milnor: *Topology from the differentiable viewpoint.* Princeton Landmarks in Mathematics. Princeton, N.J.: Princeton University Press, 1997.

[15] L. Nirenberg: *Topics in nonlinear functional analysis* (with a chapter by E. Zehnder and notes by R.A. Artino). Courant Lecture Notes in Mathematics **6**. New York: Courant Institute of Mathematical Sciences, American Mathematical Society, 1974.

[16] E. Outerelo, J.M. Ruiz: *Topología Diferencial.* Madrid: Addison-Wesley, 1998.

[17] P.H. Rabinowitz: *Théorie du Degré Topologique et applications à des problèmes aux limites non lineaires (redigé par H. Berestycki).* Lecture Notes Analyse Numérique Fonctionelle. Paris: Université Paris VI, 1975.

[18] M. Spivak: *Calculus on manifolds: A modern approach to classical theorems of advanced calculus.* Boulder: Westview Press, 1971.

Symbols

Index

Titles in This Series

108 **Enrique Outerelo and Jesús M. Ruiz,** Mapping degree theory, 2009
107 **Jeffrey M. Lee,** Manifolds and differential geometry, 2009
106 **Robert J. Daverman and Gerard A. Venema,** Embeddings in manifolds, 2009
105 **Giovanni Leoni,** A first course in Sobolev spaces, 2009
104 **Paolo Aluffi,** Algebra: Chapter 0, 2009
103 **Branko Grünbaum,** Configurations of points and lines, 2009
102 **Mark A. Pinsky,** Introduction to Fourier analysis and wavelets, 2009
101 **Ward Cheney and Will Light,** A course in approximation theory, 2009
100 **I. Martin Isaacs,** Algebra: A graduate course, 2009
99 **Gerald Teschl,** Mathematical methods in quantum mechanics: With applications to Schrödinger operators, 2009
98 **Alexander I. Bobenko and Yuri B. Suris,** Discrete differential geometry: Integrable structure, 2008
97 **David C. Ullrich,** Complex made simple, 2008
96 **N. V. Krylov,** Lectures on elliptic and parabolic equations in Sobolev spaces, 2008
95 **Leon A. Takhtajan,** Quantum mechanics for mathematicians, 2008
94 **James E. Humphreys,** Representations of semisimple Lie algebras in the BGG category $\mathcal{O}$, 2008
93 **Peter W. Michor,** Topics in differential geometry, 2008
92 **I. Martin Isaacs,** Finite group theory, 2008
91 **Louis Halle Rowen,** Graduate algebra: Noncommutative view, 2008
90 **Larry J. Gerstein,** Basic quadratic forms, 2008
89 **Anthony Bonato,** A course on the web graph, 2008
88 **Nathanial P. Brown and Narutaka Ozawa,** C^*-algebras and finite-dimensional approximations, 2008
87 **Srikanth B. Iyengar, Graham J. Leuschke, Anton Leykin, Claudia Miller, Ezra Miller, Anurag K. Singh, and Uli Walther,** Twenty-four hours of local cohomology, 2007
86 **Yulij Ilyashenko and Sergei Yakovenko,** Lectures on analytic differential equations, 2007
85 **John M. Alongi and Gail S. Nelson,** Recurrence and topology, 2007
84 **Charalambos D. Aliprantis and Rabee Tourky,** Cones and duality, 2007
83 **Wolfgang Ebeling,** Functions of several complex variables and their singularities (translated by Philip G. Spain), 2007
82 **Serge Alinhac and Patrick Gérard,** Pseudo-differential operators and the Nash–Moser theorem (translated by Stephen S. Wilson), 2007
81 **V. V. Prasolov,** Elements of homology theory, 2007
80 **Davar Khoshnevisan,** Probability, 2007
79 **William Stein,** Modular forms, a computational approach (with an appendix by Paul E. Gunnells), 2007
78 **Harry Dym,** Linear algebra in action, 2007
77 **Bennett Chow, Peng Lu, and Lei Ni,** Hamilton's Ricci flow, 2006
76 **Michael E. Taylor,** Measure theory and integration, 2006
75 **Peter D. Miller,** Applied asymptotic analysis, 2006
74 **V. V. Prasolov,** Elements of combinatorial and differential topology, 2006
73 **Louis Halle Rowen,** Graduate algebra: Commutative view, 2006
72 **R. J. Williams,** Introduction the the mathematics of finance, 2006
71 **S. P. Novikov and I. A. Taimanov,** Modern geometric structures and fields, 2006

70 **Seán Dineen,** Probability theory in finance, 2005
69 **Sebastián Montiel and Antonio Ros,** Curves and surfaces, 2005
68 **Luis Caffarelli and Sandro Salsa,** A geometric approach to free boundary problems, 2005
67 **T.Y. Lam,** Introduction to quadratic forms over fields, 2004
66 **Yuli Eidelman, Vitali Milman, and Antonis Tsolomitis,** Functional analysis, An introduction, 2004
65 **S. Ramanan,** Global calculus, 2004
64 **A. A. Kirillov,** Lectures on the orbit method, 2004
63 **Steven Dale Cutkosky,** Resolution of singularities, 2004
62 **T. W. Körner,** A companion to analysis: A second first and first second course in analysis, 2004
61 **Thomas A. Ivey and J. M. Landsberg,** Cartan for beginners: Differential geometry via moving frames and exterior differential systems, 2003
60 **Alberto Candel and Lawrence Conlon,** Foliations II, 2003
59 **Steven H. Weintraub,** Representation theory of finite groups: algebra and arithmetic, 2003
58 **Cédric Villani,** Topics in optimal transportation, 2003
57 **Robert Plato,** Concise numerical mathematics, 2003
56 **E. B. Vinberg,** A course in algebra, 2003
55 **C. Herbert Clemens,** A scrapbook of complex curve theory, second edition, 2003
54 **Alexander Barvinok,** A course in convexity, 2002
53 **Henryk Iwaniec,** Spectral methods of automorphic forms, 2002
52 **Ilka Agricola and Thomas Friedrich,** Global analysis: Differential forms in analysis, geometry and physics, 2002
51 **Y. A. Abramovich and C. D. Aliprantis,** Problems in operator theory, 2002
50 **Y. A. Abramovich and C. D. Aliprantis,** An invitation to operator theory, 2002
49 **John R. Harper,** Secondary cohomology operations, 2002
48 **Y. Eliashberg and N. Mishachev,** Introduction to the h-principle, 2002
47 **A. Yu. Kitaev, A. H. Shen, and M. N. Vyalyi,** Classical and quantum computation, 2002
46 **Joseph L. Taylor,** Several complex variables with connections to algebraic geometry and Lie groups, 2002
45 **Inder K. Rana,** An introduction to measure and integration, second edition, 2002
44 **Jim Agler and John E. McCarthy,** Pick interpolation and Hilbert function spaces, 2002
43 **N. V. Krylov,** Introduction to the theory of random processes, 2002
42 **Jin Hong and Seok-Jin Kang,** Introduction to quantum groups and crystal bases, 2002
41 **Georgi V. Smirnov,** Introduction to the theory of differential inclusions, 2002
40 **Robert E. Greene and Steven G. Krantz,** Function theory of one complex variable, third edition, 2006
39 **Larry C. Grove,** Classical groups and geometric algebra, 2002
38 **Elton P. Hsu,** Stochastic analysis on manifolds, 2002
37 **Hershel M. Farkas and Irwin Kra,** Theta constants, Riemann surfaces and the modular group, 2001

For a complete list of titles in this series, visit the
AMS Bookstore at **www.ams.org/bookstore/**.